Photosynthesis Bibliography

volume 10/2

Cumulative indexes to volumes 6–10

Editors Z. Šesták & J. Čatský

Springer-Science+Business Media, B.V. 1982

Contributors:

Z. Šesták

J. Čatský

I. Tichá

J. Pospišilová

J. Solárová

D. Hodáňová

J. Zima

ISBN 978-90-6193-050-1   ISBN 978-94-017-2628-3 (eBook)
DOI 10.1007/978-94-017-2628-3

CUMULATIVE INDEXES

These cumulative indexes to Volumes 6 to 10 contain references to papers published in the years 1975 - 1979 and also to papers published in the years 1965 - 1974 and included as addenda in Volumes 6 to 10.

The **Authors' Index** contains all names of authors, co-authors and editors. Authors' names are presented in the form in which they appear in the respective publication. The names from papers published in Cyrillic characters are transcribed as shown on p. III of this volume, part 1. Alternative spellings and forms of the name of the same author are usually cross-indexed. The numbers in *italics* refer to publications in which the author acts as editor.

The **Subject Index** contains a selection of primary items chosen according to their importance in photosynthesis research and to their relevance and occurrence. The word "Photosynthesis" is not regarded as a main theme, but partial processes, photosynthetic parameters and the factors influencing photosynthesis are listed. The processes and other characteristics are summarized into several main themes when presented in combination with individual factors, *e.g.* carbon fixation pathways, electron transport chain, chlorophyll, carotenoids, gas exchange, ecosystem and plant productivity (*including photosynthate distribution and canopy organization and functioning*), photorespiration, resistances to CO_2 and water vapour transfer, *etc.* Several items from branches related to photosynthesis research were also chosen for convenience, *e.g.* dealing with respiration, plant growth and development, water relations, anatomy, bioclimatology, *etc.* These items contain only references to papers within the scope of this bibliography.

The **Plant Index** contains a selection of plant genera and types interesting as experimental material for physiological, ecological and agricultural studies. Latin scientific names of plant genera and English names of plant groups and types are the main items which present the reference numbers.

In Subject and Plant Indexes, the references from the individual volumes are distinguished by a semicolon (;) or a paragraph.

The Editors

ERRATA

(cf. also Volume 10/1, p. IV)

Page	For	Read
Volume 6		
262 / line 6 from bottom	24215	24275
279 / line 5 from bottom	23154	23454
285 / Model of leaf ...	24963	24863
287 / line 2 from top	22344	22544
287 / Ontogeny of algae and gas ...	25021	25121
288 / line 18 from bottom	22732	23732
293 / line 21 from bottom	2244	22244
294 / line 10 from bottom	28041	23041
295 / line 5 from bottom	221117	22117
299 / line 18 from top	24704	24764
299 / Seasonal changes in electron ...	28672	23672
299 / line 20 from bottom	23771-2	23571-2
302 / line 9 from top	2173	21773
302 / Thylakoid ...	22450	22415
303 / Uncouplers ...	25929	25029
Volume 7		
51 / ref. 26068	210	209
106 / ref. 27032	159	159a
216	NAUSCH, Ya.	NAUSH, Ya.
240 / line 8 from bottom	25832	26832
249 / line 15 from bottom	27612	27812
265 / line 2 from top	25890	25690
275 / line 2 from top	26792	27692
Volume 8		
333 / line 3 from top	20128	30128
333 / Precipitation ... resistances		*delete*
Volume 9		
297 / line 12 from bottom	36400-1	36500-1
309 / line 21 from bottom	32695	32605
312 / line 13 from bottom	33820	33920
319 / line 12 from bottom	33947	33847
332 / line 12 from top	34625	34925
335 / line 9 from top	32681	36281
335 / line 18 from top	34987	34907
339 / Photophosphorylation, cyclic	33353	33853
349 / line 18 from top	34334	34834
350 / lines 21, 26 from top	Seasonal	Soil
352 / line 15 from top	32772-8	32772-3
Volume 10/1		
Contributors	J. Zime	J. Zima
145 / ref. 39064	fissazione	fissazionc
145 / ref. 39064	*1979*	*108* (*34* N.S.)
145 / ref. 39064	[In Ital.]	[In Ital., ab: E.]

This cumulative authors' index contains references to volumes 6 to 10. Authors' names are presented in the form in which they appear in the respective publication. The names from papers published in Cyrillic characters are transcribed as shown on p. III of this volume, part 1. Alternative spellings and forms of the name of the same author are usually cross-indexed. The numbers in *italics* refer to publications in which the author acts as editor.

A

AARNES, H. 32555
AARONSON, S. 21505
AASE, J.K. 28496, 32556
ABAD-ZAPATERO, C. 29782, 32557
ABARSUA, A.Z. 26709
ABAYCHI, J.K. 36588
ABDELLA, P.M. 37731
ABDEL-RAHMAN, M. 25162
ABDEL-WAHAB, O.A.L. 39923
ABDOURAKHMANOV, I.A. 36589
     see ABDURAKHMANOV, I.A.
ABDULAEV, N.G. 31167, 32558-9, 35199
ABDULLAEV, H.A. 24847
     see ABDULLAEV, Kh.A.
ABDULLAEV, Kh.A. 24848, 28497, 32222,
     32560-1, 36590
     see ABDULLAEV, H.A.
ABDULLAEV, M.A. 36591
ABDULLAEV, N.G. 39011
ABDULLAEVA, S.K. 22534, 29662
ABDULRAHMAN, F.S. 28498, 32562
ABDURAKHMANOV, A.A. 32865
ABDURAKHMANOV, I.A. 32563
     see ABDOURAKHMANOV, I.A.
ABE, A. 36517, 39589
ABEL, K.M. 28567, 36679
ABELIOVICH, A. 25163, 30873, 36592
ABER, J.D. 36593-4
ABERNETHY, R.H. 25164
ABILOV, Z.K. 21506, 22514, 36595,
     37610
ABO-EL LEL, G. 33449
ABOUL-ELA, M.M. 28619
ABRAHAMSSON, J. 33333
ABRAHAMSSON, S. *32733*
ABRAMCHIK, L.M. 28281
ABRAMOVA, I.V. 38365
ABRAMYAN, L.Kh. 21700, 27716
ABRANYI, A. 36596
ABRAVANEL, G. 36597
ABREU, I. 27380
ABROS'KINA, L.S. 35191, 36305, 36598
ABRUÑA, F. 30648
ABUL-FATIH, H.A. 36599
ABU-SHAKRA, S.S. 32832-3, 36600
ABUTALYBOV, M.G. 21506
ACEVEDO, E. 26437-8, 33527, 36601
ACKEFORS, H. 25165, 32564-5
ACKER, G. 30342

ACKER, S. 21507, 21922, 24041, 25166,
     28499, 31629-30, 32566, 35670,
     37584, 39453
ACKERMAN, T.L. 25303
ACKERSON, L.C. 22408, 27326
ACKERSON, R.C. 21508, 28500-1, 32567,
     40249
ACOCK, B. 25167-8, 28502, 29086-7,
     32568-9, 36000, 36602
ACQUA, F.M. dall' see DALL'ACQUA, F.M.
ACTON, D.W. 37654
ADABRA-MICHANOL, Y. 21509
ADACHI, S. 39818
ADACHI, T. 37533
ADALSTEINSSON, H. 38070
ADAMCHIK, G.G. 28253, 32266
ADAMCZAK, B. 23646
ADAME, E.G. 38271
ADAMOVA, N.P. 26731, 27273
ADAMS, C.A. 36603
ADAMS, C.J. 28503
ADAMS, G.E. *23292*
ADAMS, G.M.W. 25524
ADAMS, J.A. 28504
ADAMS, J.E. 25169, 28505
ADAMS, M. 22321
ADAMS, M.S. 21510, 23596, 23745,
     23992, 24773, 26754, 32570, 33441,
     36604, 38079, 38444, 39932-4
ADAMS, M.W.W. 28506, 32571, 36605-7
ADAMS, S. 32570
ADAMS, S.M. 36608, 39886
ADAMSON, H. 32572
ADARI, H. 35569, 39369
ADDY, N.D. 23582
ADEDIPE, N.O. 21511-3, 25170-1, 32059,
     36080
     see ADEPIPE, N.O.
ADELANA, B.O. 25172
ADELUSI, S.A. 34533
ADEM, L. 39636
ADEPIPE, N.O. 36609
     see ADEDIPE, N.O.
ADHIKARY, S.P. 36610, 39400
ADJEI, G.B. 37440
ADJEI-TWUM, D.C. 25173
ADLER, K. 25174, 36611
ADMAN, E.T. 36612
ADMIRAAL, W. 28507

                                (continued)

ALVIN, R.   32758
AL'ZHANOVA, R.M.   21561
AMADA, M.   38010
AMANO, H.   32617, 38917
AMANO, S.   28546, 29944
AMANOV, M.A.   23561
AMASAGA, T.   36647
AMBARD-BRETTEVILLE, F.   22264
AMBASHT, R.S.   27893-4, 34498
AMBERGER, A.   B36648
AMBLER, R.P.   21562, 23635, 28232,
     36649-50
AMBROSAŬ, A.L.   21563
     see AMBROSOV, A.L.
AMBROSE, M.J.   37845
AMBROSOV, A.L.   21564
     see AMBROSAŬ, A.L.
AMBY, R.P.   39606
AMELLA, A.   30619
AMEMIYA, A.   23352
AMEND, J.   32754
AMERKHANOVA, M.B.   21565
AMES, I.H.   32618
AMESZ, J.   21566-8, 24903, 25202,
     27653, 28547-51, 29263-5, 31511,
     31534, 32619-20, 35526, 36651-2,
     39318-9, 39822
AMEZAGA,   A. de   24766, 26179, 31988,
     32621, 33722
AMILENI, A.R.   22083
AMIRDZHANOV, A.G.   21569, 25203, 28552
AMIRI, Z.   25204
AMLA, D.V.   36653-4
AMLING, H.J.   36785
AMMA, B.S.K.   see SUMATHY KUTTY AMMA,B.
AMOUR, G.T.S.   32956
AMPOFO, S.T.   25205-7
AMUNDSON, R.G.   26895, 30563
ANAGNOSTAKIS, S.   26336
ANAN'EV, G.M.   32622, 36540, 40330
ANANYAN, A.A.   21570-1
ANDEREGG, R.J.   37625, 38197
ANDERSAG, R.   25208
ANDERSEN, A.   36655, 37217
ANDERSEN, A.S.   25209
ANDERSEN, D.C.   36656
ANDERSEN, F.Ø.   25210
ANDERSEN, J.L.   28553
ANDERSEN, J.M.   21825, 36657
ANDERSEN, K.   32623
ANDERSEN, N.R.   29365
ANDERSEN, R.A.   32624
ANDERSEN, W.   21572
ANDERSEN, W.R.   21573, 30278-9, 32373,
     36658, 39927
ANDERSON, C.E.   24748
ANDERSON, G.C.   25053
ANDERSON, I.C.   24444, 37157
ANDERSON, J.B.   36665
ANDERSON, J.D.   36614
ANDERSON, J.E.   32625, B33281, 36659
ANDERSON, J.F.   *33936*

ANDERSON, J.M.   21574-5, 28554, 28698-9,
     32626, 32889-90, 36660, 37366-7,
     40122
ANDERSON, J.W.   28555-6, 30853, 32627,
     34141, 36661-2, 38889
ANDERSON, L.E.   21576-80, 22274, 22977,
     23949, 25211-4, 28557-60, 29772,
     32628, 33809, 36663-7, 36725,
     36849
ANDERSON, L.L.   22069-70, 25713,
     33156-7
ANDERSON, M.C.   21581
ANDERSON, M.M.   27816
ANDERSON, O.R.   21582, 23527-8, 25215
ANDERSON, R.   21510, 23992
ANDERSON, R.E.   21583-4
ANDERSON, R.J.   21585, 27663
ANDERSON, R.R.   39900
ANDERSON, R.S.   28561
ANDERSON, W.K.   32629
ANDERSON, W.R.   25216
ANDERSSON, B.   25180, 25217, 28514,
     28562, 30527, B32630, 32631-2,
     36625, 38049
ANDERSSON, L.   25918, 36668
ANDO, A.   27720
ANDO, T.   24645
ANDONOVA, P.   36669
ANDRE, C.   21586, 25218
ANDRÉ, M.   23376, 28007, 32633, 36670-2,
     37624
ANDREENKO, T.I.   27535, 33114
ANDREEV, L.V.   30313
ANDREEV, N.G.   32634
ANDREEVA, A.N.   35202
ANDREEVA, A.S.   36673-4
ANDREEVA, N.E.   25219-20, 28563, 32635,
     36675
ANDREEVA, T.F.   21587, 25221, 36676
ANDREO, C.S.   21588, 24856, 25222,
     26967, 28225, 28564, 31461, 32233-5,
     32636, 36677
ANDREOLI, C.   23505
ANDREOLI, M.G.   28565
ANDREU, J.M.   32637
ANDREW, M.H.   36678
ANDREW, P.W.   21589, 29021
ANDREWS, A.K.   21590, 25223
ANDREWS, D.J.   23037
ANDREWS, H.T.   25080
ANDREWS, P.   28566
ANDREWS, P.T.   21591
ANDREWS, R.   25753
ANDREWS, R.A.   26713
ANDREWS, T.J.   21592, 23455, 26333-4,
     26958, 28567, 30636, 32638, 36679
ANDRIANOV, V.K.   25589, 27889, 33329,
     37285
ANDRIANOVA, Yu.E.   32057
     see ANDRIYANOVA, Yu.E.
ANDRIYANOVA, Yu. E.   24715
     see ANDRIANOVA, Yu.E.

ARMSTRONG, F.A.   36707
ARMSTRONG, J.E.   32669
ARMSTRONG, W.   21609, 25780, 36708
ARNASON, T.   25242-3, 31835, 35879
ARNAUTOVA, A.I.   23082, 34318, 38194
ARNESEN, U.   36709
ARNESON, R.D.   21610
ARNHEIM, K.   32670
ARNOLD, C.-G.   27962, 32671, 32872
ARNOLD, K.E.   38499
ARNOLD, M.H.   *33928*
ARNOLD, W.   25244, 32672
ARNOLD, W.A.   28591
ARNON, D.I.   21611, 25104, 26380, 27104,
     28184, 28592-4, 29038, 29070, 29949-
     -50, 32495, 32673-4, 36514, 36710,
     37884
ARNOTT, R.A.   39683
ARNTZEN, C.   38469
ARNTZEN, C.J.   21607-8, 21612-3, 22193,
     24910-1, 25240-1, 25245-6, 25819,
     28587, 28590, 28595, 28884-6,
     28935-6, 28991, 32675, 32934,
     33024-6, 34909-10, 34990, 35947,
     35987-8, 36711, 36993, 37055-6,
     38832-3, 39123-5, 39226, 39311,
     39745, 39762-3
ARO, E.-M.   28222, 28596, 32676, 36234,
     36712
ARONOFF, S.   21614-5, 28597, 32677,
     37057
ARPIN, N.   25247
ARRABACA, M.C.   37028
ARRON, G.P.   21616, 25529, 28910-1,
     36713-4
ARRUDA, J.A.   36715
ARTECA, R.N.   32678, 36716-7
ARTEM'EV, P.N.   32266
ARTIS, P.   28866
ARTYUKH, A.D.   21845, 25480
ARUGA, Y.   24713, 28598
ARYA, R.S.   36266
ARYA, S.S.   36718
ASADA, K.   21617, 23837, 24685, 25248,
     26730, 28084, 28599, *28599*, 32042,
     32679-82, 34229-30, 39441
ASADOV, A.A.   31805, 35846, 39613
ASAHI, T.   36168
ASAI, H.   21597
ASAMI, S.   21618-9, 25249, 28600,
     32585, 32683-5, 34658, 36719-20
ASANA, R.D.   32686, 36721
ASANOV, A.N.   21620, 32687
ASANUMA, K.   38258, 38859-60
ASAY, K.H.   23795-6, 31021, 31433,
     31870, 32912, 36722
ASBELL, C.W.   29793
ASCENCIO, J.   36723
ASCENSO, J.C.   25250
ASCOLINI, A.   29494
ASENSI, A.   28601, 33312
ASGHAR, A.   32688
ASHCROFT, G.L.   33854

ASHCROFT, W.J.   36724
ASHENDEN, T.W.   21621, 28602
ASHIDA, K.   28603, 38201
ASHLEY, D.A.   22489, 26093, 28005, 28785,
     30530, 31606
ASHOUR, N.I.   25297
ASHTON, A.R.   36725
ASHTON, D.H.   36726
ASHTON, F.M.   28604, 29296-8, 36727,
     39232
ASHUROVA, O.B.   25117
ASHWANI   36752
ASHWORTH, E.N.   21622
ASKAROV, M.   32816
ASLAM, M.   28605-6, 32689, 36728
ASLIN, R.   22139
ASLIN, R.G.   33984
ASLING, Kh.S.   21623
ASLYNG, H.C.   21624, 36729
ASNANI, V.L.   39652
ASOEVA, L.M.   22534, 29662
ASPERGES, M.   28607
ASPINALL, D.   25460
ASSAF, R.   21625
ASSCHE, C.J. van see VAN ASSCHE, C.J.
ASSCHE, F. van see VAN ASSCHE, F.
ASSELIN, A.   28608
ASTAUROVA, O.B.   28609-10, 36730
ASTIER, C.   34199, 36731
ASTON, A.R.   33395
ASTON, M.J.   25251, 32690, 33790, 36732
ATABEKYAN, E.A.   27110
ATALLAH, K.   28611
ATANASIU, L.   21626, 25252, 28612, 33491
ATCHISON, B.A.   30171
ATKINS, C.A.   21627, 28613-4, 29509,
     31228, 32691, 38431, 39066
ATKINS, P.W.   *29972*
ATKINSON, D.E.   *37841*
ATKINSON, M.A.   32692
ATLAS, R.M.   21628
ATRASHENOK, N.V.   36733
ATSMON, D.   22978, 33489
ATTIWILL, P.M.   36734
ATWOOD, D.K.   28615
AUBERT, S.   29538
AUBUCHON, R.   22139
AUBUCHON, R.R.   32693, 33984
AUCHTERLONIE, C.C.   31963
AUCLAIR, A.N.D.   25253-4
AUCLAIR, D.   21629, 25255, 28616-7,
     36735
AUCLAIR, J.-C.   31869, 39669
AUDUS, L.J.   *27184*, *28244*, 33270
AUERBACH, S.   32383, 32694
AUFHAMMER, W.   21630, 25256, 28618
AUGSTEN, H.   22319, 29385-6, 34060
AUGUSTIN, P.   25257-8, 32695
AUGUSTINE, J.J.   25259-61, 36736
AUGUSTO, O.   36737
AULD, B.A.   32696
AUNE, J.   28619
AUNG, L.H.   27283, 28620

(continued)

BOWES, J.M.   28903, 32954, 36968-9
BOWMAN, G.E.   37912
BOWMAN, M.K.   39910
BOWN, A.   25516
BOWN, A.W.   29363, 33981, 35925, 39689
BOWREN, K.E.   38930
BOWYER, J.   33251
BOWYER, J.R.   28904, 32955
BOX, E.   21876, 38485
BOXER, S.G.   25517, 35300, 36970-2
BOYADZHIEV, P.Kh.   30487
BOYARSHINOVA, G.A.   25310
BOYCE, C.O.L.   28905
BOYD, C.E.   21877, 25518
BOYER, J.   25519
BOYER, J.S.   21878, 22415, 22830-1,
    22974, 23673, 25520-1, 26007,
    27163, 30820-1, 33524, 34206,
    36526, 38718, 40321-2
BOYER, M.G.   28996
BOYER, P.D.   21879, 25522, *27088*,
    27661, 27933-4, 28906-8, 36282-3,
    37763
BOYER, Y.   21880-2
BOYLE, J.E.   21883
BOYLEN, C.W.   24440, 25523, 31766,
    35797
BOYNTON, J.   25395, 25438
BOYNTON, J.E.   22117, 24962, 25524,
    26144, 26295, 29839, 32361,
    32473, 33686, 36793, 39571
BØYUM, A.   25525
BOZCUK, S.   25526
BOZHKOVA, M.   33352
BOZHKOVA, M.D.   25527
B.PAPP, L.   36973
BRAAMS, R.   33307
BRACH, E.J.   32956, 39126, 39928
BRACKENHOFER, H.   23422
BRACKER, C.E.   26463, 37961
BRADBEEER, J.W.   28912
    see BRADBEER, J.W.
BRADBEER, J.W.   21616, 21884-5, 22594,
    23689, 25528-30, 26361, 27176,
    28330, 28909-11, 29922, 32957-9
    see BRADBEEER, J.W.
BRADBURY, I.K.   22584, 25531, 28913-4,
    32960
BRADFORD, A.M.   31921-2
BRADFORD, J.M.   32961
BRADFORD, K.J.   36974
BRADSHAW, R.A.   30258
BRADY, C.J.   21886, 28915-6, 29795,
    32469, 36121, 39127
BRADY, R.A.   21887
BRAEGELMANN, P.K.   38995
BRAIKOV, D.   27388, 30480, 39968
BRAIN, P.   37847
BRAMANE, A.   28465
    see BRAMANE, A.Ė.
BRAMANE, A.Ė.   28464
    see BRAMANE, A.
BRAMLAGE, W.J.   38805

BRAMMER, E.   35281
BRANCA, C.   28743
BRAND, J.J.   21888-9, 22164, 25532,
    28917, 31438, 36975-7
BRANDÃO, I.   35625
BRÄNDÉN, C.I.   32963, 36978
BRÄNDÉN, R.   32962-3
BRANDERS, C.   39673
BRÄNDLE, E.P.O.   21890
BRANDLE, J.R.   28918-9
BRANDMÜLLER, J.   *28985*
BRANDON, P.C.   28920
BRANDT, A.B.   21891-4, 25533-6, 28921-2
BRANDT, P.   28923-4, 36979
BRANGEON, J.   28925, 29835, 32964,
    35011, 36980
BRANSON, F.A.   25537-8
BRANTON, D.   21895
BRAR, G.   28926
BRAR, G.S.   32965
BRAR, J.S.   32722
BRASSEL, D.   29477
BRASSEUR, F.   25539
BRATT, G.C.   39381
BRAUM, J.G.   21896
BRAUMANN, T.   32966, 36981
BRAUN, G.   28927
BRAUN, H.J.   25540
BRAUN, J.G.   25541
BRAUNE, W.   36982
BRAUTIGAN, D.L.   27965
BRAVDO, B.   21625, 21897, 26246,
    28928-9, 32967, 36983
BRAVDO, B.-A.   28930-1, 34683, 38542
BRAVO, A.   31250
BRAY, E.   36984
BRAYMAN, A.   36985
BREAZEALE, V.D.   27772, 31699, 32968
BREBNER, J.   23538
BRECH, J. Le   37324
BRECHT, E.   36611, 38691
BREDE, J.M.   32520
BREEN, P.J.   35371, 39173
BREEZE, V.   32969
BREIDENBACH, R.W.   25259-61, 25542,
    36736
BREIDERT, D.   32970, 36986
BREITENBERGER, C.A.   36987-8
BREMNER, P.M.   27594, 31463, 32971
BRENDLER, S.   28535
BRENES, E.   33074
BRENNAN, E.   35507
BRENNER, D.   25543
BRENNER, M.L.   30912, 34936-7, 36984,
    37848, 39546
BRESENHAM, G.L.   34814
BRESKOVSKA, Ts.P.   29709
BRESSAN, R.A.   36437
BREST, D.E.   38358
BRETHERTON, G.   36989
BRETON, J.   21718, 21898-900, 22394,
    25353, 25544, 26100, 27072,
                                 (continued)

BRETON, J. (continued)
    28258, 28932-3, 29580, 29595,
    29619-21, 31186, 32280, 36046,
    36272, 36990-1, 37584, 39027-8
BRETON-PROVENCHER, M. 36992
BRETTUM, P. 39854
BRETZ, N. 26324
BREUZÉ, G. 21901, 22748
BREWER, P.E. 36711, 36993
BREWINGTON, G.T. 31631-2, 35669
BREWSTER, J.L. 23864, 36994
BREZEANU, A. 21902, 25545
BŘEZINA, V. 22362
BRIANTAIS, J.-M. 21612-3, 24910-1,
    25240, 28260, 28595, 28867, 28934-
    -6, 30950, 36995, 37706, 40068
BRIAT, B. 39326
BRICK, M.A. 25546
BRICKER, T. 37235
BRICKER, T.M. 36996, 37262
BRIDGEN, J. 25547
BRIDGES, S. 25548
BRIDGWATER, F.E. 33066
BRIGGS, R.E. 29007, 31252
BRIGGS, S.V. 28937
BRIGGS, W.R. 25559-60, 28825, 30145,
    31499, 32873, 35410, 37691-2
BRILL, A.S. 32972
BRIM, C.A. 37059
BRIN, G.P. 23201-2, 26762, 28938,
    29081, 34419, 35075-6
BRINCKMANN, E. 36997, 40213
BRINKHUIS, B.H. 21903, 25549, 28939-40
BRINKLEY, B.R. *28554*, *28849*
BRINKMAN, F.A. 33564
BRINKMAN, M.A. 28941, 32973
BRINKMANN, G. 28942, 32974-5, 36998
BRINKMANN, K. 25550
BRISBIN, I.L. Jr. *34885*
BRISKER, H.E. 25551
BRISSON, J.D. 24216
BRISTOW, J.M. 30623
BRITH-LINDER, M. 28943, 29227, 35565,
    36999
BRITTON, C.M. 25552, 32976
BRITTON, G. 21904, 22616, 25553-4,
    28944-9, 29205, 31349-50, 32441,
    36447, 37000, 37433, 37741
BRITZ, S.J. 21905-7, 25557-60, 28950,
    37001
BRIX, H. 37002
BROADHURST, M.K. 33090
BROCH-DUE, M. 32977-8
BROCK, J.T. 32941
BROCK, T.D. 21707, 21908-11, 23501,
    24998, 25561, 27000, 30696-8,
    30944, 32979, 34394-5, 34971,
    37466, 40338
BROCKINGTON, N.R. 32980
BROCKINGTON, P.J. 33404
BROCKMANN, H. Jr. 22549, 25562-3,
    27631, 32981, 37003-4, 39322
BRODA, E. 21912-3, B21914, *23478*, *24257*,

25567, *25888*, *26053*, 28951-2,
    32982-3, 37005, *40366*
BRODA, H. 37006
BRODERSEN, P. 25564
BRODY, M. 28953-4
BRODY, S.S. 22063, 25565, 25704, 28954,
    34001, 37007, 37879
BROECKER, H.-C. 32984
BROGÅRDH, T. 21916
BROLIN, S.E. 25004
BRONISZ, D. 23646
BRONNER, F. *22233*, *27390*
BROOKER, G. 28108
BROOKS, A.S. 28955, 37008
BROOKSBANK, P. 22577
BROSSEAU, J.D. 36536
BROUÉ, P. 25566
BROUERS, M. 21918, 25567, 28956, 32985,
    37009
BROUGHTON, W.J. 37010-1
BROUWER, R. 21919, 32986
BROUWER, Y.M. 24585
BROVCHENKO, M.I. 25568, 28957, 32987,
    37012
BROWN, A.D. 37013
BROWN, A.P. 22405, 28958
BROWN, A.S. 21920, 24809-10, 28959-60,
    36186, 37014, 39964-5
BROWN, B.O. 37433
BROWN, C.A. Jr. 32988
BROWN, D.A. 27707
BROWN, D.C. 32544
BROWN, D.C.W. 29351
BROWN, D.H. 21699, 28961-2, 32989-90
BROWN, D.M. 26159, 29152
BROWN, G. 27628
BROWN, G.N. 22224, 23616-7, 24213,
    27629, 28919, 28963, 29299-300,
    30833-5, 37301, 38695
BROWN, G.R. *33023*
BROWN, H.E. 21921
BROWN, H.M. 32991, 33779, 37015-6
BROWN, J. 21922, 28964, 36244, 37016
BROWN, J.A. 21923
BROWN, J.C. 37017-8
BROWN, J.H. 28496
BROWN, J.K. 25569
BROWN, J.M.A. 28975, 37024-5
BROWN, J.S. 21924-5, 22450-1, 25570,
    28965-9, 32992-4, 34688, 36245,
    37019, 38548
BROWN, K.M. 34506
BROWN, K.W. 22964, 25571-2, 28123,
    31793, 35827
BROWN, L.F. 28970-2
BROWN, L.M. 32995
BROWN, M.A. 31931
BROWN, P.R. 40329
BROWN, R.G. 39198
BROWN, R.H. 21926, 22110, 22489,
    23072, 23452, 25404-7, 25573-4,
    26093, 26181, 27566, 27737, 28005,
    28785, 29581, 31606, 32867, 32913
                                    (continued)

BULYCHEV, A.A.   25589-91, 28292, 33329,
   37047-8, 37285
BULYCHEVA, T.M.   26977
BUNCE, J. .28342, 32364
BUNCE, J.A.   25592, 27415, 28986-9,
   31230, 33019-21, 37049
BUNCE, N.J.   21954
BUNNELL, F.L.   23654
BUNT, J.S.   21955, 32940
BUNTING, A.H.   25593
BUNTING, E.S.   25594
BUNUS, F.T.   25595
BURBA, M.   34941
BURBAEV, D.Sh.   32885, 37050
BURCH, G.J.   33022
BURCHARTZ, N.   36366
BURCZYK, J.   37051
BURDEN, T.S.
   see SCOTT BURDEN, T.
BURDETT, A.N.   37052
BURDGE, E.L.   37053
BURDICK, D.   25679
BUREN', V.M.   28674
BURENIN, V.I.   22487
BURENKOV, V.I.   30389
BURET, M.   25667
BURGESS, M.D.   21956
BURGH, M.E. de   29460
BURGHOFFER, C.   22135, 25779, 36816,
   37203
BURIAN, K.   25596, 37054
BURIEL, J.F.   25597, 28990
BURINGH, P.   33023
BURKARD, G.   35991
BURKE, J.   38833
BURKE, J.J.   21957, 28595, 28886,
    28991, 33024-6, 35987-8, 37055-6,
   39763
BURKE, S.   37057
BURKETT, R.D.   21958
BURKEY, K.   37731
BURKHARD, C.   31677-8, 33960, 37858
BURKHART, H.E.   38568
BURKHOLDER, P.R.   25598
BURLAKINA, A.V.   22296
BURLEIGH, R.   25599
BURNETT, J.H.   25600, 32850
BURNHAM, B.F.   23583
BURNHAM, J.C.   25601
BURNISON, B.K.   38437
BURNS, A.R.   39878-9
BURNS, C.W.   38752
BURNS, D.D.   28992
BURNS, E.E.   38849
BURNS, N.M.   27206
BURNS, R.M.   29035
BURRELL, V.G.   30735
BURRIS, J.E.   25448, 25602, 28993-4,
   37058, 40158
BURRIS, R.H.   *25359, 25437, 25443-6,*
   25573, B25603, 25640, 25711,
   25787, 25913, 26315, 26526,

26692, 26954, 27097, 27190, 27313,
27359, 27360, *27540, 27780, 28143,*
28154, 28316, 28477, 31278, *32867,*
33027, 34051, 35315, 39725
BURROWS, F.J.   25604
BURSOVÁ, M.   30694
BÜRSTELL, H.   25869
BURSTRÖM, H.G.   25605
BURSZEWSKA-SAMIEC, H.   25155
BURT, R.L.   29377
BURTON, G.W.   33855, 35560
BURTON, H.R.   37219
BURTON, J.H.   38842
BURTON, J.W.   37059
BURTON, W.G.   39430
BURTSEVA, R.A.   25727
BURWELL, C.C.   37060
BUSBY, J.R.   33028-9
BUSCH, G.E.   25606
BUSCH, L.   24216-7
BUSCH, L.V.   22662, 26993
BUSCHBOM, U.   22379-80, 23015, 23297-
    -300, 24376-8, 25970, 26580,
   27768, 29443, 34239, 34512, 38132
BUSCHMANN, C.   21959, 24032, 28995,
   30595, 33030-1, 34606, 37061-2
BUSE, G.   33032
BUSEY, P.   37063
BUSH, T.F.   33033
BUSHEVA, M.K.   38109
BUSHKIT, V.D.   36226
BUSS, G.R.   21960
BUSTRILLOS, A.R.   27382
BUTA, J.G.   21961
BUTCHER, H.   33034
BUTCHER, J.E.   28996
BUTCHER, T.B.   25607
BUTENKO, R.G.   22205, 25265, 29283
BUTLER, D.R.   23295, 32120
BUTLER, E.I.   32887
BUTLER, J.   33307
BUTLER, L.S.   30164-5, 32405
BUTLER, R.G.   22931
BUTLER, W.L.   21664, 21962-5, 23102-3,
   24631, 25293, 25295, 26676-7,
   26914, 27324, 27724, 28027-9,
   28657, 28997-9001, 30586-8, 31625,
   31979-82, 33035-7, 35665-6, 36024,
   37064-5, 38638
BUTT, V.S.   26218, 29002, 29747
BUTTERY, B.R.   29003-4
BÜTTNER, R.   25608, 29005, 31591
BÜTÜN, G.   29121
BUURT, G. van see VAN BUURT, G.
BUVET, R.   *28540, 28670-1, 28750, 28858,*
   B29006, 29018, 29444, 29618, 29796,
   30155, 31174, 31177, 31304, 31341,
   31860, 32157, 32295, 32327, 32444
BUXTON, D.R.   29007, 31252, 34040
BUZAS, Z.   37225
BUZZARD, G.H.   27886
BUZZELL, R.I.   29003-4

BYALKO, A.V. 23996
BYERRUM, R.U. 21966
BYERS, B. 22667
BYERS, T.J. *26415, 27694, 28139, 28385*
BYKAVA, L.M. 24890
BYKOV, O.D. 21967-8, 22487, 25609-11,
    29008, 33038-40, 37066-8
BYNOV, F.A. 39411
BYOTT, G.S. 25612
BYRNE, G.F. 25613
BYSTROVA, M.I. 21969, 25614, 30418,
    32208, 37069
BYSTRYKH, E.E. 21970, 24928, 28282,
    32310-1, 36300, 40097-9
BYSTRZEJEWSKA, G. 33041
BYSZEWSKI, W. 29009
BYTEVA, I.M. 29010
BYTNIEWSKA, K. 33042

C

CABANETTES, A. 33043
CABANGBANG, R.P. 25615
CABANYES, J. de 21971
CADÉE, G.C. 21972, 29011, 33044-5, 37070
CADILHAC, B. 29538
CAGIOTTI, M.R. 32645
CAHEN, D. 25616, 29012, 31105, 33046-
    -8, 33636, 33810, 35143, 38595
CAHET, G. 25617
CAI, J.-P. 34598
CAI, K. 37071
CAILLIAU, L. 36597
CAILLIAU-COMMANAY, L. 32290
CAIRNS, J. Jr. 27940, 29013, 33049,
    33183, 35546
CALDER, J.A. 32669
CALDERONI, A. *33804, 35586*
CALDWELL, C.D. 37072
CALDWELL, D.E. 29014
CALDWELL, M. 21973
CALDWELL, M.M. 22209-10, 22950, 26532,
    27899, 28918, 29015-6, 31853,
    33050-1
CALÈ, M.T. 21974, 24781, 25618, 34100
CALE, W.G.Jr. 37073
CALHOUN, W.T. 24390
CALKINS, J. 31886
CALLAGHAN, C.H. 33052
CALLAGHAN, T.V. 24961, 26912, 33052
CALLAHAN, A.B. *37113, 38410*
CALLENDER, R.H. 37911
CALLIS, J. 33053
CALLOW, J.A. 37074
CALLOW, M.E. 37074
CALMÉS, J. 21975, 24915, 25619-20,
    34410, 34524
CALVAYRAC, R. 21976, 22264, 25621-2,
    29017, 30536, 37075-7

CALVERT, A. 21977, 25623
CALVERT, H.E. 21978, 25624-5
CALVET, P. 30989
CALVIN, G.J. 33054
CALVIN, M. B21979, 21980, 24651,
    25626-7, 29018-9, 31048, 33054-6,
    40229
CAMA, H.R. 27510, *35786, 36885*
CAMACHO-B, S.E. 22650, 29020
CAMARA, B. 33057
CAMBELL, L. 34247
CAMBURN, K.E. 25080
CAMERON, D.R. 29022
CAMERON, G.N. 32950
CAMERON, S.I. 32196
CAMM, E. 37078
CAMM, E.L. 33058
CAMMACK, R. 21583, 21591, 21981-3,
    22374-6, 22658-9, 22979, 24723,
    25628-9, 25963, 25968, 26310,
    26637, 26978, 29021, 29440, 32069,
    34081, 37079-81, 39326, 39385
CAMP, L.B. 29016
CAMP, M.E. 26560
CAMP, R.R. 21984
CAMPBELL, C.A. 29022, 37082
CAMPBELL, C.C.M. 25676
CAMPBELL, D. 37872
CAMPBELL, D.C. 33059
CAMPBELL, D.E. 33060
CAMPBELL, G.S. 25742, B29023
CAMPBELL, L.C. 31244
CAMPBELL, L.E. 21985
CAMPBELL, P. 37029
CAMPBELL, P.J. 33061
CAMPBELL, W.F. 28918, 29024
CAMPBELL, W.H. 33062
CAMPILLO, A.J. 21986, 23153, 24421,
    25630-2, 29025-7, 29620, 33063,
    33649, 35782
CAMPION, A. 29028-9
CAMPIONE, A. 22225
CAMPO, F.F. del see DEL CAMPO, F.F.
CAMURRI, L. 22417, 25633, 26012
CANAANI, O.D. 29030-1, 33064
CANALE, R.P. *26655, B33065, 33357*
CANCEL, L.E. 25634
CANDAU, P. 25635, 27048, 34745
CANELLAKIS, E.S. 25636
CANIĆ, V. 24829
CANNELL, G.H. *38082, 38100*
CANNELL, M.G.R. *25593, B25637, 25780,
    25784, 25989, 26161, 26188, 26348,
    26856, 26883, 26987, 28141, 28474,
    33066*
CANNELL, R.Q. 37083
CANNISTRARO, S. 33067
CANNY, M.J. 23634, 29052
CANTERFORD, R.L. 22014
CANTLEY, L.C. Jr. 21987-9, 25638-9
CANVIN, D. 36983
    see CANVIN, D.T.

CANVIN, D.T.   21627, 21897, 24386,
     25640, 29032, 29328, 29514,
     30622-4, 33068-9, 37084, 37515,
     38663
     see CANVIN, D.
CAPALDI, R.A. *38972*
CAPBLANCQ, J.   21990, 33070-1
CAPDEPON, M.   31482
CAPEL, M.S.   28897, 29033
CAPERON, J.   26867, 33072
CAPIEL, M.   33073-4
CAPLAN, S.R.   22179, 25296, 25804,
     25927, 27828-9, 29034, 29227-9,
     29387-9, 29606-8,   31773,
     32727, *32795*, 33046, 33075,
     33075, 33099, 33277, *33277*,
     33326-7, 33443-4, *33444*, 33636-8,
     33636-8, 33682, 33689, 33949,
     33973, 34028, 34263, 34345,
     34345, 34667-8, 34977, 35208,
     35253, 35312, 35565, *35722*,
     35805, 35913, *35913, 35951*,
     36007, 36321, 36364, 36375,
     37404-5, 37608, 39670-1
CAPLE, M.   29035
CAPLE, M.B.   37085
CAPONE, D.G.   25777, 37086
CAPPELLETTI, E.M.P.
     see PAGANELLI CAPPELLETTI, E.M.
CAPPELLI, C.   23687, 27793
CAPPY, J.J.   31071, 33076
CAPRON, T.M.   25641, 29036
CAPUT, C.   28617
CARAFOLI, E. *36892, 37800, 38938*
CARAPELLUCCI, P.A.   21991
CARBON, B.A.   26213, 37087
CARBONNEAU, A.   25642-3, 37088
CARD, K.A.   24806-7, 28406, 32439
CARDE, J.-P.   33077
CÁRDENAS, J.   29037, 31151-3, 31907
CARDINAL, A.   36992
CARDOZO, G.H.   25644
CARELL, E.F.   21992
CARELLI, M.L.C.   25652
CAREY, P.R.   31583
CARITHERS, R.P.   21993, 25645, 29038,
     32495
CARL, G.   33687
CARLES, J.   32296
CARLIER, G.   21828, 29343
CARLIER, M.-F.   37089-90
CARLIN, S.   32784
CARLING, D.E.   29039
CARLSEN, B.   29040
CARLSON, G.E.   31894, 32582, 33894,
     35933, 36623
CARLSON, J.   34379
CARLSON, P.S.   21994, 23254
CARLSON, R.E.   26318, 29041, 33916,
     37829, 38779
CARLSON, R.W.   21708, 21995, 25646,
     29042, 36817-8

CARLSSON, R.   29043, 37091
CARMELI, C.   21996-7, 22520, 22773-4,
     26388, 29044, 29630-1, 29959-61,
     33078, 33661, 37092, 37894, 39021
CARMER, S.G.   29045
CARMI, A.   29046, 33079, 37093-4
CARMON, M.   26225
CARNEVALE, J.   37095
CARNEY, C.F.   22340
CAROLIN, R.C.   29047, 33080
CAROLLO, A. *33804, 35586*
CAROW, B.   21998
CARPENA, O.   23781, 29048
CARPENTER, D.J.   25647, 37096
CARPENTER, E.J.   22717, 25543, 33081,
     37097, 38659
CARPENTER, S.B.   36637, *36837*
CARPENTER, S.M.   25647, 37096
CARPENTER, W.J.   33082
CARPENTIER, F.   38463
CARPINETTI, R.M.   31563
CARR, J.L.   29049
CARR, J.W.   38239, 38241
CARR, M.K.V.   27736
CARR, N.G.   22363-4, 33482-3, 34199,
     37452-3
CARRATORE, G. del
     see DEL CARRATORE, G.
CARRAYOL, E.   22135, 22213, 25779
CARRETO, J.I.   25648
CARRIER, J.M.   29050-1, 37098-9
CARSON, J.L.   33083
CARSON, S.D.   25649
CARSON, W.Z.   30735
CARSTAIRS, A.G.   33084
CARTER, D.P.   39746
CARTER, E.D.   39636
CARTER, J.L.   21999, 22502
CARTER, J.V.   37949
CARTER, N.W.   35399
CARTIER, L.D.   24956
CARTWRIGHT, B.   28899
CARTWRIGHT, S.C.   29052
CARVALHO, F.I.F. de   33085
CARVER, M.A.   30066
CARY, J.W.   29053-4
CASADIO, R.   23615, 28640, 30831-2,
     33086, 34841
CASADORO, G.   29055-6, 33087-9,
     37100-1, 39260
CASCIO, B.L.   see LO CASCIO, B.
CASEY, R.P.   35728
CASEY, S.   36510
CASHMORE, A.R.   25650, 33090
CASPER, S.J.   33091
CASPERS, N.   22000, 29057
CASSEL, D.K.   33092, 38123
CASSELLS, A.C.   33093
CASSELTON, P.J.   25016-7
CASSIER, A.   31292
CASSIM, J.Y.   21714-6, 25351, 35228-9,
     38817

CLOUGH, B.F.   22097
CLOUGH, J.M.   33195, 37171-3
CLOUTIER, Y.   40013
CLUTTERBUCK, B.J.   33196
CLYMO, R.S.   37174
COATS, G.E.   25744
COBB, A.H.   24997, 25745, 29140-1,
    33197-8
COBBS, M.R.   24672
COBBY, J.M.   27687, 39565
COCHRANE, L.A.   33199
COCITO, C.   37175
COCK, J.H.   30565, 37176
COCKBURN, W.   21885, 22098-9, 29142,
    30937, 37177
COCKING, E.C.   *25912*, 29612, 33643-5
COCQUEMPOT, M.F.   37178
COCUCCI, S.   28730
CODD, G.A.   23453, 24293, 26866, 28009,
    29143-6, 29836-7, 31590, 31963-4,
    33200-1, 33883, 36002, 37179,
    38955
CODDINGTON, A.B.   37594
COE, E.H. Jr.   40121
COEN, D.M.   29147, 32902, 33202, 33773,
    34618-20
COFFELT, R.J.   22008
COGDELL, R.   36272
COGDELL, R.J.   22100-1, 23965-6, 24227,
    25051, 25076, 25746, 27173,
    29148-9, 33203-4, 33252, 34687,
    36484, 37180-1
COGGINS, C.W. Jr.   22102, 37775
COGGINS, J.R.   34196
COGHI, E.   28743
COGNIAUX, F.   37306-7
COHEN, A.S.   37182, 39168
COHEN, C.E.   23291, 25132, 25747-8,
    26842, 29150, 33205, 34327, 35480,
    37183, 39279
COHEN, D.   2 1713, 25749, 26141
COHEN, W.S.   22103, 22106-7, 25750,
    33206-7, 35264, 37042, 37186,
    39075
COHEN, Y.   22104-5, 25430, 29151, 30575,
    38086, 38790
COHEN-BAZIRE, G.   22545, 29152, 31937,
    32054, 37030
COHN, D.E.   22103, 22106-7
COÏC, Y.   29153
COJENEANU, N.   33208
COKE, L.   37184
COKER, G.   37185
COLBEAU, A.   38709, 39078
COLBOW, K.   24371, 25751, 27763
COLDEA, G.   33209
COLE, D.F.   22108
COLE, D.R.   25752
COLE, E.R.   37095
COLE, H. Jr.   24256
COLE, J.   33210
COLE, K.   33211, 35793

COLE, R.M.   37186
COLEMAN, D.C.   25753, 35243
COLEMAN, J.R.   33218
COLEMAN, W.K.   25754
COLES, S.L.   29154
COLIJN, F.   22109
COLIN, H.   33829
COLLARD, R.C.   29155
COLLATZ, G.J.   28645, 29156, 30919,
    33212-4, 34945, 37187
COLLATZ, J.   25755
COLLETTI, C.   39834, 40181
COLLIER, B.D.   23654
COLLINS, C.R.   33215
COLLINS, D.   33216
COLLINS, M.L.P.   25756-7
COLLINS, N.   22110-1
COLLINS, N.J.   27302, 33052
COLLINS, O.D.G.   29157
COLLINS, W.B.   25758, 29158
COLLINS, W.J.   28566
COLLINSON, M.E.   22016
COLLOS, Y.   31869, 35911, 37188, 39669
COLMAN, B.   22609, 22876, 25759, 26219,
    26478, 27377, 28808, 29159-60,
    30797, 33217-9, 36897, 37189,
    38655
COLMAN, P.M.   33220
COLMAN, R.L.   22112, 37190
COLOMBETTI, G.   25688
COLOMBO, P.M.
    see MARIANI COLOMBO, P.
COLOWICK, S.P.   *21577, 21937, 22562,
    22699, 22713, 22802, 23598, 24492,
    25033, 26248, 34680, 35978, 38404,
    38495, 39504*
COLQUHOUN, A.J.   22113-4
COLVILL, K.E.   35176
COLWELL, G.L.   25760
COLWELL, R.R.   *24073*, 32258
COLWICK, R.F.   22749
COMBE, L.   37191
COMBS, W.S. Jr.   38686
COMERIO, G.   33533
COMPTON, W.A.   31273
CONCIN, R.   33221
CONDE, L.F.   22116
CONDE, M.F.   22117
CONDON, B.N.   26998
CONE, R.A.   36380, *37167, 37854, 39782*
CONESA, A.P.   37192
CONJEAUD, H.   25761-2, 29161, 37193,
    38639
CONKLIN, P.J.   25763
CONNER, A.J.   37194
CONNOR, D.J.   22118, 24826, 29162
CONNORS, R.E.   22088, 25733-5, 29124,
    33185
CONOVER, C.A.   25764, 29155, 29163,
    38065
CONRAD, C.D.W.   30578
CONRAD, D.D.W.   25793

(continued)

CZYGAN, F.-C. 22170, 25795, 29256,
    34457

D

DABADGHAO, P.M. 31748
DA CRUZ, G.S. 33270
DADAY, A. 29211, 38393
DADYKIN, V.P. 22171, 28275, 29212
DAFT, M.J. 31965
D'AGOSTINO, G. 36696
DAGUENET, A. 32633, 36670
DAHL, B.E. 40233
DAHL, J.S. 24034
DAHLHOFF, J.A. 21551
DAHLSTEDT, L. 31528
DAILEY, F. 33271, 33273
DAILEY, L. 35958
DAĬNEKO, M.M. 24314
DAINTY, J. *28681, 28934, 29758,* 30653-
    -5, 32319, 32477, 34033
DAITO, H. 28546
DAKWA, J.T. 29213
DAL', E.S. 23511, 27026
DALE, A.D. 22172
DALE, J.E. 29485, 30844, 40216
DALE, R.F. 23426, 34622, 35055,
    39799
DALEY, L. 33271
    see DALEY, L.S.
DALEY, L.S. 25796-7, 29214-7, 33272-3,
    37234
    see DALEY, L.
DALEY, R.J. 27988, 32999, 33274
DALGARN, D. 29218, 37235
    see DALGARN, D.S.
DALGARN, D.S. 37236
    see DALGARN, D.
DALGARN, M.C. 22173
DALLACKER, F. 28165
DALL'ACQUA, F.M. 38071
DALLE, J.F. 38463
DALLING, M.J. 31263, 35294, 39098
DALL'OLIO, G. 25995, 29465, 36249
DALTON, C. 29219
DALTON, C.C. 25798, 2922 0
DALY, J.M. 21783, 25799
DAM, K. van see VAN DAM, K.
DAM, R.J. 22174, 25800
DAMASCHUN, G. 36041
DAMASCHUN, H. 36041
DAMBYN, M. 33830
DAMISCH, W. 25801, 29221-3
DAMON, P.E. 29224
DAMSZ, B. 22175-6, 25802, 37237
DANCSHÁZY, Zs. 25803, 29225, 30239,
    33275-6, 35185
DANDONNEAU, Y. 29226, 29916, 37238
DANELYUS, R.V. 28515, 32588
    see DANIELIUS, R.V.

DANFORTH, W.F. 22177
DANG XUYEN NHU 35061
DANIEL, M. 36649
DANIEL, R.S. 24533, 31856
DANIELIUS, R. 36626
DANIELIUS, R.V. 32587, 36627
    see DANELYUS, R.V.
    see DANIELIUS, R.
DANILOVA, T.A. 32603
DANKANITS, E. 22178
DANKOV, T. 38768
DANNIGKEIT, W. 37239
DANNOWSKI, M. 39461
DANON, A. 22179, 25804, 29227-9, 33277
DANTUMA, G. 37240
DANYLUK, R.P. 25751
D'AOUST, A.L. 33278
D'AOUST, B.G. 28573
DARBYSHIRE, B. 37241, 39761
DARBYSHIRE, J. 29230
DARKANBAEVA, G.T. 32307
DARLEY, W.M. 22791, 25805-6, 40184
DARLING, M.S. 25807
DARMANADEN, J. 21872
DART, P.J. 31996
DARWINKEL, A. 33279
DAS, B.K. 35759
DAS, G. 22180-1
DAS, K. 24141
DAS, M. 22182
DAS, T.P. 29080, 33120
DAS, V.S.R. 22183, 24139, 24159-63,
    27550-6, 27580-1, 27585, 29231-2,
    30695, 31416-21, 33280, 35426-34,
    35442, 37242-4, 39237-8, 39431,
    40058
DAS GUPTA, D.K. 25808-9, 29233-5
DASH, M.C. 22184, 25716, 25810
DASHKEVICH, A.M. 22185
DASHKEVICH, E.M. 23226
DA SILVA, J.F. see SILVA, J.F. da
DA SILVA, J.V. see VIEIRA DA SILVA, J.
DASS, A.D. 26245
DASTE, P. 23808, 26127
DATIASHVILI, N. 22073
DATTA, S.K. de 27325
DAUBENMIRE, R. 25811
DAUDET, F.-A. 26618
DAUN, J.K. 25812
DAUNICHT, H.J. 23371, 29236
DAUSSANT, J. 23321
DAUTA, A. 33070
DAUTKULOV, A.D. 29237
DAVENPORT, D.C. 22186-7, 25813, B33281
DAVENPORT, H.E. 29238
DAVENPORT, J.W. 37245
DAVEY, C.B. 31032
DAVEY, M.R. 25814, 26632-3, 28398,
    38033
DÁVID, A. 25947
DAVID, K.A.V. 29239-40
DAVIDOVICH, M.A. 34361
DAVIDSON, B. 23295
DAVIDSON, D. 26931

DILOV, Kh. 33351-2
DILOVA, S. 32502, 33353, 37309, 37998
DILOVA, S.A. *B39164*
DILUNG, I.I. 23012, 30117
DILWORTH, M.F. 37310
DI MARCO, G. see MARCO, G. di
DIMITRIJEVIĆ, J. 22918
DIMITROV, G. 37620
DIMITROV, Kh. 22234, 29306
DIMITROVA, A. 39995
DIMITROVA, M. 29306
DIMITROVA, O. 28219, 37311
DIMITROVA, O.D. 39166-7
DIMON, B. 22521, 37312
DIMOV, A. 34443
DINANT, M. 22235, 32411, 39862
DINAR, M. 37313
DINER, B. 22236, 25861, 29307
DINER, B.A. 29308-9, 33354, 37314-6
DINGER, B.E. 31470, 34690
DINUR, U. 37911
DI PASQUALE, G. see PASQUALE, G. di
DI PIETRO, A. see PIETRO, A. di
DIRKS, W. 22237, 35515
DI SEMPIO, C. see SEMPIO, C. di
DISKIN, Y. 24063
DISMUKES, C. 29310
DISMUKES, G.C. 33355-6, 37546, 37565
DISTLER, E. 32937, 33442
DI TORO, D.M. 22238, 24741, 29311,
    33357
DITTO, C. 39124
DITTO, C.H. 25245-6
DITTO, C.L. 28991, 33024, 36711,
    38833
DITTRICH, P. 22239, 25862-3, 29312-3,
    31442, 37317-9
DIVAKAR, N.G. 38511
DIXON, K.R. 33358-9
DIXON, T. 36347
DJAVADI-OHANIANCE, L. *25267*
DJELEPOV, K. 25864
DMITRIEV, A.P. 33360, 37320
DMITRIEV, V.P. 36628, 37321
DMITROVSKIĬ, L.G. 29314, 32886, 33714-5,
    37666
    see DMITROVSKY, L.G.
DMITROVSKY, L.G. 28833
    see DMITROVSKIĬ, L.G.
DOBBEN, W.H. van see VAN DOBBEN, W.H.
DOBBERSTEIN, B. 29315
DÖBEREINER, J. *32867*
DOBIE, J.B. 22008
DOBIS, E. 30888
DOBREN'KOVA, L.G. 36213
DOBRENZ, A.K. 25546, 26055, 38438
DOBRETSOV, G.E. 36739
DOBREVA, S. 24619, 28018-9
DOBREVA, S.I. 28020
DOBRINSKIĬ, L.N. 33361
DOBROVOL'SKAYA, M.G. 32058
DOBRYNIN, S.A. 34778

DOCKERTY, A. 29316-8, 37322
DODA, D.D. 33362
DODD, J.D. 25552, 32976
DODD, J.L. 37323, 38423
DODELET, J.P. 37324
DODGE, A.D. 22240, 22789, 27387, 27637,
    31194-5, 39038-9, 40319-20
DODGE, J.D. 22241
DODSON, A.N. 36117
DOEHLERT, D.C. 33363, 37325
DOEMEL, W. 24998
DOEMEL, W.A. 21911
DOERING, E.J. 37518
DOHÁNYOS, M. 25865
DOHERTY, A. 37326
DOHERTY, J.H. 33010
DÖHLER, G. 25866-70, 37327
DOI, M. 37967-8
DOI, Y. 24851
DOIKOVA, M. 25871
DOKIĆ, P. 25872
DOKULIL, M. 28561, 33364-7, 37328-30
DOLAN, E. 30255, 39614-5
DOLEY, D. 25873, 29319, 33368
DOLGIKH, T.A. 38697
DOLGORUKOVA, T.V. 26928
DOLINER, L.H. 37331
DOLL, S. 25874
DOLPH, G.E. 29320
DOLPHIN, D. 22242, *32981*, *38048*
DOLZHIKOVA, N.M. 29321
DOMAN, N.G. 22054-5, 23754, 25699,
    25875, 26233, 26725, 27007, 29098,
    29322, 30990-1, 32075, 33148-9,
    33807, 35592-3, 38633, 39526,
    39876
DOMAN'SKA, G. 29323
DOMANSKAYA, I.N. 33578
DOMBROVSKIĬ, Yu.A. 37332
DOMES, W. 28623
DOMEY, J. 38575
DOMINGUEZ, C. 33369
DOMINIJANNI, R. 28055
DOMINY, P.J. 33370, 37333
DOMMERGUES, Y.R. 21666
DOMNINSKIĬ, D.A. 37980
DÖMÖTÖR, Z. 26201
DON, R. 36431
DONAGHAY, P.L. 33371
DONCHEV, Kh. 29324
DONDERSKI, W. 23646
DONDEYNE, P. 25846
DONDO, V.V. 31855
DONE, J. 28555-6, 30853, 32627
DONKIN, P. 25876
DONOV, V. 29325
DONOVAN, P.A. 32699
DONTSCHEV, T. 25877
DONTSOV, V.V. 32640
DONTSOVA, I.G. 31991, 39805
DOO, A. 39689
DOODSON, J.K. 33372

DOOHAN, M.E.   25878, 29326
DOOLITTLE, W.F.   21846, 30531
DOORLEY, P.F.   33838
DOR, I.   29327
DORAISWAMY, P.C.   37334, 37898
DORAVARI, S.   29328
DÖRING, G.   22243-4, 25879-80, 27619
DÖRING, H.W.   23372, 23568
DORNE, A.M.   30534
DÖRNEMANN, D.   34340, 38225
DORNESCU, D.   39143
DORNHOFF, G.M.   25881
DOROBANŢU, N.   22245
DOROGOSTAISKAYA, E.V.   22053
DOROKHOV, B.L.   22246-7,   30090, 39563
DOROZHKINA, L.A.   27918
DORPEMA, J.W.   22733
DORRIS, T.   32423
DORSKY, D.   22307
DORSMAN, A.   29329
DORTCH, O.   23947
DOSE, K.   35705, 38825, 39483
DOSPEKHOV, B.A.   *30396*
DOSTAL, H.C.   24071, 27148, 27159,
      30883
DOSTER, W.   35486
DOTLOV, M.M.   35779
DOTSON, L.   29330
DOUCE, R.   22967, 24049, 25701, 26553-4,
      29101, 30177, 31034-5, 33373, 33938,
      34975, 35854, 37140, 37203, 37335-7,
      37849, 38092-3, 38812, 39623
DOUCHA, J.   22248, 25121, 25882,
      38344
DOUGHERTY, C.T.   26409, 26847, 31706
DOUGHERTY, P.   22139
DOUGHERTY, P.M.   22761, 26857, 37338,
      37874
DOUGLAS, F.   24712
DOUGLAS, I.N.   39326
DOUILLARD, R.   33374, 37339-41
DOUSSINAULT, G.   38463
DOVBYSH, E.F.   31991
DOVBYSH, K.P.   24643, 28039, 36032,
      39805
DOVNAR, V.S.   27830, 33375, 37342
DOW, C.S.   32409
DOWLING, J.E.   *37167, 37854, 39782*
DOWNS, J.D.   32470
DOWNS, R.J.   24748
DOWNTON, W.J.S.   22249-52, 23219, 25883,
      26776, 29331, 39524
DOYLE, A.D.   37343
DOZIER, B.J.   24090
DRABER, W.   36175, 37344, 39956-7
DRACHEV, L.A.   22253, 25884, 32558,
      33275-6, 33376, 37345, 40094
DRAFFAN, A.G.   35596
DRĂGHICI, L.   34076
DRAKE, B.G.   25885-6, 29332-3, 37269
DRAKE, G.M.   39649
DRAKE, S.H.   34444

DRAPKIN, V.Z.   27683-4
DRASKOVITS, R.M.   22254, 25887, 37346
DRAYCOTT, A.P.   33398
DRECHSLER, Z.   21744-5, 25375, 38888
DRENNAN, D.S.H.   29334
DRESSEL, H.   30440
DREW, A.P.   22255, 26884, 29335, 30556,
      33377, 37347
DREW, E.A.   22256-8, 29336-7, 33378-9,
      37348
DREW, M.C.   29338, 37349
DREWA, G.   25129
DREWS, G.   22230, 22259, 22499, 23825-6,
      23880, 25888-9, 26660, 29339,
      29505, 33380-1, 33516, 35725,
      35968, 37350, 37485, 37675, 39502,
      40157
DREZNER, W.   22260
DRIESEL, A.J.   31674, 33256
DRIESSCHE, R. van den
      see VAN DEN DRIESSCHE, R.
DRIESSCHE, T. VANDEN
      see VANDEN DRIESSCHE, T.
DRING, M.J.   22261, 23475, 29340
DRINKWATER, A.   29485
DRISSLER, F.   22642, 33824
DROBA, M.   25890-1, 33382-3, 36405
DROBNY, J.   34074
DROMGOOLE, F.I.   28975, 33384-5, 37024-
      -5, 37351
DROOP, M.R.   *38374*
DROPPA, M.   30016, 31049, 33499, 34043-
      -4, 37280, 37723, 37919
DROSTE, T.   37746
DROUET, A.G.   29846
DROUIN, R.   39900
DROZD, H.   23456
DROZDOV, S.N.   35547, 36156, 37352
DROZDOVA, I.S.   24933, 25892-3, 32315,
      32317
DROZDOVA, N.N.   22262, 23203, 33386
DRUART, J.-C.   33535
DRUCKER, H.   *33446*
DRUCKMANN, S.   35817, 37353
DRUET, P.   *25421*
DRUMM, H.   21933, 29567, 33012
DRURY, L.N.   30882
DRUZHKO, A.B.   24211, 27626
DRYAGINA, I.V.   37519
DU Yuan-Shou   36343
DUAN Xu-Chuang   36581
DUBACQ, J.-P.   30028, 33387, 34799,
      35400, 37371, 37925, 39197, 39958
DUBAY, C.I.   37354
DUBBE, D.R.   33388, 33505
DUBÉ, P.A.   22263
DUBERT, F.   38616
DUBERTRET, G.   21976, 22264-5, 29017,
      29341, 30536, 32416, 33389-90,
      37076
DUBEY, P.S.   22266, 31395, 35454
DUBININ, N.P.   *26831*

DWYER, D.D.   24044
D'YACHENKO, A.P.   33410
D'YACHENKO, O.Z.   28883
D'YACHKIN, I.I.   22296
DYCK, L.A.   37379
DYER, A.F.   22147, 25908
DYER, M.I.   37290-1
DYKES, M.G.   37379
DYKYJOVÁ, D.   22297, 25909, 29362,
      33411-2, *33411-2, B33413, 33536,
      33703, 34387, 34486-7, 34768,
      35168-70, 35173-4, 35500, 35607*
DYLYANOK, L.A.   38190
DYMINA, G.D.   25910
DYMOCK, I.   25516
DYMOCK, I.J.   29363
DYNESIUS, R.A.   29364
DYRSSEN, D.   *29365*
DYURIAN, I.   35191
      see GYURJÁN, I.
DYUR'AN, I.   30991
      see GYURJÁN, I.
DYUSOZH, K.   29658
      see DUSOGE, K.
DYUTIN, K.E.   37380
DZAGNIDZE, D.K.   29366-7
DZEVYATAŬ, A.S.   29368, 37381
DZHABIEV, T.S.   30399
DZHAFAROV, M.I.   39138
DZHAGAROV, B.M.   29369, 33414
DZHAIANI, G.I.   27705
DZHANUMOV, D.A.   22482, 28840, 32893,
      34348, 36923, 38231
DZHUMANKULOV, Kh.D.   39138
DZIĘCIOL, U.   33415
DZNELADZE, A.A.   22298-300
DZOTSENIDZE, Ts. G.   31599
DZYARUGINA, T.F.   27248
DZYBOV, D.S.   37382
DZYUBENKO, V.S.   30909-10

E

EAGLES, C.F.   25189
EAGLESHAM, A.R.J.   31996, 34895
EAKS, I.L.   29370
EARLE, E.D.   40132
EASTERBY, J.S.   27297, 31089, 31351,
      40270
EASTIN, J.D.   33192, *37383*
EASTON, H.S.   33416
EASTON, J.   37586
EASTWELL, K.C.   36801
EATON, G.W.   23406
EATON, W.A.   26401
EBARA, H.   34118
EBELING, D.E.   32544
EBRALIDZE, Sh.S.   27705

EBREY, T.   36160, 37911
EBREY, T.G.   25191, 25352, 28722, 30055,
      32781, 33752, 34077-8, 36161
EBRINGER, L.   24544, 29371, 33417-8
ECK, H.V.   37384
ECKARD, A.N.   30250
ECKARDT, F.E.   22301-2, 25911, 29372-3
ECKERT, H.-J.   35503, 37385
ECKSTEIN, F.   31987, 39795
ÉCOCHARD, R.   37386, 39080
EDDLEMON, G.K.   33675
EDELMAN, K.   35681
EDELMAN, M.   24128, 24130, 27698, 29374,
      31545, 33419, 35497-9, 40161-2
EDELMANN, K.   31986, 39471, 39795
EDENS, J.I.   35533
EDER, A.   37387, 39429
EDER, F.A.   37388
EDER, J.   33420
EDGE, E.A.   36348, 40241
EDGERTON, M.E.   33421
EDLER, L.   37606
EDMEADES, G.O.   37389-90
EDMISTON, J.   36779
EDMONDSON, W.T.   22599, 29375, 33422
EDMUNDS, L.N. Jr.   22303-4, 33423,
      38426
EDRICH, J.   21634
EDRICH, J.A.   25264, 28624
EDWARDS, G.   22833, 29376, 33424
      see EDWARDS, G.E.
EDWARDS, G.E.   22635, 22834-40, 23231,
      23586, 24164-5, 25912-3, 26249,
      26441-5, 26785-6, 27582-4, 30034-8,
      30348, 30453-7, 31449-52, 31913,
      33363, 33425-7, 33663, 33665, 34451-
      -3, 35704, 35766, 35950, 36713-4,
      37325, 37391-3, 38340-1, 39333,
      39724-6, 39809
      see EDWARDS, G.
EDWARDS, J.H.   26421
EDWARDS, K.   31498
EDWARDS, M.B.   37394
EDWARDS, M.R.   23010, 26576, 34324,
      34698
EDWARDS, P.   36510
EDWARDS, P.A.   25914, 32375
EDWARDS, R.A.   28666
EDWARDS, R.R.C.   33428
EDWARDS, R.W.   26130
EDYE, L.A.   29377
EFFERTZ, B.   24994
EFIMOV, M.V.   21948
EFIMTSEV, E.I.   22305-6, 25915-6, 26935-7,
      29378, 33429-30, 34627, 36937
EFIMTSEVA, É.P.   33430
EFREMOV, D.F.   21541
EFREMOVICH, N.V.   27676
EGAMI, F.   29379
EGAN, A.R.   *21824*
EGAN, J.M. Jr.   22307
EGARA, K.   29380

EVANS, B. 25400
EVANS, E.H. 22363-4, 25963-4, 26376,
    29438, 31391, 33252, 33482-4, 34199,
    34744, 37452-4, 37723 39198
EVANS, I.M. 33908
EVANS, J.O. 29024
EVANS, L.T. *21951*, *22282*, 22365-9,
    22367-9, *22419*, *23587*, *23695*,
    23736, *23975*, *24444*, 25765, 25965-
    -7, 28842, 30310, 31896, 33226
EVANS, L.V. 36204, 37074, 38178
EVANS, M.C.W. 21983, 22370-6, 22658,
    23799, 24706, 25963, 25968, 29439-
    -42, 29851, 29880, 30138, 31857-8,
    33482, 33485-7, 33902, 33930-1,
    36095, 26430, 36704, 37454, 37840,
    38927, 39386-8
EVANS, N. 22377
EVANS, P.K. 37455
EVANS, P.S. 33488
EVANS, T.A. 22378
EVANS, W.C. 33401
EVANS, W.R. 27440
EVDOKIMOV, V.M. 28199
EVDOKIMOVA, I.V. 25969, 26740, 30395,
    34399-400
EVENARI, M. 21695, 22379-80, 23015,
    23297-300, 24376-8, 25970, 26580,
    27768, 28407, 29443, 34239, 34512,
    38132
EVEN-CHEN, Z. 33489
EVENSON, J.P. 25971
EVENSON, P.D. 37456
EVERS, A. 37457
EVERS, A.K. 22381
EVERSON, E.H. 29823
EVERSON, R.G. 25448
EVERT, F. 25972
EVERT, R.F. 24831, 33975, 40146-7
EVSTIGNEEV, V.B. 21725-6, 22382-6,
    22567-8, 22621-2, 23089, 23100,
    23720, 24622, 24936-7, 25361, 25973-
    -6, 26614, 26657-8, 26670-3, 26751,
    28021, 28290, 28737, 29390, 29444-
    -5, 29694-5, 31794, 32785, 33490,
    33800, 34336, 35046, 35832, 35839,
    36040, 36311, 36836, B37458,
    37459-60, 37682, 37743, 38108,
    38204, 38309, 38328-9, 38876,
    39610
EVSTIGNEEVA, R.P. 34900
EVSTRATOV, A.V. 39975
EWEL, K.C. 38754
EYCK, L.F. ten see TEN EYCK, L.F.
EYLES, J.C. 22387
EYTAN, E. 31028-9, 35053, 38882
EZE, J.M.O. 29446
EZHOVA, T.A. 29447, 37461

F

FABBRI, E. 21649-50

FABBRI, F. 21850, 22388, 28870, 36946
FABIAN, A. 29277, 36284
FABIAN-GALAN, G. 28612, 33491
FABRI, R. 29448
FABRIS, G.L. 22389
FADAYOMI, R.O. 27880
FADEEVA, L.G. 37462-3
FADEEVA, L.M. 29449
FADIA, V.P. 29450
FADRUS, H. 33492
FAGERBERG, W.R. 29451
FAGGIANI, R. 39202
FAHLBUSCH, K. 37873
FAHY, P.C. 36184
FAIR, P. 24734, 33493
FAIREY, N.A. 37390, 37464-5
FAĬZIEV, Sh. 36777
FAJER, J. 22242, 22390, 25488, 29452-4,
    32114, 33494, 33609, 36118, 37254-5,
    38095
FAJSZI, Cs. 22161, 25790
FAKOREDE, M.A.B. 29455, 33495
FAKOUSSA, R.M. 35131
FALK, H. 22391, 22888, 24368, 25977,
    27759, 28401, 31685
FALK, J.H. 25978
FALK, R.H. 25979
FALKENBORG, D.H. *22592*, *24456*, *24992*,
    25537
FALKENFLUG, V. 34954
FALKNER, G. 25980-1
FALKOWSKI, M. 29456-7
FALKOWSKI, P.G. 22392, 33496, 35204,
    39321
FALLER, N. 25982
FALLON, R.D. 37466
FAL'TSMAN, A.V. 33497
FALUDI-DÁNIEL, Á. 22393-5, 22815-6,
    23746, 25838, 26101, 27211, 28792,
    29458, 29578, 30016, 31049, 31551,
    33143, 33498-9, 34043-5, 34155,
    37919
FAM TKHAN' KHO 29459, 31975
FAN, C. B37906
FAN I-JI 25983-5, 37467
    see FÀN YÌ-ji
FAN, L.T. 24454, 31789
FÀN YÌ-ji 25986
    see FAN I-JI
FANG, H.L.-B. 39907
FANG, S.C. 26889, 33000, 38588-9
FANICA-GAIGNIER, M. 22093-4
FANIMOKUN, V.O. 34533
FANKBONER, P.V. 29460
FANSHAWE, N.C. 25499
FANTINET, M. 29461
FARAH, S.M. 22396
FARARD, D.M. 37468
FAREED, M. 33050
FARES, Y. 33500
FARINEAU, J. 22397-400, 25987-8, 28189,
    29462-3, 30537, 30936, 33501-2,
    34958, 38798
FARINEAU, N. 22401, 27667, 33503

GIVAN, C.V.   22540-1, 26152-3, 29393,
     33450, 37411, 38080
GIVNISH, T.J.   26154, 33695
GIZIŃSKI, A.   23646
GJERSTAD, D.H.   22226, 22542, 33346
GJESSING, Y.T.   22543
GJØNNES, B.   25956
GLACOLEVA, T.A.   33696
     see GLAGOLEVA, T.A.
GLADUNOV, I.M.   38635
GLADYSHEV, N.P.   33697-8
GLAESER, R.M.   33924
GLAGOLEVA, T.A.   22544, 25527, 26155,
     30223-4, 33699, 40332
     see GLACOLEVA, T.A.
GLAMM, A.B.   36950
GLASBERGEN, J.M.   28551
GLASE, J.C.   33700
GLÄSER, M.   25062, 26156, 31487-8,
     31666
GLASS, G.   28354
GLASS, R.W.   26157
GLASZIOU, K.T.   21951
GLATKI-JOŠT, M.   26552
GLAZER, A.N.   22545-6, 25579, 26158-60,
     29152, 29670-2, 30588, 36429,
     36504, 37651-2, 38386
GŁAŻEWSKI, S.   22547
GLAZOVA, M.V.   26853
GLEAVES, J.T.   36963
GLEAVES, T.   29858
GLEBA, Yu.Yu.   38288
GLEBOVA, N.T.   24770
GLEESON, A.C.   29164
GLENN, R.K.   28604, 29298, 36727, 39232
GLERUM, C.   26161
GLIDEWELL, S.M.   22548, 24170-2, 26162,
     31457, 37653
GLIEM, G.   27967, 33701, 35959-60
GLIME, J.M.   37654
GLIWICZ, Z.M.   26163, 37655
GLOCK, H.   31957
GLOE, A.   22549, 25563, 33702, 39322
GLOOSCHENKO, V.   22550, 29673
GLOOSCHENKO, W.   22550
GLOOSCHENKO, W.A.   22551, 29674
GLORY, M.   32242, 37656, 40024
GLOSER, J.   26164, 29675-6, 33703,
     35171-3, 37657-8
GLOVER, H.   22552, 29677
GLOVER, H.E.   25346, 34978, 37659
GLOYNA, E.F.   25112
GLUSHCHA, N.I.   37320
GLUTH, G.   33704
GLYNNE-JONES, E.   22553
GNAIGER, E.   33704
GNANAM, A.   22554, 23242, 24919-22,
     26165, 27789, 30082, 33705, 37616,
     38334, 38770-1
GNANARETHINAM, J.L.   22555, 38685
GNAUCK, A.   24989, 26166
GNAUCK, A.H.   24990, 28359

GOATLY, M.B.   22556
GÖBEL, E.   26167
GÖBEL, F.   33706-7
GÖBL, F.   32802
GOCHEV, A.D.   33708, 37660-1
GOCKE, H.   29678
GODBY, E.A.   35564, 35929
GODDEN, D.A.   24090
GODIK, V.I.   28878, 29679, 33709,
     37662-3
GODINOT, C.   29618
GODLEVSKA-LIPOVA, A.   29658
GODLEWSKA-LIPOWA, W.A.   29937
GODZIEMBA-CZYŻ, J.   22557
GOEDHEER, J.C.   22558-9, 24729, 26168,
     28234, 29680-2, 31638, 33258,
     33341, 39880
GOELLER, H.E.   28648
GOERING, C.E.   21658
GOERING, J.J.   27103, 27250, 38031
GOESCHL, J.D.   26169, 33321, 33500
GOFF, L.J.   26170
GOFF, N.M.   33509
GOFFER, J.   21744
GOGEL, G.E.   32149
GOGOTOV, I.N.   23202, 26171, 29683-4,
     30535, 31274, 31730, 33710-1,
     34405, 35304-5, 35456, 36578,
     38424, 39102
GOGOTOVA, G.I.   25478, 34131
GOH, S.H.   39866
GOÏSA, N.I.   27149, 38753
GOÏTSA, N.I.   23665
GÖKÇEOĞLU, M.   29685
GOLBECK, J.H.   26172-4, 29686-8, 33712-
     -3, 37664
GOLD, K.   31326
GOL'D, V.M.   22560, 26175, 37665
GOL'DANSKII, V.I.   36860
GOLDBERG, D.   32796
GOLDBERG, E.D.   24946, 29311, 31316
GOLDEN, M.   24047
GOLDEN, M.L.   39355
GOL'DFEL'D, M.G.   22561, 23076, 26176,
     28833, 29314, 29689-90, 32886,
     33714-9, 34311, 36312, 37666-71
     see GOLDFIELD, M.G.
GOLDFIELD, M.G.   33720
     see GOL'DFEL'D, M.G.
GOLDHABER, A.S.   26177
GOLDKORN, T.   26178
GOLDMAN, C.R.   22436, 24766, 26179,
     26407, 27376, 29511, 31988, 32136,
     32138, 32621, 33721-2, 36141
GOLDMAN, J.C.   29691, 33723, 37672-3
GOLDSCHMIDT, C.R.   26180, 29692, 32812
GOLDSCHMIDT, E.E.   25233, 25551, 27025,
     37674
GOLDSTEIN, J.   36896
GOLDSTEIN, L.   27634
GOLDSTEIN, L.D.   23072, 25445, 26181
GOLDSTEIN, N.H.   24381-2

HASHIMOTO, K.   37822
HASHIMOTO, S.   27221
HASHIMOTO, T.   34093
HASHINOKUCHI, K.   31091, 35124-5
HASHIZUME, K.   26308
HASHWA, F.   22695
HASKINS, F.A.   24235, 24584, *37383*
HASLEMORE, R.M.   32364
HASLETT, B.   22529, 28033
HASLETT, B.G.   26309-10, 27503, 27708,
    29857-8, 32948, 33907-8, 34081
HASPEL-HORVATOVIČ, E.   22471, 26311,
    29859
    see HASPELOVÁ-HORVATOVIČOVÁ, A.
HASPELOVÁ-HORVATOVIČOVÁ, A.   22696,
    22805, 26312, 33909
    see HASPEL-HORVATOVIČ, E.

HASSELT, P.R. van see VAN HASSELT, P.R.
HASUMI, H.   33910, 37823
HATA, M.   26313
HATA, Y.   29433
HATAKENAKA, J.   24392
HATAKEYAMA, I.   35021, 36195-6, 38854,
    39977
HATANAKA, A.   28572, 35748
HATANO, M.   38766
HATANO, S.   33911-2
HATCH, A.   37824
HATCH, C.R.   36218, 37825
HATCH, M.D.   22697-9, 22713, 22983,
    26314-7, 27621, 29085, 29860-3,
    30191, 33913-4, 37120, 37392-3,
    37826-7
    see KHETCH, M.D.
HATCHER, B.G.   29864, 33915
HATCHIKIAN, E.C.   34974
HATEFI, Y.   *25267*
HATER, G.R.   34885, 37478
HATFIELD, J.L.   26318, 33916, 34090,
    37828-9, 37974
HATFIELD, S.G.S.   38242
HATTERSLEY, P.W.   22700, 26319-20,
    29865
HATTORI, A.   35148, 38818
HATZIOS, K.K.   33917-8, 37830
HAUG, A.   22795, 28741, 32791
HAUGHNESS, J.A.   34247
HAUGSTAD, M.   38903
HAUN, J.R.   22701
HAUPT, W.   22702-3, 24366, 26321-4,
    37001, 37764, B37831, 37832-3,
    39259, 39449, 39488,*39529, 39825*
HAURY, J.F.   29866
HAUSHILD, W.L.   24763-4
HAUSKA, G.   22704, 22706-7, 24298,
    24790, 25030, 29867-70, 32710,
    33919, 37578-9
HAUSKA, G.A.   22705, 24095, 24097,
    27787, 36749
HAUSWIRTH, N.   22973, 26557
HAVEL, J.J.   25607
HAVELANGE, A.   29871, 36926
HAVELKA, U.D.   22684-5, 24125, 29832,
    33871-3

HAVEMAN, J.   22708-9, 23574, 26325,
    28236, 30781-2
HAVRANEK, W.   25371
HAVRANEK, W.M.   33920
HAWCROFT, D.M.   22710-1
HAWKE, J.C.   22712, 34820, 38671
HAWKER, J.S.   22250, 22713, 34749
HAWKES, S.P.   25200
HAWKINS, S.E.   *21622, 21982, 22370-2,*
    22655-6, 22710, 25235
HAWKRIDGE, F.M.   26615, 29872, 30255-6,
    30509, 35516
HAWXBY, K.   29873
HAWXBY, K.W.   38688
HAXO, F.T.   22930, 23611, 25055-6,
    26326, 26678, 27512-3, 27952,
    31812, 32442, 36448
HAY, R.L.   22714
HAYAISHI, O.   *28599, 32680*
HAYAKAWA, S.   39593
HAYAMA, T.   38202
HAYASHI, F.   31522
HAYASHI, H.   23891, 30934
HAYASHI, K.   36500-1
HAYASHI, K.-I.   29874, 33921
HAYASHI, Y.   38908
HAYASHIDA, F.   33922
HAYASHIYA, K.   22715, 37834
HAYDEN, D.B.   22799, 26327, 29875-6,
    30006
HAYDOCK, K.P.   22716
HAYES, J.D.   23944, 32431, 40225
HAYET, M.   24868
HAYHOME, B.A.   37835
HAYMAN, E.P.   26328
HAYNES, L.   31415
HAYNES, R.C.   33923, 34537
HAYS, R.L.   26329
HAYSTEAD, A.   37836
HAYWARD, S.B.   33924
HEAD, A.   26330
HEAD, R.N.   27459
HEAD, W.D.   22717
HEADY, H.F.   *25537*
HEAGLE, A.S.   37837
HEAL, O.W.   *22053, 23304,* 33925, *B33926,*
    34190, 34880, 35301, 35928, 36037,
    37174, 37709
HEANEY, S.I.   33927, 37818
HEAPY, L.A.   26331
HEARD, A.J.   30167
HEARN, A.B.   33222, 33928, 36207, 37838
HEARN, A.R.   28502
HEATH, O.V.S.   B22718
HEATH, R.   22062
HEATH, R.L.   22136, 22719, 22906, 29877,
    30561, 32726, 33929, 39777
HEATHCOTE, L.   37839
HEATHCOTE, P.   22720, 29441, 29878-80,
    33487, 33930-1, 36095, 36430, 37840,
    39388
HEATHERLY, L.G.   29881
HEATHERSHAW, A.D.   29882
HEBBERT, N.P.   35031, 35924

HERBERT, S.J.   33961-3
HERBETTE, L.   33964
HERBLAND, A.   22742, 26358, 29914-7,
    32308, 33965, 36295, 37859, 40095
HERCZEG, T.   37280, 37860-1, 38460
HERDMAN, M.   29918, 40036
HERGENRADER, G.L.   28210, 28553, 28982
HERICH, R.   26449, 38741
HERLIHY, W.C.   37625, 38197
HERM, K.   32086, 36101
HERMAN, A.W.   29919
HERMAN, E.M.   22743-4, 26359
HERMANN, T.R.   33966
HERMOSO, J.   36649
HERN, S.C.   36089
HERNÁNDEZ-GIL, R.   29920
HERNROTH, L.   25165, 32564
HERODEK, S.   22745-6
HEROLD, A.   26360, 29891, 29921, 32343,
    37862-3
HEROLD, B.   33884
HERRENDORF, V.S.
    see SORBELLO HERRENDORF, V.
HERRERA, A.   26361, 28910, 29922
HERRERA, R.   34829
HERRIDGE, D.F.   29923, 35255
HERRMANN, F.   22643, 24379
HERRMANN, F.H.   21860, 26362, 26700,
    33967
HERRMANN, G.   28042-3
HERRMANN, K.   22747
HERRMANN, R.G.   31674, 33256, 38301
HERTEL, R.   30145
HERTOGH, A.A.de see DE HERTOGH, A.A.
HERTZBERG, S.   26363, 29924, 37864
HERVEY, A.   33968
HERVISH, P.V.   22341
HERVO, G.   22748, 29925, 34687
HERWIG, K.   24127
HERZ, J.M.   39022
HERZOG, H.   33969-70
HERZOG, R.   33971
HESKETH, J.D.   21535, 22749, 22787,
    23587, 23603-4, 25184-5, 26364,
    29984, 30513, 38650
HESLOP-HARRISON, J.   22750
HESS, B.   22022, 22218, 26365, 29926,
    30390-1, 30471, 33972-3, 34398,
    35486, 37016, 37865, 38282-3
HESS, J.L.   29536, 36706
HESS, M.   29927
HESS, S.   32373
HESS, W.M.   24476
HESSE, H.   25487, 26366
HESSE, M.   26367, 29928
HESSENBERG, B.   31505
HESSLER, R.   33974
HETEŠA, J.   33971
HEUER, B.   31319
HEUNERT, H.-H.   34396
HEUPEL, A.   22345, 22347, 25939, 25942,
    33822
HEVESI, J.   22751, 32730
    see KHEVESI, Ya.

HEW, C.-S.   23787, 26368, 32458
HEWETT, R.K.   37241
HEWITT, E.J.   24334, *37216*, *37768*, *37866*,
    38434, *38496*, *39142*, *39800*
HEWITT, H.G.   22752, 26369
HEWSON, A.   25599
HEYES, L.M.   28944
HEYLAND, K.-U.   35938, 37867
HEYN, M.P.   22753, 25337, 29099, 29929,
    33327-8, 34977, 37283
HEYSER, R.   22754
HEYSER, W.   22754, 33975
HEYWOOD, P.   29930
HIATT, H.D.   36846
HIBBLE, J.   22034-5
HIBI, K.   30104-5
HICKLENTON, P.   23872
HICKLENTON, P.R.   26370, 29931, 33976
HICKMAN, J.C.   22755
HICKMAN, M.   22756, 23107, 26371, 33977,
    37868
HIDEG, K. see KHIDEG, K.
HIEBSCH, C.K.   23007, 26372-3
HIEKE, B.   26399, 29932-3, 29978, 34014,
    34582, 37869-70, 37901, 40093
    see KHIKE, B.
HIEKEL, H.-G.   22757-8
HIGASHI, R.M.   36499
HIGASHI, S.   22879
HIGASHIDA, M.   23723, 26374, 34984
HIGGINBOTHAM, K.O.   26375, 28024
HIGGINS, T.J.V.   28891, 33978, *33978*,
    35106
HIGGS, R.E.A.   28625
HIGHFIELD, P.E.   29934, 33454, 33979
HIGINBOTHAM, N.   B38544
HIGUCHI, M.   33980
HIGUTI, T.   22354, 22759, 25953
HIKICHI, M.   23588
HILDEBRAND, E.   22760, 29935, 37284,
    39470
HILE, J.L.   23979
HILL, A.   29787
HILL, A.C.   21748, 30019
HILL, B.   25516, 29363
HILL, B.C.   33981
HILL, D.J.   36695
HILL, G.D.   33962-3
HILL, J.   37871
HILL, L.D.   *25576*
HILL, R.   26376, 29936, 33982
HILL, R.D.   23764, 27881, 30804
HILLBRICHT-ILKOWSKA, A.   25455, 29937,
    32137, 37655, *37985*
    see GILL'BRIKHT-IL'KOVSKA, A.
HILLER, R.G.   29938-9, 33983, 37872
HILLMAN, J.R.   22113-4, 22663, 37227
HILLMAN, W.S.   26377
HILLMER, P.   29940-1, 37873
HILLS, F.J.   28503
HILSCHER, H.   38284
HILTON, J.L.   27184, 28012, 36171,
    37792, 39307
HILTON, J.R.   30173

HINCKLEY, T.M.   22139, 22761-2, 28125,
    28919, 28963, 29881, 32110-1,
    32693, 33984-5, 35335, 37338,
    37874, 39134
HIND, G.   *28570, 28575, 28595, 28644,*
    28720, 28804, 28905, 28964, 28997,
    29031, 29035, 29080, 29135, 29173,
    29339, 29358, 29440, 29452, 29458,
    29474, 29483, 29592, 29754, 29788,
    29896, 29942, 30051, *30189, 30248,*
    30256, 30589, 30591, 30672, 30678,
    30782, 30796, 30813, 30870, 30875,
    30975, 31046, 31097, 31138, 31219,
    31369, 31638, 31662, 31756, 31785,
    31788, 31812, 31872-3, *31930, 32113,*
    32124, 32127, 32132, 32168, 32214,
    32282, 32357, 32585,
    32623, 32768, 32822,
    32963, 33068, 33171, 33445, 33673,
    33770, 33873, 33986, *34018, 34055,*
    34065, 34072, 34173, 34178, 34509,
    34618, 34657, 34683, 34817, 34890-
    -1, *34911, 35142, 35212-3, 35372,*
    35472, 35693, 35877, 35915-8,
    36159, 36331, 36412-3, 36415,
    36452, 36658, 36754, 36978, 37132,
    37143-4, 37222, 37875, *38356,* 38733-
    -4, 39674-5
HINDÁK, F.   22763
HINDE, R.   22764, 37876
HINDE, R.W.   32469
HINDMAN, J.C.   29943, 33987-8
HINDS, W.T.   22765
HINES, G.   22530
HINKLE, P.C.   33989
HINO, A.   28546, 29944
HINSON, K.   32923
HINTON, G.C.F.   34793
HIPKIN, C.R.   28124, 29945
HIPKINS, M.F.   22766, 29946, 33990-1,
    37877
HIRAI, A.   33992
HIRAKI, E.   39998
HIRAKI, K.   33993
HIRANO, M.   36622
HIRAO, M.   36199
HIRAO, S.   38201
HIRAOKA, T.   38910
HIRASAWA, T.   34118-9, 38017-8
HIRATA, H.   25110, 40311
HIRAYAMA, O.   26378, 29947-8, 33994-5,
    37878
HIRAYAMA, T.   36198
HIROSAKI, S.   22896, 23894, 31130
HIROSE, H.   *22692,* 22767, *22767, 22803,*
    23040, 23668, 23843, 24819
HIROSE, S.   22874, 25098, 28450
HIROSE, T.   22768
HIROSE, Y.   38149
HIROTA, N.   33996-7
HIROTA, O.   22769, 33998-4000
HIRSCH, M.D.   26379
HIRSCH, R.E.   34001, 37879

HIRSCHBERG, J.G.   23146
HIRTH, L.   37513
HISCOX, J.D.   37880
HITAKA, N.   30351-2
HITCHCOCK, G.L.   34002
HITZ, W.D.   34003, 37881
HIURA, H.   34217
HIXSON, C.S.   22546, 26159, 29672,
    37651
HIYAMA, T.   26380-1, 28184, 29788, 29949-
    -50, 32674, 37767, 37882-4
HNILICA, P.   39112
HO, C.-H.   26382
HO, E.   24205
HO, K.K.   34004, 37885
HO, L.C.   22037, 22770-1, 23817, 25682,
    26383-5, 28302, 29951-4, 32339-40,
    34005-6, 36329, 37886-8, 38893
HO, P.P.   25114
HO, P.T.   23272
HO, Y.B.   37889-90
HO, Y.-K.   37891-2
HOAD, G.V.   22772
HOAGLAND, K.D.   39341
HOAGLAND, R.E.   35261
HOANG-THI-HÀ   26076
HOARAU, A.   38716-7
HOARAU, J.   22623, 23344-5, 23641, 29955-
    -6, 30854-6, 31480-1, 34007, 35502,
    38446-7
HOBART, D.   29125
    see HOBART, D.R.
HOBART, D.R.   29126, 37160-1, 38281
    see HOBART, D.
HOBBS, E.H.   23223
HOBBS, J.F.F.   27654
HOBÉ, J.H.   40039
HOBERG, W.   30532
HOBOM, G.   35730
HOBRO, R.   28568
HOBSON, G.E.   29957
HOBSON, L.A.   26386-7, 37893
HOCH, G.E.   29958, 32935-6
HOCH, J.C.   29124
HOCHHEIMER, H.J.   37765
HOCHMAN, A.   22773-4, 29959-61
HOCHMAN, Y.   22520, 26388, 37894
HOCHSTER, R.M.   *29697*
HOCK, B.   32337-8, 36328
HOCKING, D.   27029
HOCKING, P.J.   35256
HOCKLEY, D.G.   37713-4
HODÁŇOVÁ, D.   22775-6, 37895
HODDINOTT, J.   22777, 26389, 28053,
    29963-4, 36044, 37896-7
HODGES, C.F.   29965-6, 38416
HODGES, H.F.   34622, 39799
HODGES, T.   29967, 31994, 37334, 37898-9
HODGES, T.K.   *29304, 31440*
HODGKINSON, K.C.   26390
HODKINSON, I.D.   22778
HODSON, R.   32108
HODSON, R.E.   26391

IVE, J.R.   26501-2
IVERSON, R.L.   25436, 26503, 38031
IVLEV, A.A.   26504
IVNITSKAYA, I.N.   23012, 30117
IVORY, D.A.   34134-5
IWAI, S.   22869, 26505
IWAKI, H.   22895-6, 26506, 32046, 35674,
   39976
IWAKIRI, S.   22880, 22897, 34136
IWAMURA, T.   22898, 38149
IWANAGA, M.   34137
IWANIJ, V.   22899
IWANZIK, W.   36101
IWASA, T.   28150, 38032
IWASAKI, A.   36063
IWATA, F.   32662
IWATA, T.   23669
IZAGUIRRE, M.L.   30823
IZAWA, S.   22900, 25385-6, 26507, 28766,
   29697, 30118-9, 34138, 34286, 39041,
   40124
IZDEBSKI, K.   30120
IZHAR, S.   38033
IZMEST'EVA, L.R.   30121, 30407, 31531,
   34139, 38307, 39083
IZUMI, K.   34477, 38364
IZVOSHCHIKOV, V.P.   22901
IZZA, C.   21974, 24781, 25618

J

JABBEN, M.   22902, 34140
JABLONSKI, P.P.   34141
JACHUCK, P.J.   39439
JACK, T.R.   38451
JACKSON, A.H.   22377, 26508
JACKSON, C.   29286, 30122, 30922, 34142
JACKSON, D.S.   26509, 38567
JACKSON, G.A.   34143
JACKSON, J.B.   22903-4, 24315-6, 25914,
   31179, 31287-8, 32375-6, 33252,
   33400, 36363, 39023, 39118-21
JACKSON, J.E.   31199, 38034
JACKSON, M.B.   38035
JACKSON, R.D.   30070-1, 33438, 34090,
   37974
JACKSON, W.A.   39019
JACOB, J.-L.   38036
JACOBI, G.   22905, 23195-8, 24367, 27500,
   30123, B38037
JACOBS, E.   38038
JACOBS, J.M.   38039
JACOBS, N.J.   38039
JACOBS, R.P.W.M.   38040
JACOBS, S.W.L.   29047, 33080
JACOBSEN, J.V.   33978
JACOBSEN, O.S.   36657
JACOBSEN, T.R.   34144
JACOBSON, B.S.   22906
JACQMARD, A.   36926

JACQUARD, P.   22907
JACQUES, G.   23809, 26510, 29171
JACQUES, G.L.   22908-9
JACQUES, R.   25778, *26816*, 26855, *27183*,
   27541, *30951*, 39958
JACQUES, T.G.   30124
JACQUET, J.   34145-6
JACQUINOT, L.   22910
JACQUOT, J.-P.   29579, 30125, 34147,
   36274, 38041
JADHAV, S.J.   22911
JADŻYN, C.   37538
JAFFE, M.J.   39865
JAGANNATH, M.K.   23220, 26778
JAGELS, R.   32377
JAGENDORF, A.   27335, 32851, 37261
   see JAGENDORF, A.T.
JAGENDORF, A.T.   22912-3, 23906, 26210,
   27495, 27669-70, 28542, 28802,
   29730-1, 30126-7, 31667, 32616,
   32949, 36645
   see JAGENDORF, A.
JÄGER, E.   21765
JÄGER, H.-J.   30128
JAGER, J.M.de see DE JAGER, J.M.
JÄGER, R.   37559
JAGGER, J.   30129
JAGIELO, J.A.   35731
JAGOW, G.von   38042
JAHN, G.   25496
JAHN, O.L.   22914, 26511, 38043
JAHNKE, H.   34148
JAHNKE, L.S.   22915, 34149
JAHREN, A.   23105
JAKOB, K.M.   31545, 35497-9
JAKRLOVÁ, J.   22916, 37501, 38044-5
JAKUCS, P.   22917, 36973
JAMALE, B.B.   26512
JAMES, D.M.   24334
JAMES, G.B.   23295, 26515
JAMES, T.D.W.   34150
JAMESON, D.M.   33581
JAMIESON, P.D.   38046
JAMRICH, V.   30130
JANA, B.L.   34151
JANA, M.K.   21603
JANA, P.K.   22524
JANÁČEK, K.   35606
JANÁK, J.   *24409*
JANARDHAN, K.   38047
JANARDHAN, K.V.   27210, 30131, 34152-4,
   35044
JANAUER, G.A.   26513
JANAVE, M.T.   24751
JANÍČEK, J.   22829
JANJIČ, V.   30132
JANK, H.-W.   21602
JANKIEWICZ, L.S.   25328
JANK-LADWIG, R.   26366
JANKOVIČ, M.M.   22918
JANKOWICZ, M.   27744
JANKOWSKA, K.   22919
JANNASCH, H.W.   *38374*

KOZLOVA, Zh.I. 24198-9
KOZLOVSKAYA, N.G. 39612
KOZLOVSKIĬ, V.S. 32588, 36626-7
    see KOZLOVSKIJ, V.S.
KOZLOVSKIJ, V.S. 32587
    see KOZLOVSKIĬ, V.S.
KOZŁOWSKI, S. 30464
KOZŁOWSKI, T.T. *21748, 22028*, 22189-91,
    22719, 23342, 23551, *23716-7,*
    24716, 24754, 25194, 25521, 25604,
    26460, 26759, 27062, 27432-3,
    28544, B28545, 28703, 29243, *29496,*
    29524, 29790, 30699, 30764, *30916,*
    30929, 30945, 30979, 31004, 31041,
    31268-70, *31496, 31841, 32033,*
    32436, 32504, 38310, 38920,
    39035-6, *39037*
KOZMA, L. 29481
KOZUBOV, G.M. 24228
KPODAR, M.P. 34410
KRAAY, G.W. 22531, 29655, 33676,
    37635
KRAAYENHOF, R. 22426, 23192, 30409-11,
    35728
KRAHNERT, S. 21860
KRAĬNOVA, N.N. 23995, 33146
KRAKHMALEVA, I.N. 23119-20, 26687,
    27869, 27871, 30328, 31805, 35846
KRALJEVIČ-BALALIČ, M. 23193
KRALJIČ, I. 38311
KRÁLOVÁ, V. 34412
KRÁL'OVIČ, J. 23194, 30412-3, 34411-3
KRAMAR, T.I. 38312
KRAMER, D. 32439, 34414, 38237
KRAMER, H.J.M. 40040
KRÄMER, M. 34415
KRAMER, P.J. 22116, 23980, 25597,
    27415, 27897, 31439, 34416, 35283
KRAMER, T. *40103*
KRAMINER, A. 29891
KRAMPITZ, M.-J. 33561
KRANS, J.V. 25032, 38313
KRANTZ, J.K. 38065
KRANZ, J. 28623
KRAPF, G. 22905, 23195-8, 26760, 29889,
    30414
KRASAVTSEV, O.A. 38314
KRASHCHUK, L.S. 39083
KRASICHKOVA, G.V. 22534, 23199, 29662,
    30415, 38315-6
KRASLOVÁ, J. 38317
KRASNA, A.I. 30416, 34417, 36025,
    38318, 39176
KRASNOVSKIĬ, A.A. 21969, 22262, 23021,
    23118-20, 23200, 23469, 24481,
    24844, 24930, 25614, 25958, 26687,
    26761-3, 27357, 27381, 27869-70,
    29081-2, 30228, 30326, 30328,
    30417-9, 31191, 31427, 32208,
    32314, 33386, 35075-7, 36305-6,
    37032, 37069, 37448, 38232-3,
    38319-20, 39554, 40102
    see KRASNOVSKY, A.A.

KRASNOVSKIĬ, A.A. ml. 23204-6, 26764,
    26868, 30420, 34418, 34543
    see KRASNOVSKY, A.A. Jr.
KRASNOVSKY, A.A. 26765, 30229, 30327,
    30421-3, 34350, 34419, 38321
    see KRASNOVSKIĬ, A.A.
KRASNOVSKY, A.A. Jr. 30424-5, 34420,
    38322
    see KRASNOVSKIĬ, A.A. ml.
KRASZEWSKA, E. 24628
KRASZNER, É.B.
    see BERNDORFERNÉ KRASZNER, É.
    see KRASZNER-BERNDORFER, E.
KRASZNER-BERNDORFER, E. 26766
KRATKY, C. 30426
KRATOCHVÍLOVÁ, H. 30427
KRATZ, H. 29773
KRAUSCH, H.-D. 34421
KRAUSE, C. 23207, 29977
KRAUSE, G.H. 23208-9, 26767, 30428-30,
    30637-9, 34422-4, 36995, 39432
KRAUSPE, R. 23968-9, 26768, 35242
KRAUSS, R. *34818*
KRAUSSE, H.J. 32970
KRAUZE, A. 34425
KRAVCHENKO, L.V. 26499
KRAVYAZH, K. 30431
KRAWIEC, R.W. 22287
KREEB, K. 24988, 27773-4, B30432, 35727
KREEB, K.H. 39218, 39503
KREICBERGS, O. 23222, 24918
    see KREĬTSBERG, O.É.
KREĮTH, F. 32625, B33281
KREĬTSBERG, O.É. 26769, 28274, 34426,
    34439
    see KREICBERGS, O.
KREMER, B.P. 23210-4, 24984, 26770-2,
    28357, 30433-5, 30968, 31676, 34427-
    -30, 34473, 38323-7, 40210-1
KREMER, D. 28527, 39097
KREMER, D.F. 35866-7, 39637
KRENDELEVA, T.E. 23215, 25124, 25893,
    26773, 29449, 30302, 30436-7, 32315,
    32481-2, 34431-2, 35530, 36200,
    37749, 39987-8
KRENZER, E.G.Jr. 23216-7
KRESLAVSKIĬ, V.D. 38328-9
KRESOVICH, S. 34809
KRETOVICH, V.L. 35554
KRETZER, F. 24529, 26774, 31105
KREUTZ, W. 34433
KREUZ, E. 30438
KRIEBEL, A.N. 29659, 38330
KRIEBEL, K.T. 38331
KRIEDEMANN, P.E. 23218-9, 26775-6, 31627,
    34632, 40126
KRIEG, D.R. 28500-1, 39815, 40233
KRIEG, K. 30439-40
KRIETSCH, W.K.G. 34434
KRINGSTAD, R. 34298
KRINSKY, N.I. 34435, 38332-3
KRISHNAMANI, M.R.S. 26777

MAHON, J.D. 23509, 27011, 28408, 30545,
    30705-6, 38574-5
MÄHR, E. 33756
MAHR, H. 26379
MAHRO, B. 32966
MAĬ, V.V. 39972
MAĬCHEKINA, R.M. 27012, 32307
MAĬDUROVA, V.E. 27799
MAIER, M. 38576
MAIER, R. 27013-4, 30707, 38577-9
MAILLARD-SEVHONKIAN, S. 34710
MAILLEFER, C. 39135
MAĬRANOVSKIĬ, V.G. 27015
    see MAIRANOVSKY, V.G.
MAIRANOVSKY, V.G. 23510
    see MAĬRANOVSKIĬ, V.G.
MAISON, B. 30708-9
MAISON-PETERI, B. 30710
MAĬSTER, A. 37640
    see MEISTER, A.
MAITY, S.P. 22524
MAJOR, D.J. 27016, 30711, 34711-2
MAKARENKO, K.I. 23972
MAKAROV, A.D. 24107, 24253, 27017-20,
    27038-9, 27482, 27523, 27712-3,
    27979, 35548, 35579-81, 35969, 36538,
    38580, 38981, 39374-5, 40301
MAKAROV, P.P. 26853
MAKAROV, Yu.A. 27683-4
MAKAROVA, A.Ya. 21845
MAKAROVA, N.N. 30712-3
MAKARSKA, E. 38581
MAKEDONSKA, Ts. 27021, 29325, 38582-3
MAKEVNINA, M.G. 23576
MAKHAMADZHANOV, I. 27022
MAKHMADBEKOVA, L.M. 27023
    see MACHMADBEKOVA, L.M.
MAKHNEVA, Z.K. 25957, 29431-2, 33468,
    37446
MAKI, T. 34362, 34713-4, 38584
MAKINO, H. 27024, 36063
MÄKIRINTA, U. 34715
MAKOVCOVÁ, O. 30714, 34716
MAKOVEC, P. 30693, B34706
MAKOVETSKAYA, R.V. 32058
MAKOVETZKI, S. 27025
MAKOVKINA, L.E. 23150, 30982-3, 34717
MAKSIMOV, V.N. 23190, 34407, 34719, 37481,
    38585, 40101
MAKSIMOVA, È.A. 38585
MAKSIMOVA, I.V. 23511, 27026, 34718-9,
    38586-7
MAKSIMOVA, L.A. 22421, 26020, 28168-9,
    32163, 32300, 34720
MAKUNGA, O.H.D. 34721
MALAFEEV, Yu.M. 33361
MALAKONDAIAH, N. 28038, 38588-9
MALANIYA, D.G. 22299-300
MALARA, G. 27266
MALASHÉVICH, A.V. 27027
MALCHEV, E. 34722
MALCOLM, A.D.B. 34196
MALCOLM, D.C. 22584, 28914, 32960
MALDONADO, J.M. 25675

MALESZEWSKI, S. 26911, 26996, 30462,
    31908, 38122
MALET, P. 38590
MALETS, L.N. 24463
MALEVA, M.N. 25357
MALEWICZ, B. 22211
MAL'GOSHEVA, I.N. 25614, 32208, 37069
MALHOTRA, H.C. 25555-6, 28949
MALHOTRA, S.S. 27028-9, 30715, 34307,
    39438
MALIGA, P. 23512, 36529
MALIK, C.P. 23513-4, 32843, *35609*
MALIK, M.N. 38591
MALIK, N. 39890
MALIK, S.K. 27808
MALINOVSKIĬ, A.V. 27030
MALINOVSKIĬ, V.I. 38312
MALINOWSKI, K.C. 20116
MALISHEVSKIĬ, S. 22608
MALIWAL, G.L. 23950, 35221
MALKIN, R. 21712, 23515-9, 25347-8,
    26694-5, 27031-4, 28667-8, 28720,
    30337-8, 30716-7, 31976, 34357-8,
    34723-7, 36389, 36823, 37065,
    37115, 38592-4
MALKIN, S. 22683, 23477, 23520, 25616,
    27872, 27879, 29012, 29464, 30718-
    -20, 30838, 30955, 31808-9, 32071,
    33047-8, 34728-9, 35863, 37469,
    38595-7, 38816, 38982, 39622
MALKINA, L.S. 23521, 27035, 30721-2,
    34730-1, 40291
MALL, L.P. 21793-5, 25422-3, 28800-1,
    31395, 39257
MALLEY, M.M. 34969
MALLON, D.E. 27254
MALLOT, P.G. 23522, 34732
MALMBERG, G. 35181
MALNOÈ, P. 35544, 38598
MALOFEEV, V.M. 23523-4, 27036, 30723
MALOFEEVA, I.V. 23525
MALONE, T.C. 30724-5, 33635, 34733
MALONEY, T.E. *22592, 24456, 24992*
MALOOF, C. 39398
MALUEG, K.W. 27766, 38409
MALÝ, J. 33492
MAL'YAN, A.N. 25467, 27019, 27037-9,
    27712-3, 30726-7, 34734, 36538,
    38599, 38981
MALYSHEV, O.G. 23033
MAMET-BRATLEY, M. 40012
MAMKAEVA, K.A. 29752
MAMLEEVA, N.A. 27040, 30728-9, 35050,
    38600-1
MAMUSHINA, N.S. 26021, 29500, 40332
MAMYTOV, A.M. 34735
MANABE, K. 36667
MANAKOV, K.N. 38602
MANANKINA, E.E. 23642, 35836, 39598
MANANKOV, M.K. 23526
MANCHESTER, J. 39008-10
MANCINELLI, A.L. 23527-8, 27041, 34736
MANCINI, J.L. 22238, 29311
MANDAHAR, C.L. 23529, 26107, 27042,
    34737

                                              (continued)

MOHANTY, P.   23673, 25432-3, 27163,
    28817, 30892, 31356, 32862, 34919,
    36899, 38773
MOHAPATRA, P.K.   25810
MOHAPATRA, S.C.   27164
MOHR, H.   21933, 22474, 22902, 23034,
    23674, 26077, 26603, 29567-8,
    29623, 30893-5, 31096, 33012,
    33603, 34920-2, 35132-4, 38936
MOHR, W.P.   38774
MÖHWALD, H.   30657, 33941, 34679, 38536-
    -7
MOISEENKO, E.I.   22893
MOK, M.C.   23675, 27165
MOKRONOSOV, A.T.   21863, 22544, 26155,
    28649, 30896-7, 32931, 33696,
    33699, 34923-6
MOLCHANOV, A.G.   27166, 30898-9
MOLCHANOV, M.I.   23676-7, 27167-70,
    30900-1, 34864, 34927-31, 35554
MOLDAU, H.   30902-4
    see MOLDAU, Kh.A.
    see MOLDAU, Kh.
MOLDAU, Kh.   23678
    see MOLDAU, H.
    see MOLDAU, Kh.A.
MOLDAU, Kh.A.   23679, 26593, 27171,
    30905, 32015
    see MOLDAU, H.
    see MOLDAU, Kh.
MOLDOVAN, I.   29277
MOLINARI, M.T.   37512
MOLINE, H.E.   23680-1
MOLL, A.   34835, 34932-3
MOLL, R.   22654
MOLL, R.A.   30906
MOLL, W.A.W.   31874, 34934-5, 38775
MOLLENHAUER, H.H.   37961
MØLLER, B.L.   35874, 38557
MÖLLER, G.   30907
MØLLER, I.   22677
MOLNÁR, E.   27507
MOLNÁR, E.N.   38776
MOLNÁR, J.   38777
MOLNÁR, M.   22751, 30908
MOLNÁR, P.   38778
MOLOTKOVSKIĬ, Yu.G.   30909-10, 31082,
    40290
    see MOLOTKOVSKY, Yu.G.
MOLOTKOVSKY, Yu.G.   28483
    see MOLOTKOVSKIĬ, Yu.G.
MOLVINSKIKH, S.L.   37648
MOLYAKA, O.N.   23682-4, 29342, 30911
MOMEN, N.N.   38779
MOMENT, G.B.   36479
MONAGHAN, J.L.   27172
MONAKOVA, S.V.   30522
MONCHOR, D.   38780-1
MONCUR, L.   28339
MONDAL, M.H.   30912, 34936-7
MONDOVI, B.   30913, 34938
MONÉGER, R.   21509, 21653, 23324-5,
    25287-9, 26292, 26855, 28651, 30551,
    30914, 32278, 33057, 34581

MONGER, T.G.   22101, 27173, 29025, 30915,
    31219
MONHEIMER, R.H.   23685-6, 38782
MONICA, R.F. Ia 27174
MONK, C.D.   29254
MONK, R.L.   36703
MONMA, E.   38783
MONNIER, A.   *28681, 28934, 29758, 32319,*
    32477, 34033
MONOSOV, E.Z.   26727
MONSELISE, S.P.   22323, 34939, 36851
MONSERUD, R.A.   37407
MONSI, M.   27227, *34037, 34101, 34269,*
    34325, 34467, 34940, 35105, 35150-1,
    35158, 35674, 35807, 36066, 36090,
    36195-6, 36198, 36212, 36225, 40287
MONTAL, M.   32167, 37910, 39491
MONTALBINI, P.   23687, 27793, 30704,
    34941
MONTALVO, M.R. de see DE MONTALVO, M.R.
MONTECUCCO, C.   25273
MONTEITH, J.L.   21801-3, B23688, *24215,*
    24244, 24261, 24677, 24740, 24947,
    25195, 25571, 26515, 26912, 26923,
    27587, 27630, 27656, 27728, 27981,
    28197, 29289, 29413, *29413,* 30205,
    30916-8, 34942, 38500-1
MONTENY, B.   27175, 34943-4, 37696
MONTES, G.   23689, 25530, 27176, 28910
MONTGOMERY, R.E.   26137
MONTI, L.M.   29499
MONTIES, B.   23690-1, 29172, 29586
MOOK, W.G.   38167
MOON, R.   25825, 29451
    see MOON, R.E.
MOON, R.E.   25824, 27177-8, 29254, 33292
    see MOON, R.
MOONEY, H.A.   21813, 23692-3, 25902,
    25924, 27179-80, 30919, 32665-6,
    33436, 34945-6, 36853, 37397, 38784,
    39424
MOOR, H.   21895
MOORBY, J.   22076, 23694-6, 29485, 30920,
    31592, B38739-40, 40216
MOORE, A.L.   29286, 30122, 30921-2,
    34142
MOORE, C.B.   *39731*
MOORE, D.G.   28126
MOORE, D.P.   34947
MOORE, F.D.   23697, 30923-4, 34948,
    39970
MOORE, F.D. III   33897
MOORE, H.   30689
MOORE, J.E.   30925
MOORE, J.R.   32809
MOORE, K.G.   23698, 24432, 25205-7,
    27430, 35800
MOORE, M.N.   36988
MOORE, P.D.   23699-701, 38785
MOORE, P.H.   27055, 34949
MOORE, R.   38786
MOORE, R.C.   32867
MOORE, R.T.   29016, 33051
MOORE, T.A.   30926, 33421

(continued)

MURATA, Y. (continued)
     B23731, 23732-7, *23732*, *23737*,
     23893, *23925*, *24661*, 24662, *24705*,
     24822, *24841*, 26474, 30077, 30097-
     -100, 32312, 34000, 34095-7,
     34122-3, 34374-6, 36303-4, 37981-3,
     37987-9, 38020, 40324
MURAWSKI, D.A.   36093
MURAYAMA, S.   31077
MURCHISON, C.   40133
MURDOCH, S.   27208
MUREĬ, I.A.   23738, 27209, 30976, 31801,
     35001-4, 35842, 38839
MUREŞAN, T.   23739-40
MURPHY, C.E. Jr.   27888
MURPHY, D.J.   26893, 30977-8, 35005,
     38840
MURPHY, P.G.   23741, 28687
MURPHY, S.J.   33455, 37423
MURPHY, T.M.   35006
MURR, D.P.   39979
MURRAY, A.M.   37087
MURRAY, D.B.   30979, 34626
MURRAY, D.R.   29700, 30980, 33731,
     36724
MURRAY, S.N.   26933, 38499
MURTAGH, G.J.   35007
MURTHY, K.S.   35437, 40083
MURTHY, M.S.   35008
MURTY, K.S.   24142, 24609, 27210,
     30131, 34152-4, 35044, 36185,
     38539, 38872-3
MURTY, P.S.S.   38873
MURZA, L.I.   33778
MURZAEVA, S.V.   21530, 23742-3, 35009
MUSASHI, A.   38950
MUSATENKO, L.I.   24673
MUSCATINE, L.   23744
MUSGRAVE, R.B.   32294, 40077
MUSHAK, P.O.   24644
MUSHKETIK, L.S.   24188, 25099-100,
     30214, 32309, 32490, 40302
MUSHKETIK, N.S.   25101
MUSICK, J.T.   26484, 37384
MUSIENKO, N.N.   24108
MUSSELL, H.   *38075*, *38429*, *39771*, *39812*
MUSSELLMAN, R.C.   35563
MUSSELMAN, R.C.   23745, 26840, 35010
MUSTAFA, J.   38935
MUSTARDY, L.   25837, 36980
MUSTÁRDY, L.A.   22816, 23746, 27211,
     28792, 29458, 29578, 30016, 33143,
     34044-5, 34155, 35011, 38841
MUSZBEK, L.   30981
MUSZYŃSKI, S.   23747
MUTO, S.   30886
MUTSAERS, H.J.W.   27212-3
MUTUSKIN, A.A.   23150, 23748, 25145,
     26721, 30982-3, 34717, 38269,
     38656
MUUS, L.T.   *29972*
MUZAFAROV, E.N.   21531, 22894, 23709,
     23749-50, 27857, *28523*, 28524,
     30115, 30726-7, *30727*, 30984-5,
     30984-5, 35012-3, 36847, 37151

MUZAFAROVA, S.   21545
MUZAFAROVA, S.M.   27214
MUZAFFAROVA, S.M.   32561
MUZIK, T.J.   28368
MUZQUIZ, M.   21971
MUZTAR, A.J.   38842
MVÉ AKAMBA, L.   30986, 31814, 38843-5,
     39627
M'YAKUSHKO, V.K.   35014
MYDLARZ, J.   40253
MYERS, A.L.   25783
MYERS, B.J.   37063
MYERS, G.A.   30819, 34828
MYERS, J.   21889, 23751, 28008, 28328-9,
     31438, 32357-9, 35015
MYERS, S.P.   31955
MYERS, V.B.   26503
MYHRE, D.L.   29984
MYKLESTAD, S.   30987-8
MYRON, J.C.   30337-8
MYRUP, L.O.   24090
MYSHKOVETS, E.N.   34318, 38194
MYSLOVICH, V.O.   27898
MYULLER, I.   26710

N

NAABER, L.Kh.   23752-3
NABEDRYK-VIALA, E.   30989
NAD', A.   23754, 30990-1
     see NAGY, A.H.
NADAKAVUKAREN, M.J.   23755, 30992-3,
     38846, 39701
NADEEM, M.T.   32688
NĂDEJDE, M.   39943
NADLER, K.D.   32820, 36867
NAFZIGER, E.D.   27215
NAGAI, H.   22898
NAGAI, S.   39440
NAGANO, M.   35092
NAGAO, N.   31139
NAGARAJAH, S.   23756-7, 27216, 35016-7,
     38847-8
NAGASHIMA, H.   22350
NAGATO, K.   23758
NAGAYAMA, J.   33912
NAGL, W.   27217, 35018
NAGLE, B.J.   38849
NAGORNAYA, R.V.   38441
NAGY, A.   37229
NAGY, A.H.   29777, 30009, 33817, 37756
     see NAD', A.
NAGY, K.   33275
NAGY-TÓTH, F.   35309, 38850
NAÏDE, O.V. de   see   VILLEDON de NAÏDE, O.
NAÏDENOVA, Ts.   38293, 38851
NAÏDICH, V.I.   32649, 36689
NAIK, G.R.   30175, 38852
NAIK, M.S.   27218, 35676, 39401
NAIM, R.   35019
NAIMAN, R.J.   23759, 27219, 35020
NAIR, K.P.P.   24511

                                              (continued)

PLATT, S.G.   24059, 27461-2, 31313-5,
     35341-2, 39149-50
     see PLATT, S.
PLATT, T.   24060-2, 26516, 27463, 29287,
     30676, 31316-7, 32032, 35865
PLATT-ALOIA, K.   29793, 31318, 33589
PLATT-ALOIA, K. A.   39151
PLATZ, R.A.   29211
PLAUT, Z.   24059, 24063-4, 27461-2,
     27464, 31314-5, 31319-21
PLESCHER, A.   29979
PLESNIČAR, M.   27465, 27996, 30132, 30769,
     31322, 33008, 39152
PLESSER, A.   28361, 31323
PLESSIS, J.G. du   see   DU PLESSIS, J.G.
PLOEG, R.R. van der
     see   VAN DER PLOEG, R.R.
PLOHR, A.   37962
PLOTNER, A.   35894
PLOTNIKOVA, A.N.   24941, 31324
PLUMB, R.T.   30167
PLUMB-DHINDSA, P.L.   39153
PLUMLEY, F.G.   35343
PLUMLEY, R.   22960
PLUMMER, G.L.   35344
PLUS, R.   23360, 24065, 26986
POCHINOK, Kh.N.   27466
POCKER, Y.   35345
PODERGINA, T.A.   34320
PODOLÁK, M.   35346
PODOLEANU, M.   39749
PODOLNY, V.Z.   39203
     see   PODOL'NYĬ, V.Z.
PODOL'NYĬ, V.Z.   35245
     see   PODOLNY, V.Z.
PODVALKOVA, I.A.   28674
POGGIONI, S.   32647
POGOL'SKAYA, V.I.   27466
POGONCHEVA, E.   23019
POGONCHEVA, E.M.   26587
POGOSYAN, S.I.   34856
POGREBNYAK, M.P.   24477
POHJONEN, V.   33878
POHL, G.W.   23320
POHL, P.   26167
POHLHEIM, E.   27467
POHLHEIM, F.   27467
POINCELOT, R.P.   24066, 27468-9, 27546,
     31325, 38967, 39154
POIRIER, G.G.   34546
POIRIER, L.   40013
POKORNY, K.S.   31326
POKORNÝ, V.   36275
POKROVSKAYA, T.N.   27470, 31327-8, 39155
POLACCO, E.   37832
POLACCO, J.C.   21994
POLAK, J.   35347
POLAK-CHARCON, S.   23915, 24067
POLCHANINOVA, T.V.   38198
POLEK, B.   32526
POLESCU-IONĂȘESCU, L.   31329, 35348
POLETAEV, V.V.   24428

POLEVAYA, V.S.   24068, 24568, 26467,
     27471, 35932, 37975, 39699,
     40301
POLI, E.   26903
POLIKARPOVA, N.N.   28628
POLING, S.M.   24069, 27472
POLISHCHUK, A.I.   24188, 24458, 25099-
     -100, 32490
POLYAKOFF-MAYBER, A.   *22484-5, 23274,
     B24070, 24755*, 26577-8, 30218,
     35161
POLLARD, A.   32477, 39156
POLLARD, A.L.   26228
POLLES, J.S.   26042, 26713, 29521
POLLHAMER, E.   39157
POLLMER, W.G.   29565-6
POLLOCK, C.J.   27473
POLLOCK, R.B.   39158
POLNA, M.   25895
POLONSKIĬ, V.I.   24485, 27474, 31330-1,
     35349-52
POLOZOVA, T.G.   22053
POLUEKTOV, R.A.   31332
POLYÁK, B.   37225
POLYAKOV, A.M.   28287
POLYAKOV, M.A.   24932, 26600, 27475,
     27807, 28286
POLYAKOV, M.I.   33621
POMAZKOVA, C.I.   38307
POMERANTSEVA, O.M.   37750
POMEROY, W.M.   27476
PON, N.G.   35510
POND, S.   40195
PONGRATZ, P.   35353
PONNAMPERUMA, C.   *26719, 31652*
PONOMAREVA, R.P.   27477, 39159
PONOMAR'OVA, S.O.   30186
PONZI, R.   39160-1
POOL, R.   34632
POOL, R.R.   23744
POOL, R.R. Jr.   36179
POOLE, D.K.   34887, 35354-5
POOLE, R.J.   27478, 35356
POOLE, R.T.   25764, 29163
POOVAIAH, B.W.   24071, 27479, 32678,
     36716-7, 39163
POPA, F.G.   39053
POPDIMITROVA, N.   35357
POPE, D.H.   24072-3, 35570
POPE, M.   28056
POPENOE, H.L.   30945
POPESCU, D.   33013
POPESCU, I.   34876
POPESCU, V.   27480
POPIELA, C.C.   27574
POPIOŁEK, Z.   30120
POPOV, A.M.   23180
POPOV, B.A.   26742
POPOV, È.G.   36629, 37352
POPOV, G.   35972
POPOV, K.I.   *B39164*
POPOV, V.I.   21546, 26831, 27481-3,
     28074-5, 31333-6, 32029, 36054

RADEVA, R.V.   see   V"LKOVA RADEVA, R.
RADEVA, V.   28827, 31408, 39223
RADIĆ, M.   38505
RADIN, J.W.   39224
RADKOV, P.   29269
RADMER, R.   23148, 24133, 24607, 26719-
     -20, 30765, 31409-11, 35418, 39225
     see   RADMER, R.J.
RADMER, R.J.   24134, 27548, 31412,
     38623
     see   RADMER, R.
RADOSEVICH, S.R.   27549, 36775, 39125,
     39226
RADUNZ, A.   23624, 24135-6, 24351,
     26711, 30361-2, 31413-4, 31662,
     31664-6, 34372, 34850, 35419-21,
     35696-700, 38458-9, 39227
RADWAN, S.S.   24137
RADWAY, J.   31415
RADYUK, M.S.   27027
RADZEĬ, Ĭ.   32518
RADZHABOV, H.   21545
RADZIEJ, J.   36553
RAFAREL, C.R.   22034-5
RAFFERTY, C.N.   25839, 33190, 35422,
     35951, 37166, 39228-31, 39729
RAFII, Z.E.   39232
RAGAN, M.A.   39233
RAGGI, V.   24138, 35423-4, 39234
RAGHAVENDRA, A.S.   24139, 27550-6,
     27585, 29231, 31416-21, 35425-34,
     37242-3, 39235-9
RAGHI-ATRI, F.   27366, 31422, 35435
RAGHUNATHA, G.   23220, 26778
RAHAT, M.   27557
RAHMAN, F.M.M.   37039
RAHMSDORF, U.   30667, 38546
RAI, A.K.   31423
RAI, B.   39652
RAI, H.   24140, 35436
RAI, K.S.V.   see   VINAYA RAI, K.S.
RAI, M.   24141
RAI, P.   31748
RAI, R.S.V.   24142, 27210, 35437
RAI, V.   24609
RAILTON, I.D.   31424-5
RAINBIRD, R.M.   38431
RAINE, R.C.T.   40224
RAINS, D.W.   29203-4, 33261, 36728,
     39240
RAISON, J.   39241
     see   RAISON, J.K.
RAISON, J.K.   24468, 35438, *36753,*
     36898, 37524, 37713, 38680,
     39172, 39242, 40030-1
     see   RAISON, J.
RAJAGOPAL, M.D.   31202
RAJAGOPALAN, C.K.S.
     see   SOUMINI RAJAGOPALAN, C.K.
RAJAN, A.K.   24143
RAJASHEKARA, B.G.   23220, 26778
RAJENDRAN, A.   35439
RAJENDRUDU, G.   39238
RAJKI, S.   22222

RAJU, T.V.K.
     see   KARIVARATHA RAJU, T.V.
RAJYALAKSHMI, T.   24144
RAJZMANN, M.A.   26555
RAKHIMBERDIEVA, M.G.   30229, 31426-7,
     34244
RAKHIMOV, A.   24145-6
RAKHIMOV, A.R.   32815
RAKHIMOV, G.T.   24147-8, 27558, 39243
RAKHIMOV, S.   24145
RAKHMANKULOV, S.A.   22870, 27559,
     32706, 35440
RAKHMANKULOVA, M.E.   24149, 34319
RAKHMATOV, N.A.   36532
RAKHTEENKO, I.N.   24150
     see   RAKHTSEENKA, I.N.
RAKHTEENKO, L.I.   22893, 26499
     see   RAKHTSEENKA, L.I.
RAKHTSEENKA, I.N.   39244
     see   RAKHTEENKO, I.N.
RAKHTSEENKA, L.I.   23828
     see   RAKHTEENKO, L.I.
RAKITIN, L.Yu.   24151
RAKITIN, V.Yu.   24151
RAKITIN, Yu.V.   27560
RAKITINA, Z.G.   35441
RAKOVÁN, J.N.   33816-7
RAKOW, G.   27561
RALPH, R.K.   27562
RAM, G.   28662
RAMACHANDRA PRASAD, T.V.   23220
RAMACHANDRA REDDY, A.   35442
RAMA DAS, V.S.   see   DAS, V.S.R.
RAMADASAN, A.   22010, 23570, 27069
RAMADHAS, V.   31428, 31989
RAMADOSS, C.S.   35443
RAMAGE, R.   24152
RAMAKRISHNA, J.   32840-1, 35444
RAMAKRISHNA, Y.S.   31842
RAMAKRISHNAN, P.S.   35445
RAMAKRISHNAN, T.V.   39245-6
RAMAKRISHNAYYA, G.   27210
RAMALINGAM, R.S.   35446
RAMAM, S.S.   24153, 27563
RAMANATHAN, V.   39016
RAMANIS, Z.   27696-7, 27891
RAMANUJAM, S.   *36959, 38539*
RAMA RAO, S.V.   see   RAO, S.V.R.
RAMASWAMY, N.K.   27564-5, 35447, 39247
RAMATI, A.   39248
RAMEY, H.H. Jr.   28431
RAMIG, R.E.   24207, 35517
RAMINA, A.   39249, 39398
RAMÍREZ, J.M.   24858-9, 26146, 29272,
     29282, 29664, 31299, 37642
RAMÍREZ, R.   39250
RAMIREZ-PONCE, M.P.   37642
RAMKRISHNA, D.   39576
RAMON, J.   36865
RAMSAY, W.E.   35448
RAMSHAW, J.A.M.   24335, 27566, 27737,
     29555, 32948, 33220, 33590-1,
     33907, 35679

RAY, P.K.   33409
RAY, R.C.   39274
RAY, T.   25797
RAY, T.B.   23072, 25445, 26181, 27089,
    27596-8, 29215, 31465-6, 34798,
    35477-8, 39275-7
RAYBIN, R.A.   39507
RAYFIELD, G.W.   33966
RAYLE, D.L.   26536
RAYMUNDO, L.C.   27644
RAZAKOV, A.   27059
RAZJIVIN, A.P.   32587, 36627
    see RAZZHIVIN, A.P.
RAZORITELEVA, E.K.   21729
RAZUVAEV, A.I.   32545
RAZVYAKOÚ, V.A.   29368
RAZZHIVIN, A.P.   28515, 32588, 36626
    see RAZJIVIN, A.P.
READ, D.J.   22850, 24632
READ, G.   22553
READ, M.   36126
REARDON, E.M.   35479
REBANE, K.   21636, 27599
REBEIZ, C.A.   23581, 24177-80, 24551,
    25344, 25747, 27081, 27526-7,
    27927, 29150, 30790-2, 31880-1,
    32776, 34790, 35480, 36819-20,
    36837-8, 37183, 37556, 38660,
    39278-9, 39686
REBEIZ, C.C.   24178-80, 27081, 33205,
    35923
REBELLA, C.   32097
REBELLO, A.   31467
RECKERMANN, J.   35995
RECOUVREUER, M.   38963
REDDY, A.N.   39253
REDDY, A.R.
    see RAMACHANDRA REDDY, A.
REDDY, A.R.C.   39431
REDDY, C.S.   39431
REDDY, G.G.   34499
REDDY, G.M.   21596
REDDY, M.N.   28038
REDDY, M.R.   39280
REDDY, P.R.   27971
REDDY, T.P.   39281
REDGWELL, R.J.   28796, 35481
RED'KO, T.P.   34983
REDLINGER, T.   39282
REDLINGER, T.E.   31468, 35482-3
REDMANN, R.E.   24181, 24215, 27630,
    35484
REED, A.J.   31469
REED, D.W.   23998, 24182
REED, G.H.   28136
REED, K.L.   31470, 36332
REED, M.L.   25431, 31471-2, 35485,
    39283
REED, R.L.   31543
REED, S.C.   29024
REED, T.   35486
REES, A.R.   22771, 27600, 29953
REES, T. ap see AP REES, T.

REES, W.A.   25728
REEVES, S.G.   22375, 31437, 31473,
    32571, 35487
REFFYE, P. de   27601
REGAN, D.L.   39663
REGEHR, D.L.   24183, 27602
REGER, B.J.   39284
REGINATO, R.J.   30070-1, 33438, 34090,
    37974
REGITZ, G.   24184, 27603
REHDER, H.   24185, 27604, 29685
REHFELD, D.W.   21783
REHM, S.   35488
REIBACH, P.H.   27605, 31474, 35489,
    40052
REICH, H.   39323
REICH, R.   27606-8, 31475, 31740-2,
    38095, 39285
REICHE, H.   39286
REICHENBÄCHER, D.   31476, 35490
REICHLE, D.E.   24186
REICHMAN, G.A.   37518
REICOSKY, D.A.   27609, 35491-2
REID, C.P.P.   31477
REID, E.   *21651*, *37335*, *39333*
REID, F.M.H.   35493, 36117
REID, G.A. Jr.   27065
REID, J.   39366-7
REIF, G.   37833
REILLY, A.   39287
REILLY, C.   39287
REIMANN, K.   35494
REIMER, S.   22706, 24187, 24790, 27610,
    28165, 31098, 32161, 35136, 35495,
    39288-9, 39957
REIMER, T.O.   39290
REIMOLD, R.J.   34621
REINERS, W.A.   40277
REINERT, R.A.   24716
REÏNGARD, T.A.   24188, 24457-8, 25099-
    -101, 30214, 32490, 35825, 40302
    see REINGARD, T.
REINHARDT, B.   31506-7
REINHOLD, R.   *33874*
REINKE, D.C.   39291
REINMAN, S.   34758, 35496, 39292,
    39902
REÏNUS, R.M.   22544
REIS, R.R.   32544
REISEN, W.K.   27611
REISFELD, A.   24128, 24130, 29374,
    31545, 33419, 35497-9, 40161-2
REISIGL, H.   33756
REISSER, W.   27612-3
REISS-HUSSON, F.   22626, 22954, 30134,
    30671-2, 31478, 34687
REITMEIER, H.   31997
REITZEL, L.   27614
REIX, M.   23388-9
REJEWSKI, M.   23646
REJMÁNKOVÁ, E.   35500
RELIMPIO, A.M.   25675, 27048
RELTON, J.   33508, 36845

ROTEM, J. 22269, 28705
ROTH, D. 21765, 24033
ROTH, H. 26708
ROTH, H.D. 23290
ROTH-BEJERANO, N. 23014, 24247
ROTHER, J.A. 39361
ROTHSCHILD, K.J. 39362
ROTT,J. 33198
ROTT, R. 31548, 36743
ROTTENBERG, H. 24042, 25296, 29388,
    31549, 33443, 35568
ROTTENBURG, T. 23005
ROTTY, R.M. 28648
ROUGHAN, P.G. 27665, 39363-5
ROUHANI, I. 27666, 31550
ROUSSAUX, J. 22401, 27667, 33503
ROUSSEAU, B. 25904
ROUSSEAU, K. 22242
ROUX, E. 21899-900, 22624, 23361,
    24840, 29458, 30156, 30410, 31551,
    32198-9, 32443
ROWBERRY, R.G. 30674
ROWE, A. 23597
ROWE, G.T. 24248
ROWE, J. 39366-7
ROWE, M.P. 39368
ROWE, P.R. 23071
ROWELL, D.L. 31657
ROWELL, P. 27668, 31964, 32656, 36002
ROWLAND, A.O. 22727
ROWLAND, R.G. 35127, 39177
ROWLANDS, P.G. 38066
ROWLEY, J.A. 22449, 28099-100
ROY, G. 23538
ROY, H. 27669-70, 31552-3, 32128-9,
    33770-2, 35569-70, 39369
ROY, S. 39370
ROYCHOWDHURY, J. 39653
ROZEMA, J. 24249, 27671
ROZHDESTVENSKII, V.I. 27672
ROZHKO, I.I. 26221, 31554
ROZONOVA, L.N. 28468, 34317, 35571,
    36541-2
ROZOV, N.F. 27673
ROZOVA, T.L. 31555
RÓZSA, Z.Sz. 33498
RUBAN, V.V. 36733
RUBIN, A.B. 21855, 23215, 23470,
    23525, 23970-1, 24189, B24250,
    25088, 25668-9, 26731, 26773,
    26974, 27075, 27260, 27273-4,
    27535-7, 27615, 28876, 29073-4,
    30252, 30386, 30437, 31053-4,
    31072-4, 31225, 31335, 31455,
    31556, 32482, 32602, 32924, 33114,
    34393, 34431-2, 34780, 35246,
    35409, 35530-2, 35572-3, 35576,
    36739, 36860-1, 36952-4, 38538,
    38918, 38989, 39268, 39988, 40056
RUBIN, B.A. 21727, 22513, 23491,
    23831, 25362, 26118, 27674-5,
    31557, 35574, B35575, 37615
RUBIN, L.B. 23970-1, 31225, 31333,
    31335, 35246, 35576, 38191

RUBIN, P.M. 24251, 39371
RUBINSHTEIN, A.I. 39372
RUBTSOV, P.M. 27676
RUBY, R.H. 39373
RUCHTI, J. 31700
RUCKENBAUER, P. 24252
RUCKLIDGE, G.J. 37866
RUD', G.Ya. 35577-8
RUD', M.S. 24394
RUDENKO, T.I. 24253, 27018, 35579-81,
    39374-5
RUDICH, J. 31558, 37313
RÜDIGER, W. 23178-9, 24254, 27677,
    27997, 31682, 33420, 34339, 34976,
    35582, 38223, 39376, 39465
RUDNICKI, R.M. 39428
RUDOI, A.B. 24912, 27678, 35583, 35835
RUDOLPH, E.D. 36427
RUDOLPH, H. 31559
RUDOLPH, K. 39377
RUDOVA, T.S. 27790, 40089
RUDY, K.C. 35584
RUECKERT, L. 35570
RUETER, J.G. 34953
RÜFFER, U. 35585
RUFFNER, H.P. 24255
RUFNER, R. 24256
RUGGIERO, P. 34107-8
RUGGIU, D. 33804, 34474, 35586-7,
    35586, 35654, 39378
RÜHLE, W. 25029, 32418, 39379-80
RUIJGROK, T.W. 24300
RUIZ, J.B. see BONILLA RUIZ, J.
RULE, D.E. 27679
RUMBERG, B. 24257, 24374, 27680,
    31560-2
RUMI, C.P. 31563
RUMYANTSEVA, V.B. 26735, 36281
RUMYANTSEVA, V.I. 38443
RUNDEL, J.A. 39382
RUNDEL, P.W. 39381-3
RUNECKLES, V.C. 22181, 24592-3, 28757,
    35720
RUNGE, T. 38094
RUNNING, S.W. 22255, 28341, 33985
RUNQUIST, J. 30939
    see RUNQUIST, J.A.
RUNQUIST, J.A. 30626
    see RUNQUIST, J.
RUPERT, C.S. 35588-9
RUPP, H. 39384-5
RÜPPEL, H. 37016
RUPPEL, H.G. 25874, 30668, 35590,
    38183
RURAINSKI, H.J. 24258, 27681, 27747,
    31564, 35591
RUSANOV, S.Yu. 38171
RUSH, J.D. 29438, 33483, 37454
RUSHLOW, K.E. 33841
RUSINOVA, N.G. 35592-3, 39526
RUSS, P.N. 29536
RUSSEL, C. 39022
RUSSEL, W.J. 24259-60, 29881

SAIJO, R. 35123, 35619
SAIJO, Y. 23659, 26530, 33851
SAINIS, J.K. 27699, 39401
SAINT, S. 34146
SAITO, A. 33964
SAITO, H. 28087
SAITO, T. 23182, 35620-1
SAITO, Y. 21523, 22769
SAITOH, M. 24282, 31142, 35178-9
SAJI, T. 31577
SAKA, H. 24283, 31578, 35023, 38103
SAKAI, A. *33911, 34597, 35223, 35999*
SAKAI, S. 35622, 39402
SAKAI, W.S. 24943
SAKAI-IMAMURA, M. 24284
SAKAKIBARA, M. 35100
SAKALO, N.D. 39403
SAKAMOTO, E. 22884
SAKANISHI, Y.· 24285
SAKANO, K. 23256, 28385
SAKANOSHITA, A. 31781, 35813
SAKATA, K. 25953
SAKATA-SOGAWA, K. 31056
SAKER, L.R. 37349
SAKHAROVA, O.V. 24286, 31579, 34918,
    36558-9, 38769, 39404
SAKO, S. 30787
SAKS, N.M. 27700
SAKUNTS, L.E. 39114
SAKURA, T. 35178-9
SAKURAI, H. 31580
SAKURAI, T. 24282, 31142
SĂLĂGEANU, N. 24287
SĂLĂGEANU, V. 39405
SALAI, L. 23364
    see SZALAY, L.
SALAMA, F.M. 29410, 31581, 35623
SALAMAKHA, O.V. 28609, 36730
SALAMATOVA, L.V. 40089
SALAMINI, F. 38809
SALAMON, Z. 24288, 26062, 28395-6,
    28434, 31582, 32433, 39406
SALARES, V.R. 31583
SALATENKO, V.N. 24289, 31584-5, 35624
SALAZAR, C.R. 39407
SALCHEVA, G. 24290, 26131, 29726,
    39408-10, 40325
SALDANA, G. 21921
SALE, P.J.M. 24291, 27701, 31586
SALEM, L. 27702
SALEMA, R. 35625
SALEMME, F.R. 30398
SALERNO, J.C. 28832, 36919
SALIH, F.A. 31587
SALIN, M.L. 31588
SALINGER, S. 21862
SALISBURY, F.B. 31501
SALISBURY, J.L. 24292, 28250, 29512,
    31589, 35626
SALLAL, A.-K.J. 24293, 29143, 31590,
    33201
SALMINEN, R. 33878, 37808, 38174
SALMON, J.-M. 23146

SALMON, R.T. 30509
SALMONSON, B.J. 25126
SAL'NIKOV, A.I. 39411
SALNIKOW, J. 32264, 40050
SALOKHIDDINOV, K.I. 33414
SALONEN, K. 24294, 39412-4
SALONTAI, A. 25312
SALUJA, A.K. 35627
SALUNKHE, D.K. 22911
SALVADOR, G. 35072
SALVADOR, G.F. 27703, 35628-30, 39415
SALVUCCI, M. 39416
SALZER, J. 29005, 31591
SAMARAKOON, A.B. 39417
SAMARRAI, S.M. 24295
SAMBEEK, J.W. van see VAN SAMBEEK, J.W.
SAMBO, E.Y. 31592
SAMEJIMA, M. 30098, 30887, 35631
SAMEOTO, D.D. 39418
SAMI, M. 32688
SAMLEV, Kh.S. 31593
SAMISH, Y.B. 22819, 24296-7
SAMMIS, T.W. 32797
SAMMONS, D.J. 35632, 39419
SAMOĬLOVA, L.A. 21968
SAMOĬLOVA, O.P. 27704, 27950
SAMOKHVAL, E.G. 26710, 30359-60
SAMOKHVALOV, G.I. 23510
SAMORAY, D. 24298
SAMOSHINA, N.M. 31594
SAMPAIO, E.S.V.B. 30288-9
SAMPSON, E.J. 31595
SAMSONOVA, I.A. 35128, 35633
SAMSONOVA, L.P. 29212
SAMSUDDIN, Z. 25663, 31596, 33105-6,
    35634, 39420-3
SAMUELSSON, G. 31597, 32000, 35635
SAMUILOV, F.D. 35636-8, 39293-7
SAMUILOV, V.D. 21690-1, 22253, 25323,
    25884, 28515, 28700-1, 31479,
    31598, 32760, 33709, 36786-7,
    37345
SAMUNI, A. 37353
SANADA, M. 35157
SANADA, Y. 31058
SANADI, D.R. *26720*, 27390, *32675,
    32860, 32890, 33253, 33380, 33686,
    34277, 34357, 34762, 34803, B35639-
    -40, 35741, 35745*
SANADZE, G.A. 27125-6, 27705, 31599-
    -600, 32087, 34866, 35641
SANAI, S. 31601-2
SÁNCHEZ, C.S. 31603
SANCHEZ, S.M. 31604
SÁNCHEZ-DÍAZ, M.F. 39424
SANCHO, C. 33530
SANDERS, D.C. 24299
SANDERS, F.E. *23385, 24114, 24632,
    25072*
SANDERS, J.G. 39425
SANDERS, J.K.M. 25844, 27706, 31605,
    33332, 35642, 36357
SANDERS, J.L. 27707

SHARP, D.D.   24424
SHARP, J.H.   22353, 39559
SHARP, R.E.   39560-1
SHARPE, D.M.   24425
SHARPE, F.T. Jr.   39464
SHARPE, P.J.H.   24426, 33321, 38168,
    38605
SHARPE, S.A.   22126-7, 37200
SHARPLESS, R.G.   28318, 36335
SHARUPICH, V.P.   35787, 35850
SHATILOV, I.S.   24427-31, 27813, 31753,
    35788-91, 39562
SHATILOV, V.R.   40089
SHATKOVSKIĬ, T.A.   39563
SHAŬCHUK, S.M.   29593, 33630, 37602
    see SHEVCHUK, S.N.
SHAUKAT, S.S.   24432
SHAVEL', S.Kh.   38886
SHAVER, G.   34888
SHAVER, G.R.   27814, 35792
SHAVIT, N.   24416, 24471, 25177,
    27745, 27815, 29105, 31754-5,
    31799, 32736, 35838, 39608
SHAVIV, G.   28070
SHAVNIN, S.A.   39391
SHAW, A.B.   27816
SHAW, A.F.   29954, 37888
SHAW, E.K.   23909, 27340, 27342,
    28232, 31137
SHAW, E.R.   23057, 30255
SHAW, M.A.   31756
SHAW, N.H.   22716
SHAW, R.H.   35903-5, 35977, 38779,
    39660-1
SHCHEGOLEVA, N.A.   23438
SHCHERBAKOVA, I.Yu.   26143, 27817,
    37640-1, 40102
SHCHERBOVA, M.A.   21541
SHCHUTSKAYA, V.V.   21563-4
SHEAR, D.B.   24433
SHEAR, H.   24434
SHEARER, J.A.   33297, 37484
SHEARMAN, R.C.   24435
SHEATH, R.G.   27818, 31757-8, 33211,
    35793, 39564
SHEEHY, J.E.   24436-8, 27819, 31759-62,
    39565-7, 39566, 40275
SHEEN, S.J.   29762, 35794, 37219,
    39568
SHEFFER, K.M.   35795
SHEIKH, A.S.   27808
SHEIKH, K.H.   *25526, 25653*
SHEIKHOLESLAM, S.N.   24439, 31763-4
SHEĬNIN, D.M.   25823
SHEĬTANOV, Kh.   35796
SHEKHAR, V.C.   31765
SHEKHTMAN, L.M.   23486
SHELDON, R.B.   24440, 25523, 31766,
    35797
SHELDON, R.W.   35798
SHELDRAKE, A.R.   39569
SHELEF, G.   31767, 32825, 35799, 38985
SHELEMETOVA, L.I.   31154

SHELEPOVA, V.M.   39187
SHELL, G.S.G.   24441, 26844, 27820
SHELP, B.   27821, 38658, 39570
SHEMBERG, M.A.   33725
SHEMIN, D.   26248, 27822, 31005
SHEN, T.C.   27506, 39866
SHEN, Y.   40307
SHENDEROVA, L.V.   33131
SHENGELIA, K.Ya.   32904
SHEN-MILLER, J.   31768
SHEPARD, A.   28402
SHEPARD, D.V.   35800
SHEPHARD, D.C.   30923, 34948
SHEPHERD, H.S.   39571
SHEPHERD, W.D.   35801
SHEREVERYA, N.I.   39572
SHERIDAN, M.A.   39040
SHERIDAN, R.P.   27823-5, 35802, 39573
SHERIFF, D.W.   27826, 31769-71
SHERMA, J.   35803
SHERMAN, D.   32049
SHERMAN, D.B.   39850
SHERMAN, L.   23322
SHERMAN, L.A.   27827, 31772, 35058,
    35804, 39574
SHERMAN, W.V.   27828-9, 31773, 35805,
    39575, 39670-1
SHERRATT, D.   28910, 28912
SHERSTENIKINA, A.V.   23550
SHESTAK, Z.   31010
    see ŠESTÁK, Z.
SHETH, M.   39576
SHETLIK, I.   21770
    see ŠETLÍK, I.
SHEU-HWA, C.-S.   24442
SHEVCHENKO, A.I.   39577
SHEVCHENKO, V.A.   36669
SHEVCHENKO, Zh.P.   31774
SHEVCHUK, N.V.   27855
SHEVCHUK, S.N.   27854, 29594, 35806,
    37603
    see SHAŬCHUK, S.M.
SHEVELUKHA, T.A.   38198
SHEVELUKHA, V.S.   27830
SHEVYRĖVA, V.V.   31557, 34856
SHEVYRNOGOV, A.P.   37648
SHEWAN, J.M.   *31965*
SHEYTANOV, H.E.   39578
SHI, J.   39579, 40307
SHIBA, T.   33863, 39580-1
SHIBA, Y.   35807
SHIBASAKI, S.   21518
SHIBATA, H.   24443, 27831-2, 31091,
    31775, 35124-5, 38932, 39582
SHIBATA, K.   22856-7, 22883-4, 23132,
    23874-6, 24842, 26480-2, 26703-4,
    27222, 30085-6, 30355, 31128,
    31776, 34104-6, 34367, 34526,
    35139-40, 36508, 37996-7, 38249-50,
    38796, 38820
SHIBAYAMA, M.   38020
SHIBLES, R.   24444, 25881, 27833
SHIDA, S.   27834-5

SILIPRANDI, N. (continued)
    33350, 33787, 33945, 34165, 34805,
    35568, 40167
SILKIN, V.A. 39633-4
SILSBURY, J.H. 26087, 29575-6, 31822,
    33611, 34810, 39635-6
SILVA, J.F. da 27880
SILVA, J.V. da see VIEIRA DA SILVA, J.
SILVA, L.F. 40025
SILVANOVICH, M.P. 27881
SILVER, A.N. *23147*
SILVER, B.L. 23799
SILVERBERG, B.A. 24499-500
SILVERMAN, M.P. 38403, 38405
SILVERT, W. 35865
SILVERTHORNE, J. 33454
SILVIUS, J. 24501
SILVIUS, J.E. 24502, 31823, 31894,
    35866-7, 37124-5, 39637-8
SILVOLA, J. 39639
SIM, S.L. 26688
SIMARD, Y. 38454
SIMIC, M. 23850, 24612
SIMIDU, U. 33863, 39580-1
SIMINICEANU, E. 39143
SIMIONESCU, B.C. 31824, 35868
SIMIONESCU, C.I. 31824, 35868
SIMKINS, J. 26138
SIMMELSGAARD, S.E. 27882
SIMMONDS, P.G. 35869
SIMMONS, G.M. Jr. 24503, 37354
SIMMONS, R.E. 32997, 37021
SIMON, B.M. 30166, 38076
SIMON, H. 28611
ŠIMON, J. 24504, 31825-6, 35870-1
SIMON, J.C. 31827
SIMON, J.-P. 39641-4
SIMON, W. 22675
SIMONETT, D.S. *37784*
SIMONIS, W. 21859, 23165, 31828,
    38453, 39645
SIMONOVA, E.I. 31829, 34094
SIMONSEN, J.F. 35872
SIMPKINS, I. 40141
SIMPLĂCEANU, V. 33667
SIMPSON, D. 35873-4
SIMPSON, D.I. 27883, 32632
SIMPSON, D.J. 24505-6, 27884, 30028,
    30216, 30688, 31830-4, 35875-6,
    38557, 39646
SIMPSON, E. 35877
SIMPSON, E.E. 33122
SIMPSON, G.M. 33184, 36016-7
SIMPSON, J.R. 25842
SIMPSON, K. 33196, 40329
SIMPSON, K.L. 26157, 27644-5, 27885,
    29205, 37433
SIMPSON, R. 32403
SIMPSON, R.L. 36386-7, *36388*
SIMPSON, W.H. 24197
SIMS, J.L. 29762
SINADA, F. 31965
SINCLAIR, J. 25242-3, 27208, 31835,
    35878-9, 37609, 39647-8

SINCLAIR, M. 35880
SINCLAIR, T.R. 22598, 24507-8, 27886-8,
    31836, 36372, 38535, 39649-50
ŠINDELÁŘ, L. 30714, 34716
SINDEN, S.L. 39651
SINEL'NIKOVA, V.N. 28199
SINENSKY, M. 35881
SINESHCHEKOV, O.A. 27889, 31837
SINESHCHEKOV, V.A. 23432-4, 26938,
    27391, 27626, 27861, 27890, 31838
SINGER, B. 27891
SINGH, A.P. 39742
SINGH, B. 37302
SINGH, B.D. 36268
SINGH, B.K. 35225-6
SINGH, B.P. 35882
SINGH, D. 35883
SINGH, D.P. 35884
SINGH, H. 39652
SINGH, H.G. 27074
SINGH, H.P. 35883
SINGH, J.S. 24509, 25753, 35243,
    35885
SINGH, K.P. 31839
SINGH, L. 31840
SINGH, L.B. 31841
SINGH, M. 27892, 34437, 39741
SINGH, M.K.. 35886
SINGH, M.P. 35887, 39655
SINGH, N. 24510, 36881
SINGH, O.S. 33680
SINGH, P. 35883, 39401
SINGH, P.K. 39803
SINGH, R. 34834, 39653-4
SINGH, R.B. 36268, 39654
SINGH, R.K. 21904, 25555-6, 28948-9,
    38512
SINGH, R.M. 36268
SINGH, R.P. 24511, 31839-40, 31842
SINGH, S. 39659
SINGH, S.P. 35888, 35966, 39344
SINGH, U.N. 27893-4
SINGH, V.B. 27389, 35225
SINGH, V.P. 21795, 24512, 25424, 27895,
    32212, 39256-7
SINGHAL, G. 28817
SINGHAL, G.S. 28789, 31356, 35889,
    36899, 37007
SINGHAL, N.C. 39655
SINHA, B.D. 35890
SINHA, S.K. 25299, 25845, 27896, 35891-3,
    36760, 39656
SINITSYNA, Z.A. 24018
SINK, K.C. Jr. 30346
SINYAKOVA, R.S. 25122, 36541
SIONIT, N. 27897, 37184
SIPOS, G. 31843
ŠÍPOSOVÁ, M. 39727
SÎRBU, M. 38725
SIRCAR, S.M. 25686, 26806, 27384-6,
    39033-4
SIRECI, J.E. 35894
SIRENKO, L.A. 24513-22, 27898

TROXLER, R.F.   21920, 24809-10, 26558,
   28959-60, 32173, 33234, 36186-7,
   37014, 39964-6
TRUBACHEV, I.N.   21735, 31331, 35351
TRUBACHOV, I.N.   26151
TRUBAČÍK, S.   33418
TRUBY, Y.   29451
TRUDEL, M.J.   28608
TRUKHAN, É.M.   24811-3, 37288
TRUMP, C.K.   37278
TRUMP, K.   22702
TRUNOV, I.A.   39967
TRUONG QUANG TAN 23258
   see TAN, T.Q.
TRÜPER, H.G.   24814, 28178, 30079,
   31576, 36188-9
TRUSCOTT, T.G.   26832, 32812
TRUSOVA, V.M.   23676-7, 27167-70,
   30900-1, 34864, 34928, 34930-1
TRUTNÉVA, I.A.   23460-1
TSAGALOFF, A.   *31103*
TSAKIRIS, S.   28586, 32595, 32661
TSANG, M.L.-S.   24815-6
TSANKOV, B.   27388, 30480, 39968
TSAO, M.S.   24205
TSAPIN, A.I.   22561, 26176, 29690,
   32885, 33719, 36312
TSCHAKALOVA, E.   28179-80
   see TSCHAKALOWA, E.
TSCHAKALOWA, E.   28181
   see TSCHAKALOVA, E.
TSCHÄPE, M.   28182
TSCHERMAK-WOESS, E.   39969
TSCHISMADIA, I.   30924, 39970
TSCHUMI, P.A.   32174
TSEL'NIKER, Yu.L.   24817-8, 32175,
   36190, B36191, 39971-2
TSENG, Y.-W.   38815
TSENOVA, E.   28220, 32262, 39973
TSENOVA, E.N.   28183, 31042, 32176,
   36192
TSENOVA, M.   32177, 39974
TSENOVA, M.P.   32178-9
TSIMASHÉNKO, M.K.   31209
TSIMILI-MICHAEL, M.   32595
TSIMILLI-MICHAEL, M.   23957, 28520,
   36193
TSISKARISHVILI, L.P.   33152
TSIVION, Y.   32516, 36194
TSOGLIN, L.N.   32180, 39534, 39975
TSOKOS, C.P.   38057
TSONEV, Ts.   39751, 39995
TSUBO, Y.   24819, 38950
TSUCHITANI, Y.   26717
TSUCHIYA, M.   36070
TSUCHIYA, T.   39976
TSUJI, H.   33302, 35021, 36195-6,
   38853-4, 39977
TSUJI, K.   39978
TSUJIMOTO, H.Y.   27104, 28184, 32674,
   37884
TSUJITA, M.J.   39979
TSUK, R.M.   25938

TSUKADA, O.   36197
TSUKAMOTO, A.   23261
TSUKIHARA, T.   31100
TSUNEWAKI, K.   34137
TSUNO, Y.   24323-4, 24820-1, 32181-3,
   36198-9
TSUNODA, S.   24822, 34308-10, 35660,
   35886, 36064, 38783, 39054, 39980-1
TSUSHIMOTO, G.   24823
TSUTSUI, Y.   35137
TSUTSUMI, M.   25097
TSUZUKI, E.   27834-5
TSUZUKI, M.   39982
TSVETKOVA, A.M.   21726, 23996
TSVETKOVA, I.V.   34545
TSVETKOVA, S.   35973
TSVIRKO, M.P.   22585, 31618, 33407,
   33763, 37374-6
TSVIRKO, M.Ts.   33408
TSVYLEV, O.P.   32184
TSYABUT, L.F.   24548
TSYARÉNTS'EÜ, V.M.   28185, 30400
TSYDENDAMBAEV, V.D.   27806
TSYGANKOVA, I.G.   26670
TSYGANKOVA, T.A.   28186
TSYTSARIN, G.V.   30843
TU, J.C.   24824, 39983
TUAL, Y.   37192
TUAN, H.   40367
TUBA, Z.   29480, 32185
TUBEA, B.   29873
TUCKER, C.J.   28187-8, 32186-9, B37906,
   39984-5
TUGARINOV, V.V.   23272, 30488, 32190
TUGNAWAT, R.K.   32191
TUGULEA, L.   33667
TUKENDORF, A.   21698, 28708, 32769,
   36803, 39986
TULBU, G.V.   26773, 30436, 36200,
   39987-8
TULEY, J.   37816
TULLY, R.E.   37803, 39989-90
TUMANYAN, É.R.   26232
TUMERMAN, L.   24825
TUMIDAJOWICZ, D.   36201
TUNG, H.F.   21886
TUNGAROV, G.   36021
TUNSTALL, B.R.   24826
TUPPY, H.   31915
TUPYK, N.D.   24644
TUQUET, C.   22627, 24827-8, 28189, 29765,
   32192, 33805, 37747
TUR, N.M.   23661
TURČIĆ, M.   24829
TURCOTTE, E.L.   36202
TURCZYŃSKA, J.   28190, 32137
TURGEON, A.J.   28191
TURGEON, R.   24830-1
TURI, A.   30888
TURI, B.   30745
TURISCHEVA, M.S.   35128
   see THURISHCHEVA, M.S.
   see TURISHCHEVA, M.S.

(continued)

382

VUNKOVA-RADEVA, R. 39410
    see VANKOVA-RADEVA, R.
    see V"LKOVA RADEVA, R.
VUNKOVA-RADEVA, R.V. 40113
VUOKKO, R. 34291
VU VAN VU 23777, 23779-80
VYARK, É. 26620
    see VYARK, É.Ya.
VYARK, É.Ya. 40114
    see VYARK, É.
VYAS, A.B. 32324
VYAS, L.N. 24942, 28294, 32325,
    36319, 40115
VYAS, N.L. 24942, 28294, 32325
VYAS, O.P. 24310
VYSHKVARTSEV, D.I. 40116
VYSKOT, B. 28216, 32326

W

WAAL, F.E.B. de see DE WAAL, F.E.B.
WAALAND, J.R. 28295, 34853, 38703
WAALS, J.H. van der
    see VAN DER WAALS, J.H.
WABER, J. 24943
WACKER, G. 40117
WACHTER, E. 39518
WADA, K. 22693, 24648, 24944-5,
    26305-6, 29852-3, 31100, 33905,
    36505, 38630
WADE, C.G. *28437*
WADSWORTH, R.M. 25943
WAFAR, M.V.M. 31202, 35439, 36320
WAFFORD, J.D. 28296
WAGENER, K. 24946, 31467, 32327
WAGENMANN, R. 33420
WAGGONER, P.E. 24947, 32328, 40118
WAGHMODE, A.P. 40119
WAGNER, E. 28297, 29267, 32329
WAGNER, F. 24924
WAGNER, G. 25720, 28298-9, 32330,
    34033, 35135, 36321-3
WAGNER, G.H. 23229
WAGNER, G.J. 24948
WAGNER, G.L. 32544
WAGNER, H. 33457
WAGNER, J. 26854
WAGNER, R. 32331-2
WAGNER, W. 32333, 36324, 39816
WAGO, K. 31000
WAHUA, T.A.T. 36325
WAIDYANATHA, U.P.de S. 24949-50, 30288
WAIN, R.L. 35384
WAINWRIGHT, S.J. 28510
WAISEL, Y. 21722, 24469, 27860, 28728,
    31798, 36831, 39248, 39603-5
WAITZ, G. 21554
WAJDA, L. 36803
WAKABAYASHI, S. 29854, 33902-5
WAKAMATSU, K. 22995, 24455, 24951,
    40120

WAL, H.N. van der
    see VAN DER WAL, H.N.
WALACH, M. 35336
WALBOT, V. 32334-6, 33053, 40121
WALCKHOFF, B. 35718
WALCOTT, J. 23696
WALCOTT, J.J. 28300
WALCZAK, T. 33615, 36326
WALDEN, D.B. 22799, 30007
WALDEN, R. 36327
WALDRON, J.C. 28301, 32626, 40122
WALI, M.K. 26724
WALK, R.-A. 24952, 32337-8, 36328
WALKER, A.J. 28302, 32339-41, 36329
WALKER, D. 33424
    see WALKER, D.A.
WALKER, D.A. 23415-6, 24442, 25833,
    26360, 26376, 27906-9, 28303-7,
    29891, 31939, 32342-3, 33310,
    33426-7, 36330-1, 37844, 37863,
    38275, 38452, 38487, 39333-7,
    40123
    see WALKER, D.
WALKER, G.H. 40124
WALKER, I.D. 25547
WALKER, J.N. 28369-70
WALKER, J.R.L. 28308, 40125
WALKER, K.F. 24953
WALKER, L.L. 26431, 30025
WALKER, N.A. 24954, 32344-5
WALKER, R.B. 21705, 26895, 28475,
    30563, 36332
WALKER, R.R. 40126
WALKER, R.W. 40283-4
WALKER, T.E. 38271
WALL, B.H. 26502
WALL, J.D. 24955, 30762, 32346, 38570
WALLACE, A. 21670, 23112-3, 26684-5,
    28309-13, B30325, 31535, 32347
WALLACE, B. *35167*
WALLACE, D.G. 28314-5
WALLACE, D.H. 28316, 28382, 31165,
    31250-1, 38346
WALLACE, H.R. 34851
WALLACE, J.S. 25430, 37592
WALLACE, L.L. 36333
WALLACE, R.B. 24205
WALLBANK, B. 21591
WALLEN, D.G. 24956
WALLENTINUS, I. 24957, 28317, 32348,
    33813, 36334
WALLER, G.R. *39007*
WALLER, S.S. 40127
WALLES, B. 23224, 24958-9
WALLIHAN, E.F. 28318, 36335
WALLIN, R. 36336
WALLIS, A.C. 34191-2
WALLSGROVE, R.M. 30853, 40128
WALNE, P.L. 27383, 29364, 32349-50
WALSBY, A.E. 24434, 29729, 34395,
    36337, 37097
WALSH, G.E. 22790, 24960

                                                            (continued)

This cumulative index contains a selection of primary items chosen according to their importance in photosynthesis research and to their relevance and occurrence. The word "Photosynthesis" is not regarded as a main theme, but partial processes, photosynthetic parameters and the factors affecting photosynthesis are listed. The processes and other characteristics are summarized into several main themes when presented in combination with individual factors, e. g. carbon fixation pathways, electron transport chain, chlorophyll, gas exchange, ecosystem and plant productivity (*including photosynthate distribution and translocation, and canopy organization and functioning*), photorespiration, resistances to CO_2 and water vapour transfer, *etc.*
    Several items from branches related to photosynthesis research were also chosen for convenience, e. g. dealing with respiration, plant growth and development, water relations, anatomy, bioclimatology, *etc.* These items contain only references to papers within the scope of this bibliography.
    References from the individual volumes are distinguished by a semicolon (;) or a paragraph.

A

Abscisic acid   see   Growth regulators ...

Absorbance in canopy   see   Canopy, radiation profile

Accumulation of dry matter   see   Biomass distribution ...; Dry-matter production ...;
        Ecosystem production ...

Achlorophyllous cells and organs, respiration   see   Respiration of achlorophyllous
        tissues in light, light inhibition of respiration

Action spectra   see   Irradiance, spectral composition ...

Adenosine triphosphate   see   ATP

Aerodynamic methods, bioclimatological methods (sampling, measurement of wind, rain,
        dew, *etc.*)
        21801, 21803, 22302, 22505, 24005, 24677, 24782, 24907; 25195, 25279, 27434,
        27912, 28010-1, 28475; 31272; 36880, 37992, 38064, 38302-3, 38442, 39984,
        40194

Age of algae, leaf, plant   see   Ontogeny ...; Canopy, leaf age

Agrotechnics and carbon fixation pathways   37862

Agrotechnics and carotenoids   34707

Agrotechnics and chlorophyll   23053, 24019; 27980, 28156; 34707; 39532

Agrotechnics and ecosystem and plant productivity   21686, 22428-9, 22503, 22532, 23405,
        23426, 23589, 23790, 23923, B24070, 24289, 24429, 24470; 25454, 25613, 25997,
        26030, 26874, 27094, 27740, 27982, 28212, 28381; 28813, 29072, 29492, 29566,
        30648, 30953, 31208, 31952, 32545; 32713, 32738, 32761, 33532, 34500, 35101,
        35514, 35620, 35871, 36086, 36388, 36433, 36497, 36500-1, 37192, 37302,
        37441, 37488, 37807, 38082, 38100, 38123, 38627, 38666, 38790, 38823, 39140,
        39280, 39667, B39903, 40250

Agrotechnics and electron transport chain   34316

Agrotechnics and gas exchange   23053, 23194, 24431; 25613, 26014, 26182, 26373, 26591,
        28020; 28844, 29366-7, 29491, 30930, 31946; 32713, 33532, 33677, 33845,
        34707, 35201, 36086, 36133, 36497; 37381, 37862, 38790, 38968

Agrotechnics and resistances to CO_2 and water vapour transfer   30930; 32713, 35610; 36637, 37441, 38790

Agrotechnics and respiration   36497

Air-conditioning in photosynthesis measurement   see   Gasometric system, conditioning of air

Air-flow rate   see   Wind ...

Albedo, canopy   see   Canopy, radiation distribution

Algae and photosynthetic bacteria, cultivation (*cf.* also Algae mass cultures productivity)
   B22052, 22931, 23142, 24287, 24498, 24513, 24587; 26357, B28494; 29230, 29913, 30185, 30601, 31159, 31299, 31329, 31772; 32623, 34143, 34191-2, 34363, 35109, 35193, 35195, 36001, 36248; 36620, 37195, 37672-3, 37800, 38344, B38477, 38484, 38938, 38985, 39196, 39405, 39576, 39882

Algae and secondary production of reservoirs
   22155, 22436, 22454, 22578, 23294, 23301, 23314, 23316, 23646, 23844, 24599- -600, 24687, 24766; 25743, 26130, 26242, 26371, 26953, 27281, 27392, 27766, 28076, 28394, 28465; 28985, 29311, 29426, 29588, 29658, 29739, 30407, 30605- -6, 30735, 30776, 30807, 31891, 32413; 32564, 32941, 33115, 33169, 33466, 33675, 33896, B34248, 34293, 34517, 34763, 34818, 35298-9, 35436, 35462, 35552-3, 35597, 35715, 36141; 36886, 37478, 37606, 37655, 38031, 38076, 38306-7, 38521, 38752, 39060, 39255, 39781, 39823, 39886

Algae, blue-green, chromatophores in   see   Chromatophore ...

Algae carotenoids   see   Xanthophylls of algae

Algae chlorophylls   see   Chlorophylls c, d

Algae, CO_2 and O_2 exchange   see   Gas exchange in algae

Algae, depth distribution in reservoirs
   21682, 21726, 21742, 21805, B21829, 21936, 21972, 22155, 22158, 22256-7, 22268, 22389, 22417, 22446, 22453, 22488, 22493-4, 22517, 22531, 22578, 22628, 22646, 22742, 22745-6, 22866, 22921, 22943-4, 23016, 23065, 23087, 23098, 23133, 23269, 23314, 23377, 23387, 23441, 23456, 23593, 23606, 23824, 23931-2, 24051, 24058, 24061, 24140, 24144, 24194, 24200, 24248, 24349, 24440, 24572, 24578-9, 24616, 24618, 24680, 24708, 24743, 24763, 24766, 24857, 24896, 24906, 24953, 25053, 25130, 25157
   25210, 25282, 25390-1, 25426, 25455, 25485, 25525, 25588, 25617, 25633, 25694, 25841, 25849-50, 25894, 26002, 26051, 26163, 26179, 26217, 26242, 26273, 26330, 26358, 26371, 26407, 26420, 26518, 26529, 26540, 26571, 26582, 26656, 26823, 26849, 26953, 27057, 27143-4, 27225, 27227, 27376, 27400, 27459, 27470, 27534, 27567-8, 27633, 27648, 27741, 27838, 27913, 27938, 27956-8, 27960-1, 27970, 27986, 27988, 28004, 28118, 28129, 28140, 28162, 28174, 28394, 28471-2, B28494
   28536, 28553, 28561, 28740, 28797, 28955, 28977, 28982, 28984, 29041, 29151, 29171, 29185, 29226, 29242, 29288, 29365, 29375, 29426, 29428-9, 29511, 29588, 29655, 29674, 29716, 29778, 29802, 29809, 29916-7, 29919, 30032, 30075, 30121, 30147-8, 30196, 30225, 30281, ¯30305, 30356-7, 30397, 30407, 30645, 30724, 30863-4, 31216, 31278, 31311-2, 31316, 31328, 31373, 31396, 31430, 31531, 31555, 31714, 31766, 31965, 31988, 32032, 32043, 32136-8, 32174, 32308, 32368, 32380
   32565, 32598, 32705, 32712, 32825, 32999, 33018, 33070-1, 33152, 33238-9, 33274, 33365, 33367, 33379, 33466, 33531, 33586, 33656, 33676, 33721-2, 33804, 33847, 33851, 33923, 33977, 34139, 34177, 34216, 34240, 34359, 34395, 34407, 34474, 34519, 34528, 34537, 34567, 34571, 34592, 34793, 34878, 34994, 35070, 35196, 35254, 35299, 35347, 35407, 35436, 35462, 35587, 35713, 35734, 35799, 35841, 35864, 35880, 35896, 35908, 35946, 35965, 36061, 36079, 36123, 36141, 36155, 36216, 36271, 36295, 36512
                                                                              (continued)

Algae, depth distribution in reservoirs (continued)
        36634-5, 36750, 36835, 36868, 37110, 37205, 37226, 37238, 37258, 37354,
        37369, 37484, 37540, 37635, 37809, 37818, 37859, 37993, 38031, 38067, 38070,
        38086, 38124, 38251, 38307, 38359, 38374, 38451, 38659, 38686, 38693-4,
        38752, 38973, 39025, 39108, 39209, 39314, 39327, 39346-8, 39378, 39418,
        39455, 39509, 39530, 39696, 39722, 39836, 39854, 39871, 39919, 39924, 40084,
        40095, 40153, 40232, 40252, 40327

Algae in cosmonautics B22052

Algae in sediments
        32998-9, 33320, 33531, 33594, 33877, 34184; 37258, 37443, 37977, 38826, 39992

Algae in sewage cleaning 26267, 26420; 30925; 34362, 35799; 38490, 38585, 39457

Algae life cycles see Ontogeny of algae ...

Algae mass cultures productivity (*cf*. also Algae and photosynthetic bacteria, culti-
        vation)
        21735, 21789, 23229, 24119, 24287, 24454, 24702; 26702, 26882, 27652, 28295;
        28763, 29230, 29276, 29691, 30185, 30757, 30836, 31162, 31203, 31603; 33728,
        34143, 34363, 34404, 34538, 35109, 35195, 35995, 36197; B36859, 37195,
        37672-3, 38503, 38985, 39634, 39658, 39962, 40163

Algae photosynthesis and production
        21707, 21742, 21772, 21805, B21829, 21952-3, 22000, 22132, 22155, 22211,
        22248, 22256-8, 22268, 22287, 22292-3, 22494, 22538, 22628, 22898, 22943-4,
        23074, 23188-9, 23225, 23278, 23301, 23377, 23686, 23824, 24118, 24144,
        24308, 24349, 24420, 24440, 24562, 24586, 24644, 24687, 24708, 24786, 24838,
        24906, 24957, 24960
        25210, 25272, 25286, 25367, 25426, 25452, 25525, 25541, 25588, 25633, 25739,
        25763, 25806, 25847, 25849-50, 26002, 26089, 26110, 26163, 26179, 26214,
        26217, 26242, 26287, 26304, 26344, 26371, 26458, 26516, 26529, 26540, 26563,
        26568, 26570-1, 26752, 26838, 27227, 27231-2, 27377, 27392-6, 27418, 27463,
        27470, 27534, 27567, 27633, 27650, 27662, 27859, 27913, 27956, 27959-60,
        27970, 28026, 28076, 28088, 28129, 28152, 28162, 28174, 28317, 28325, 28394
        28536, 28553, 28561, 28687, 28776, 28797, 28830, 28985, 28996, 29089, 29151,
        29242, 29653, 29809, 29821, 29825, 29842, 29914, 29917, 30072, 30075, 30147,
        30151, 30196, 30281, 30356, 30484, 30486, 30501, 30522, 30605-6, 30645, 30724,
        30732, 30757, 30843, 30873, 30906, 31183-4, 31202, 31278, 31311, 31317, 31327-
        -8, 31382, 31396, 31514, 31555, 31700, 31912, 31988, 32032, 32091, 32137-8,
        32174, 32184, 32203, 32269, 32348, 32356, 32494, 32544, 32554
        32564-5, 32570, 32598, 32669, 32694, 32705, 32712, 32814, 32818, 32825,
        32887, 33061, 33070-2, 33081; 33091, 33152, 33163, 33169, 33243, 33334,
        33340, 33365, 33367, 33422, 33428, 33511-2, 33586, 33588, 33619, 33635,
        33656, 33676, 33704, 33722, 33851, 33888, 33896, 33915, 33923, 33965, 33977,
        34016, 34030, 34052, 34058, 34143, 34177, 34184, 34222, 34359, 34395, 34407,
        34436, 34489, 34515, 34555, 34571, 34741, 34811, 34818, 34840, 34870, 34884,
        34951, 34985, 34991-2, 34994, 35070, 35196, 35217, 35299, 35313, 35462,
        35584, 35586, 35597, 35672, 35733, 35797-8, 35872, 35880, 35896, 35908,
        35946, 36061, 36079, 36088, 36117, 36123, 36141, 36153, 36271, 36298, 36477,
        36512
        36657, 36715, 36750, 36758, 36778, 36868, 36886, 36992, 37008, 37054, 37058,
        37086, 37106, 37152, 37226, 37246, 37279, 37327, 37330, 37354, 37435-6,
        37450, 37466, 37540, 37606, 37673, 37868, 37945, 38050, 38060, 38067, 38079,
        38086, 38277, 38304, 38336-7, 38359, 38375, 38402, 38414, 38473, 38514,
        38559, 38585, 38686, 38692-4, 38752, 38782, 38807, 38826, 39025, 39083,
        39155, 39209-10, 39256-7, 39314, 39325, 39327, 39348, 39361, 39378, 39455,
        39509, 39530, 39668, 39696, 39714-5, 39718, B39759, 39768, 39781, 39787,
        39836, 39854, 39871-2, 39908, 39919, 39924, 39931, 39937, 39960, 39962,
        39992, 40089, 40116, 40153, 40172, 40175, 40224, 40234, 40266

Algae, primary productivity in reservoirs (*cf*. also Chlorophyll and production of
       algae and water reservoirs)
       21707, 21731, 21955, 22169, 22604, 22634, 23164, 23268, 23414, 23534, 23569,
       23644, 23685, 23765, 24062, 24447, 24503, 24608, 24617, 24699-700, 24736,
       24969, 25003, 25005, 25121
       25341-3, 25788, 25848, 26072, 26344, 26705, 26875, 27371, 27811, 27940
       29346, 29448, 29469, 30585, 32136, 32356
       33852, 34016, 34099, 34387, 34436, 34571, 34818, 34840, 34955, 35127, 35227,
       35244, 35407, 35735, 35965, 36079, 36097, 36141, 36295, 36369
       36592, 36715, 37796, 38124, 38755, 38807, 38973, 38998, 39055, 39108, 39186,
       39290, 39317, 39325, 39405, 39576, 39634, 39908, 39959, 39960, 40163, 40175

Algae, primary productivity, methods {*cf*. also O_2 determination (other than O_2 elec-
       trode); O_2 electrode}
       21742, 21952, 21955, 21958, 22143, 22177, 22577, 22592, 22654, 22757-8,
       22921, 22944, 23414, 24087, 24294, 24456, 24514, 24732, 24786, 24957, 25005
       25389, 25647, 25724, 25805, 25847, 25851, 26108, 26129, 26195, 26330, 26387,
       26393, 26563, 26697-8, 27067, 27112, 27144, 27266, 27376, 27400, 27418,
       27642, 27785, 27859, 28076, 28126, 28284, 28354
       28568, 28615, 28687, 28983, 29242, 29511, 29588, 29716, 29802, 29915, 29919,
       29988, 30072, 30147, 30166, 30196, 30247, 30517, 30815, 31127, 31183, 31382,
       31422, 31532, 31597, 31869, 31944, 32090-1, 32137, 32174, 32184, 32274, 32396
       32809, 32830, 32887, 32941, 32979, 33115, 33163, 33367, 33371, 33404, 33492,
       33535, 33721, 33813, 34247, 34386, 35216, 35546, 35635, 35672, 35797-8, 36149,
       36271, 36286, 36474
       36750, 36835, 37086, 37096, 37581, 37635, 37809-10, 38076, 38337, 38437,
       38462, B38477, 39096, 39101, 39412-4, 39669, 39715, 39718, 39962, 39996,
       40095, 40137, 40168, 40224, 40234, 40305

Algae synchronous cultures see Algae and photosynthetic bacteria, cultivation;
       Ontogeny of algae ...

Altitude see Pressure, altitude ...

Amino acids see Proteins, amino acids, nucleic acids ...

δ-Aminolaevulinic acid see Chlorophyll biosynthesis ...

Amphistomatous leaf, gas exchange in (*cf*. also Leaf epidermis, stomata) 22960;
       28623, 30721; 37313

Anaerobic atmosphere see N_2, anaerobic atmosphere ...

Antibiotics and biliproteins 37622

Antibiotics and carbon fixation pathways
       21616, 21702, 21724, 22018, 22117, 22192, 23451, 23480, 23715, 24016;
       25199, 25672, 25730, 26144, 27790, 28303; 28910-1, 29942, 30962, 31973,
       32176, 32262, 32291; 33518, 34065, 35088, 36533; 36784, 37557, 37650,
       37973, 38606, 39885

Antibiotics and carotenoids
       21642, 24778; 27972; 29266, 32014, 32018; 33177, 33592, 33684, 34111; 36763,
       39119-20, 39316

Antibiotics and chlorophyll
       21505, 21642, 21679, 21724, 21741, 22107, 22225, 22317, 22338, 22469, 22889,
       23025, 23095, 23108-9, 23224, 23272, 23396, 23528, 23851, 23882, 23969,
       24016, 24076, 24397, 24443, 24462, 24464, 24534, 24544, 24745, 24778, 25019,
       25075

(continued)

Antibiotics and chlorophyll (continued)
        25200, 25433, 25504, 25730, 26094, 26143, 26253, 26310, 26415-6, 26594,
        26885, 27025, 27041, 27076, 27081, 27140, 27394, 27775, 27817, 27831, 27853,
        27972
        28583, 28628, 28680, 29101, 29361, 29458, 29636, 29662, 29773, 29939, 29946,
        30201, 30488, 30594, 30935, 31064, 31288, 31687-8, 31795, 32001, 32098, 32414,
        32421, 32451
        32854, 32864, 33026, 33177, 33592, 33625, 33639, 33684, 33727, 34032, 34111,
        34207, 34883, 34891, 34934, 35093, 35144, 35283, 35651, 35657, 35671, 35814,
        35835, 35917, 36045, 36181, 36249
        36763, 36979, 36987,   37557-8, 37703,   37860, B38118, 38130, 38460, 38466,
        38651, 38819, 39415, 39587, 39598, 39707-8, 39766, 39988, 40112, 40135

Antibiotics and chloroplast (chromatophore)
        21545, 21863, 21931, 22083, 22117, 22469, 22729, 23224, 23637, 24076, 24348,
        24354, 24462
        25463, 25558, 25730, 25853, 26144, 26252, 26950, 27076, 27214, 27249, 27407,
        28292
        28897, 28923, 29465, 29824, 29839, 30036, 30334, 30682, 31056, 31155, 31204,
        31541, 31757, 32414, 32421
        32561, 32671, 32859, 32920, 33170, 33282, 33417, 33592, 33597, 33625, 33678,
        34883, 35104, 35383, 35657-8, 35705, 35728, 36249, 36284, 36488, 36529, 36544
        36645, 36793, 36895, 36979, 36987, 37055, 37135, 37175, 37557, 37641, 37860,
        38466, 39300, 39742-3, 39797-8, 40162, 40314

Antibiotics and electron transport chain
        21611, 21669, 21859, 21938, 21997, 22018, 22154, 22179, 22499, 22501, 22536,
        22612-3, 22773, 22809, 22894, 22903-4, 22923, 22970, 23124, 23127, 23199,
        23362, 23627, 23704, 23718, 23721, 23851, 24193, 24241, 24315-6, 24354,
        24757, 24869-70, 25090
        25590, 25804, 26194, 26253, 26310, 26415, 26442, 26968, 27023, 27076, 27262,
        27275, 27353-5, 27394, 27448, 27491, 27856, 28307, 28330, 28483
        28592-3, 28672, 28812, 28904, 29107, 29192, 29541, 29939, 29946, 29959, 30034-
        -5, 30114, 30183, 30307, 30361, 30437, 30594, 30599, 30831, 30982-3, 31049,
        31123, 31147, 31179, 31288, 31479, 31487, 31646, 31664, 31683, 31872-4,
        32145, 32303, 32321, 32375, 32408, 32414, 32421, 32496, 32498
        32657, 32673, 32703, 32733, 32755, 32853, 32864, 32949, 32955, 33047, 33086,
        33251-2, 33402, 33690, 33786, 33886, 33901, 33932, 33986, 34032, 34059,
        34259, 34282, 34356, 34372, 34674, 34789, 34841-2, 34866, 34883, 34891,
        35104, 35568, 35705, 35752, 35915, 35917-8, 35983, 36282, 36363, 36484
        36699, 36710, 36745, 36852, 37326, 37345, 37830, 37860, 37875, 37891, 37907,
        37942, 37956, 38030, 38453, 38620, 38647, 38656, 38824-5, 39023, 39118-20,
        39289, 39296, 39673, 39675, 39702, 39708, 39730, 39842, 39885, 40132, 40182,
        40298

Antibiotics and gas exchange
        22018, 22834, 22946, 23641, 23718; 25853, 25913, 26583, 26585, 27972; 29101,
        29775, 29942, 30026, 30034, 30223, 30935, 30962, 32093, 32145, 32414; 33169,
        33482, 34059, 34891; 37140, 37557, 37737, 37860, 38460, 39885, 40162

Antibiotics and resistances to CO_2 and water vapour transfer   29076

Antibiotics and respiration 37860

Antigens see Electron transport chain, serological analysis

Antitranspirants
        22171, 22187, 22190-1, 23703, 24842, 25140; B25775, 25813, 27693, 27983;
        29072, 29212, 29486, 30507, 30930, 31379, 31416, 32352; 32625, 33096, 33165,
        B33281, 33288, 33532, 34743, 34814, B34849, 35019, 35049, 35476, 35939,
        36011, 36371; 37184, 37243, 38134, 38345, 38532, 38609, B38722, 38790,
        40141

Architecture of canopy see Canopy ...

Assimilates see Photosynthates ...

Assimilation chamber
       21693, 21858, 22109, 22302, 22830, 22845, 22901, 22939, 23306, 23443, 23457,
       23608, 23727, 23782, 23834, 24029, 24045-6, 24259, 24291, 24341, 24429,
       24532, 24592, 24626, 24734, 24835, 24889
       25251, 25257, 25264, 25331, 25418, 25502, 25549, 26027, 26044, B26120, 26164,
       26372, 26375, 26506, 26743, 26745, 26895, 27280, 27421, 27461, 27826, 28085,
       28087, 28366, 28380, 28475
       28502, 28540, 28571, 28728, 28939, 29065, 29372, 29400, 29491, 29675, 29722,
       29864, 30263, B30325, 30343, 30358, 30394, B30432, 30490, 30545, 30563,
       30706, 30860, 30946, 31020, 31301, 31711, 31943, 32015
       32941, 33278, 33384-5, 33500, 33529, 33703, 33796, 34511, 34809-10, 35081,
       35341, 35827, 36069, 36137, 36406, 36434, 36515
       36687, 36735, 36802, 36804, 36900, 36985, 37023, 37360, 37658, 37761, 37808,
       37837, 38140, 38388, 38575, 38869, 38934, 39097, 39367, 39501, 39649, 39741,
       39756, 39946, 40033

ATP   21530, 21561, 21650, 21744, 21780, 21847, 21859, 21879, 21913, 21934, 22017-
       -8, 22051, 22064, 22134, 22201, 22215, 22233, 22287, 22392, 22426, 22474,
       22515, 22528, 22567, 22586, 22644, 22689-90, 22704-5, 22720-2, 22724, 22742,
       22785-6, 22813-4, 22824, 22904, 22912-3, 22923, 22935, 22946, 23005, 23060,
       23078, 23126-7, 23150, 23170, 23207, 23231, 23274, 23301, 23363, 23370,
       23484, 23491, 23506, 23586, 23641, 23721-3, 23809, 23847, 23888-9, 23899,
       23949, 23989, 24035, 24081, 24107, 24160, 24169, 24172, 24206, 24233, 24255,
       24264, 24312, 24333, 24351, 24384, 24416-8, 24484, 24568, 24638, 24652,
       24724, 24730, 24742, 24768, 24789, 24804, 24859-60, 24999, 25057, 25124,
       25159
       25177, 25184, 25226, 25244, 25261, 25305, 25317, 25320, 25401, 25487, 25508-9,
       25522, 25528, 25671, 25913, 25949-50, 25983-6, 26148-9, 26193, 26198,
       26238, 26268, 26291, 26322, 26332-3, 26341-2, 26441, 26472-3, 26500, 26517,
       26566, 26574, 26714, 26726, 26756, 26767, 26773, 26835, 26872, 26890, 26980,
       26997, 27006, 27017-9, 27037, 27039, 27064, 27091-2, 27112, 27131, 27194,
       27204, 27244, 27255, 27353, 27374, 27376, 27504, 27553, 27589-92,
       27597-8, 27640-1, 27652, 27661, 27665, 27712-3, 27745, 27783, 27788-9, 27809,
       27815, 27859, 27898, 27907-9, 27933-4, 28035, 28037, 28047, 28227, 28297,
       28303-4, 28307, 28327, 28354, 28411, 28445
       28524, 28581, 28592-4, 28633, 28636, 28672-3, 28755, 28777, 28784, 28879,
       28906, 28957, 28983, 29011, 29070, 29075-6, 29101, 29151, 29304, 29314, 29353,
       29385, 29449, 29533, 29583-4, 29596, 29668, 29695, 29837, 29847, 29870, 29885,
       29942, 29961, 29974, 29977, 30048, 30062, 30126, 30136, 30166, 30197, 30215,
       30219, 30223, 30259, 30280, 30288, 30297, 30322, 30336, 30339-41, 30369,
       30420, 30429, 30470, 30528, 30567, 30572, 30690, 30744, 30805-6, 30830, 30845,
       30853, 30855, 30859, 30868, 30892, 30952, 30963-5, 30985, 31096, 31149, 31247,
       31299, 31320, 31324, 31343, 31380, 31402, 31412, 31456, 31458, 31489, 31518,
       31522-3, 31525-6, 31549, 31560, 31693, 31702, 31713, 31715-6, 31852, 31872,
       31939, 32043, 32050, 32068, 32092, 32107, 32128-9, 32134, 32217, 32223,
       32233, 32244-6, 32274, 32316, 32330, 32366, 32376, B32410, 32437, 32445,
       32489, 32496, 32507, 32514, B32515
       32656, 32673, 32703, 32733-4, 32736, 32839, 32843, 32860, 32886, 33004,
       33086, 33116-7, 33159-61, 33557, 33606, 33647, 33691, 33714-5,
       33744, 33754, 33787, 33820, 33842, 33911, 34013, 34025-6, 34104, 34247,
       34250, 34255, 34316, 34355, 34424, 34520-1, 34575-6, 34589, 34661, 34674,
       34802, 34841-2, 34862, 34897, 34904, 34982, 34996, 35009, 35111, 35148,
       35187-8, 35216-7, 35356, 35396, 35410, 35475, 35487, 35523, 35566, 35644,
       35676, 35682, 35732, 35798, 35838, 35863, 35908, 35926, 35975, 36173-4,
       36200, 36282-3, 36392, 36398, 36435, 36513
       36710, 36745, B36768, 36796, 36812, 36842, 26868, 36884, 36975, 37048, 37103,
       37137, 37140, 37186, 37245, 37372, 37423, 37442, 37445, 37467, 37533, 37561,
       37566, 37665-6, 37682, 37704, 37708, 37763, 37793, 37800, 37815, 37841, 37844,
       38030, 38042, 38076, 38080, 38169, 38202, 38243, 38249-50, 38276, 38335,
       38352, 38394, 38439, 38454, 38487, 38526, 38658-9, 38661, 38709, 38716, 38757,
       38888, 38901, 38936, 39020, 39034, 39041, 39078, 39119-21, 39152, 39194,
                                                 (continued)

ATP (continued)
       39213, 39250, 39336, 39351-2, 39393, 39397, 39433, 39498, 39607, 39629,
       39675, 39702, 39776, 39795-6, 39912, 39957, 40024, 40080, 40123, 40132,
       40247, 40307, 40311, 40314, 40321, 40332, 40338

ATP, methods
       21934, 22172, 25004; 25305, 25509, 25777, 26391, 26590, 28047; 30166, 33017,
       33183, 34247, 34434, 36283; 38938, 38981, 39101

ATPase, coupling factor 1
       21514, 21559, 21575, 21588, 21613, 21649-50, 21688, 21752-4, 21766, 21777-8,
       21879, 21915, 21987-9, 21996-7, 22017, 22513, 22520, 22539, 22575, 22594,
       22611-2, 22688-90, 22704, 22874, 22912-3, 22923, 22936, 22947, 22967, 22972,
       22996, 23032, 23077, 23126-7, 23192, 23321, 23507, 23513, 23591, 23662,
       23711, 23721, 23800, 23906, B23933, 24034, 24081, 24174, 24182, 24188, 24221,
       24311, 24316, 24355, 24367, 24416, 24471, 24871, 24891, 24973, 25006, 25098,
       25105, 25110, 25159
       25177, 25292, 25317, 25387, 25399, 25417, 25420, 25427, 25461, 25465, 25467,
       25638-9, 25830, 25861, 25888, 25891, 25914, 26103-4, 26210, 26230, 26301,
       26352, 26354, 26366, 26374, 26388, 26418, 26523, 26554, 26631, 26665, 26968-
       -9, 27004-6, 27017, 27037-8, 27113, 27138, 27194, 27204, 27243-5, 27335,
       27338, 27390, 27482, 27486, 27495, 27554, 27661, 27712-3, 27745, 27788,
       27816, 27829, 27881, 27898, 27907, 27910, 27916, 27992, 28034-5, 28045,
       28164, 28411, 28450, 28483
       28554, 28564, 28633, 28660, 28670-1, 28697, 28727, 28730, 28758, 28778-9,
       28789, 28802, 28817, 28906, 28908, 28992, 29044, 29076, 29105, 29114, 29216-7,
       29356, 29404, 29406, 29584, 29618, 29630-1, 29634, 29642, 29667-8, 29696,
       29700, 29719-20, 29730-1, 29749-51, 29781, 29837-8, 29844, 29856, 29934,
       29956, 29961-2, 29980-1, 29999, 30027, 30039, 30066, 30115, 30126-7, 30135,
       30153, 30178, 30192, 30292, 30340, 30409, 30411, 30498, 30548, 30726-7, 30744,
       30805-6, 30875, 30884, 30966, 30969, 30989, 31024, 31026, 31055, 31123-4,
       31147, 31234, 31295-6, 31335, 31365, 31381, 31420, 31461, 31523, 31561, 31580,
       31610, 31634; 31667, 31671-2, 31683, 31688, 31702, 31754-5, 31799, 31818,
       31871, 31915, 31924, 31930, 31942, 31955, 31985-7, 32128-9, 32157, 32233,
       32235, 32245, 32260, 32267, 32275, 32331-2, 32375-6, 32384, 32386-7, 32393,
       32438, 32445, 32480, 32507-8
       32582, 32632, 32636-7, 32661, 32675, 32710, 32733, 32736, 32811, 32839,
       32843, 32851, 32862, 32930, 32949, 33078, 33206, 33452-3, 33647, 33661,
       33690, 33744, 33753, 33769, 33773, 33827, 33853, 33882, 33886, 33906, 33964,
       34203, 34266, 34280, 34299, 34314, 34372, 34488, 34587, 34598, 34734, 34802-
       -5, 34822, 34890, 35053-4, 35100, 35231, 35252, 35382, 35396, 35420, 35433,
       35487, 35495, 35511, 35705, 35716-7, 35752, 35838, 35859, 35934, 35981-3,
       35985, 36035, 36041, 36236, 36283, 36363, 36377, 36382, 36404, 36449, 36493,
       36526, 36538
       36611, 36677, 36749, 36755-6, 36764, 36829, 36852, 36891, 36894, 36957,
       37089-90, 37092, 37261, 37305, 37335-6, B37493, 37629, 37737, 37815, 37856,
       37891-2, 37894, 37944, 38042, 38361, 38541, 38599, 38661-2, 38674, 38734,
       38746, 38797, 38808, 38820, 38824-5, 38833, 38835, 38864, 38868, 38882,
       38888, 38983, 39017, 39051, 39221, 39289, 39471, 39483, 39518, 39521, 39531,
       39607-8, 39629-30, 39704, 39796, 39810-1, 39845, 40090, 40247, 40321-2

ATPase, methods
       21656, 21777, 22217, 22996, 23321, 25105; 25492, 26301, 26969, 27243, 27330;
       29044, 29357, 29631, 29750, 29837, 30153, 30411, 30981, 31024, 31296, 31580,
       32508; 32851, 33661, 33690, 34372, 34546, 34598, 34680, 34803, 35396, 35984,
       36538; 36852, 36891, B37458, 38495, 38746, 38983, 39051, 39704

Autotrophy see Carbon metabolism types ...

B

Bacteria, photosynthetic see Photosynthetic bacteria ...

Bacteriochlorophylls (*cf*. also Chlorophylls,*Chlorobium*)
        21690, 21703, 21707, 21726, 21843, 21993, 22020, 22090-1, 22093-4, 22126,
        22198, 22230, 22253, 22259, 22262, 22267, 22288-9, 22378, 22390, 22408,
        22416, 22443, 22499, 22549, 22580-1, 22626, 22642, 22733, 22773, 22887,
        22954, 23022, 23046, 23089, 23120, 23203, 23206, 23312, 23347, 23436-7,
        23825-6, 23857, 23909, 23965-6, 23998, 24095, 24109, 24189, 24205, B24250,
        24305, 24315, 24327, 24459, 24550, 24560, 24690, 24696, 24796, 24844, 24937,
        25051
        25263, 25323, 25488, 25562, 25606, 25733-4, 25741, 25746, 25756-7, 25773,
        25791, 25884, 25889, 25903, 25957, B26185, 26205, B26220, 26247, 26395,
        26401, 26404, 26406, 26410, 26486, 26508, 26545, 26609-10, 26614, 26658,
        26672, 26681, 26687, 26727, 26731, 26761, 26763, 26808, 26942-3, 26974,
        26985, 27000, 27173, 27254, 27273, 27282, 27303-4, 27340-2, 27381, 27504-5,
        27518-9, 27536, 27615, 27631, 27653, 27726, 27731, 27780-1, 27822, 27867-9,
        27871, 27879, 27904, 27945, 28036, 28131, 28135-6, 28155, 28232, 28237-8,
        28259, 28290, 28335, 28409, 28455
        28515, 28548, 28570, 28582-3, 28644, 28672, 28700-1, 28822, 28876, 28878,
        28892, 28905, 29014, 29025-6, 29035, 29074, 29080, 29098, 29124, 29126,
        29134, 29149, 29174, 29197-200, 29263-5, 29282, 29339, 29358, 29369, 29431-2,
        29452-4, 29474, 29483-4, 29505, 29679, 29682, 29710, 29712, 29738, 29749,
        29818, 29828, 29878-9, 29943, 29947, 29969, 29973, 29996-7, 30050, 30079,
        30134, 30138, 30179, 30227, 30248, 30328, 30385-6, 30418, 30420, 30625-6,
        30671-2, 30696-7, 30718, 30742, 30780, 30789, 30915, 30934, 30939, 31033,
        31046, 31060, 31073-4, 31094, 31137-8, 31219, 31225, 31288, 31299, 31308,
        31354-5, 31369-71, 31415, 31534, 31548, 31556, 31741, 31756, 31805, 31845,
        31892-3, 31900, 31902, 31929, 32045, 32048, 32114, 32123-4, 32127, 32149,
        32168-9, 32208, 32251, 32280-2, 32367, 32409, 32549
        32563, 32600, 32619, 32657, 32670, 32685, 32710, 32744, 32754, 32756-7, 32760,
        32857-9, 32890, 32925, 32981, 33120, 33185-90, 33203-4, 33237, 33241, 33259,
        33337, 33380-1, 33386, 33403, 33468-9, 33484, 33494, 33515-6, 33552, 33574-5,
        33581, 33640, 33702, 33707-9, 33735, 33745, 33779, 33863, 33980, 33987, 34009-
        -10, 34020, 34022-4, 34039, 34194, 34237, 34270, 34281, 34338, 34347, 34363,
        34393, 34469, 34522, 34537, 34566, 34761, 34791, 34795, 34842, 34862-3, 34957,
        34965, 34968-9, 35066, 35078, 35085, 35107-8, 35166-7, 35193, 35236, 35238,
        35246, 35274, 35292, 35311, 35323, 35333, 35376, 35386-7, 35422, 35572, 35642,
        35668, 35683, 35724-5, 35745, 35758, 35821, 35846-7, 35855, 35899, 35901,
        36118-9, 36247, 36292, 36352, 36357, 36399, 36471, 36484-5, 36546
        36620, 36626, 36699, 36771-2, 36786-7, 36860, 36911, 36943-4, 36971, 37117,
        37165-7, 37180-1, 37206, 37255, 37350, 37370, 37446-7, 37485, B37493, 37545-
        -6, 37598, 37661-3, 37675, 37759-60, 37773, 37800, 37914, 37954-5, 37957-8,
        38039, 38086, 38111, 38153-4, 38204, 38240, 38320, 38322, 38333, 38351,
        B38415, 38571, 38613, 38637, 38648, 38682-4, 38766, 38795-6, 38895, 38905-6,
        38918, 38957, 38972, 39095, 39121, 39180, 39222, 39229-31, 39297, 39303,
        39322, 39440, 39443, 39451-2, 39491, 39502, 39580-1, 39613, 39626, 39666,
        39681, 39700, 39822, 39901, 39909, 40129, 40186, 40190, 40346

Bacteriorhodopsin see · *Halobacterium* photosynthesis

Bibliographies of photosynthesis, biographies
        B23816, 24410; B27801-2, 27803; B28522; B33583, 35769-70, B35771-2; B39164,
        39544, B39545

Biliproteins see also Phycocyanins; Phycoerythrins

Biliproteins absorption spectra *in vitro*
        21725, 21836, 22460, 22545, 23010, 23152, 24254, 24288, 24450, 24697-8;
        25361, 25579, 26061, 26992, 27299; 28960, 29152, 29672, 30373, 30588, 31011,
        31423, 32071; 32617, 33134, 34381-2, 34696, 35804, 36071; 37535-7, 38223,
        38248, 38348, 38945, 38947, 39744

Biliproteins absorption spectra *in vivo*
22461, 22642, 24345, 24450, 24495-6; 25867, 25869, 25958, 26158, 26207, 26250, 26992, 27299, 27615, 28008; 28598, 28737, 29378, 29525, 29670, 30552, 30678, 31206, 31505; 33746-7, 33758, 34594, 34698, 35015, 35478, 35684, 36071, 36575; 36836, 37030, 37448, 37599, 37601, 38201, 38248, 38270, 38948, 39516, 39610, 39994

Biliproteins and production of algae and water reservoirs
29498, 30226, 32368; 33535, 34394; 40153, 40163

Biliproteins biosynthesis, precursors
21836, 23843, 23869, 24254, 24809; 25439, B26185; 28719, 29866, 30311, 30531, 32102, 32173; 32779, 33234, 33657, 34468, 34846, 35015, 35308, 35706; 37622, 37652, 38212, 38223, 38947, 39484, B39693, 39964, 39966

Biliproteins chemical structure
21836, 22545, 22949, 23178-9, 23182, 23247, 24254, 24642, 24698, 24810; 25402, 26159, B26185, B26220, 27299; 28959, 29670, 32173; 34339; 37014, 38347-8, 39465, B39693, 39901

Biliproteins complexes *in vitro*   23182; 38248

Biliproteins complexes *in vivo*
21836, 22590, 24909; 26158, B26185, 26207, 27299, 27615; 29590, 29670, 30372, 30552, 30877, 30940, 31901, 32540; 33133, 33257, 33584, 36429; 37014, 38248, 38270, 39744, 39901, 39965, 39994

Biliproteins degradation   23178-9; 30299, 30531; 33133-4, 34881, 35308, 36186; 38948

Biliproteins, delayed light emission, luminescence *in vitro*   25223; 34382

Biliproteins determination, column chromatography
21920, 22546, 24409, 24695, 24810; 25579, 25631, 26160; 28960, 29672; 33234, 34381, 36575; 36608, 38348, B39693

Biliproteins determination, electrophoresis and other methods
21920, 22545, 23152, 24515; 25402, 25579, 26098, 26160, 26250; 29152, 29591, 29782, 30373, 30588; 33234, 33420, 33776, 34324, 34698, 36504, 36575; 37601, 37652, 38201, 38270, 38295

Biliproteins determination, paper chromatography, thin-layer chromatography
30475; 33776, 34382; B38477

Biliproteins determination, spectral methods
21920, 23327, 24518-20, 24697, 24810; 27086, 27804; 29152; 36575; 38446, B39693, 40305

Biliproteins energetic states *in vitro*
23532, 24288, 24573; 25394, 26082; 28675; 33576, 34697; 37535, 37537, 38248

Biliproteins energetic states *in vivo*
23187, 24496; 28968, 29590, 30525, 30970, 31108, 31709, 32357-8, 32540; 33258, 33757-8, 34832, 35366, 35738, 35804, 36575; 38248, 39497, 39516

Biliproteins, enzymes of synthetic and degradation processes   34468; 36887, 39964

Biliproteins fluorescence *in vitro*
21725, 21836, 23010, 23152, 24254; 25431, 25973; 28504, 28675, 30373, 30588, 32071, 32368; 34381-2, 34629, 34696, 34740; 37535-7, 38248

Biliproteins fluorescence *in vivo*
22458, 22495, 22590, 23728; 25958, 26098, 26168, 26207, 27319; 28737, 29014, 29378, 29590, 29592, 29723, 29836, 30372, 30588, 30678, 30940-1, 32357, 32540; 33257-8, 33757-8, 33884, 34629, 34698, 34832, 35366-7, 35478, 36504,
(continued)

Biliproteins fluorescence *in vivo* (continued)
   36575; 36706, 36836, 37030, 37448, 37526, 37599-601, 37622, 37652, 38248,
   38270, 39110, 39198, 39359, 39497, 39499, 39516, 39994

Biliproteins in model systems
   21725, 23532; 25361, 25394, 26062; 28737, 29545; 32785, 33576; 37134,
   37535

Biliproteins in mutants see Mutants, biliproteins in

Biliproteins in photosynthesis mechanism
   21725, 22046, 22495, 23043, 23187, 24450, B25064; 25973, 26160, B26185,
   26718, 27319; 28675, 28737, 29589-90, 29670, 29723, 29836, 30135, 30323,
   30442, 30525, 30877-8, 30970, 31108, 31485, 32071, 32357-8, 32540; 32744,
   32785, 33258, 33569, 33746-7, 33883, 34697, 34832, 34893, 35366, 36575;
   B36769, 36944, 36982, 37314, 37764, 37785, 38150, 38160, 38325, 39198,
   39276, 39344, 39516, 39685, 39901, 40178

Biliproteins in physiology of photosynthesis 26608; 28613, 32269

Bioclimatological methods see Aerodynamic methods ...

Biological clock see Diurnal changes ...

Biomass distribution and redistribution in plant
   21525, 21556, 21686, 21795, 21802, 21832, 21853-4, 21862, 21878, 21949-50,
   B21979, 22007, 22015, 22034-5, 22074, 22118, 22184, 22210, 22282, 22301,
   22331, 22343, 22365, 22369, 22380, 22428-9, 22447-8, 22510, 22524, 22563,
   22593, 22676, 22725, 22765, 22768, 22793, 22817, 22829, 22844, 22850, 22909-
   -10, 22917, 22937, 23037-9, 23125, 23145, 23181, 23222-3, 23250-1, B23262,
   23372, B23407, 23443, 23466, 23499-500, 23523, 23578, 23604, 23646, 23665,
   23678, 23698, 23712, 23736, 23739, 23741, 23758, 23763, 23776-7, 23779,
   23790, 23802, 23818, 23863, 23871, 23897, 23975, 24000, 24007, 24020, 24052,
   24075-6, 24085, 24115, 24123, 24127, 24181, 24210, 24249, 24260, 24267,
   24275, 24291, 24314, 24429-30, 24438, 24472, 24504, 24511, 24534, 34557-8,
   24625, 24633, 24661, 24666, 24716, 24748, 24765, 24770, 24773, 24780, 24841,
   24961, 24969-72, 24985-6, 25047, 25049, 25052, 25059, 25103 ·
   25164, 25171-2, 25230, 25232, 25264, 25299, 25319, 25435, 25469, 25473,
   25531, 25566, 25594, 25660, 25666, 25725, 25737, 25753, 25758, 25771, 25810,
   25845, 25857, 25917, 25970, 25997, 26017, 26028-30, 26055, 26073, 26085,
   26096, 26110, 26132, 26197, 26202, 26204, 26213, 26246, 26263, 26267, 26274,
   26361, 26383, 26400, 26427, 26438, 26461, 26465, 26474, 26484, 26501, 26506,
   26530, 26533, 26537, 26607, 26643, 26688, 26776, 26788, 26798, 26803-4,
   26812, 26827-8, 26843, 26847, 26904, 26906, 26912, 26952, 26954, 27013-4,
   27062, 27094, 27101, 27108, 27139, 27209, 27212, 27216, 27228, 27235, 27248,
   27253, 27327, 27333, 27343-4, 27362, 27409, 27430, 27437, 27457, 27466,
   27521, 27538, 27540, 27563, 27647, 27664, 27719-20, 27725, 27734-6, 27748,
   27812, 27882, 27931, 27983-4, 27991, 28023, 28050, 28086, 28099, 28134,
   28157, 28175, 28207, 28212, 28300, 28312, 28337, 28347, 28381, 28392, 28400,
   28408, 28427, 28470, 28477, 28495
   28516-7, 28533, 28544, 28566, 28578, 28624, 28648, 28703, 28782, 28791, 28800,
   28809, 28899, 28918, 28926, 28941, 28972-3, 29007, 29016, 29022, 29042, 29053,
   29068, 29086, 29103-4, 29136, 29153, 29157, 29164, 29193, 29203, 29221, 29254,
   29290, 29331, 29335, 29351, 29366, 29373, 29443, 29509, 29524, 29550, 29566,
   29576, 29605, 29628, 29652, 29656, 29771, 29794, 29819-20, 29835, 29883, 29923,
   29968, 29975, 29977, 30011, 30044, 30053, 30195, 30203, 30272, 30353, 30412,
   30427, 30451, 30461, 30476, 30478, 30483, 30505-6, 30532, 30571, 30581, 30642,
   30680, 30691, 30699, 30711, 30733, 30820, 30824, 30891, 30932, 30957, 30967,
   30979, 31013, 31015, 31021, 31032, 31067, 31069, 31086, 31107, 31116, 31130-1,
   31148, 31163, 31180, 31201, 31228, 31242, 31272-3, 31303, 31305, 31307, 31331,
   31348, 31376-7, 31392, 31399-400, 31408, 31439, 31463, 31492, 31535, 31584-5,
   31592, 31595, 31608, 31695, 31761, 31792, 31822-3, 31825-6, 31868, 31916,
   31922, 31933-4, 31941, 31978, 31996, B32004, 32046, 32059, 32062, 32065, 32076,
(continued)

Biomass distribution and redistribution in plant (continued)
        32115, 32117, 32133, 32151, 32170-1, 32231, 32243, 32257, 32322, 32325, 32347,
        32364, 32403, 32412, 32431, 32434, 32436, 32483, 32504, B32528, 32530, 32542-3
        32612, 32629, 32633, 32648, 32652, 32678, 32752, 32762, 32832-3, 32836,
        32965, 32980, 33021, 33066, 33074, 33079, 33082, 33101, 33107, 33162, 33184,
        33196, 33209, 33222, 33226, 33229, 33232, 33255, 33260, 33262, 33279, 33293,
        33306, 33331, 33358-9, 33368-9, 33375, 33377, 33394, 33398, 33411, 33495,
        33507, 33513, 33523, 33536, 33547, 33570, 33611, 33614, 33624, 33655, 33668-
        -9, 33703, 33730, 33740, 33759, 33797, 33833, 33859-60, 33894, 33898-90,
        33921, 33925, 33951, 34037, 34076, 34091-2, 34095-6, 34129, 34137, 34176,
        34190, 34287, 34291, 34308, 34330, 34439, 34453, 34486-7, 34499, 34503,
        34507, 34611-2, 34621, 34635, 34721, 34735, 34754, 34787, 34810, B34849,
        34854, 34873-6, 34880, 34887, 34895, 34933, 34946, 35004, 35014, 35017,
        35021, 35025, 35034, 35039-40, 35047, 35079, 35091, 35117-8, 35154, 35169,
        35174, 35194, 35202-3, 35205, 35221, 35225, 35243, 35255, 35261, 35263,
        35268, 35293, 35301, 35314, 35321, 35335, 35350, 35394, 35448, 35458, 35463,
        35513-4, 35529, 35541-3, 35561, 35574, 35598, 35607, 35610, 35662, 35691,
        35721, 35781, 35785, 35787, 35808, 35815, 35870, 35882, 35884, 35907, 35928,
        35933, 35971, 36010, 36014, 36036, 36068, 36070, 36076, 36112, 36116, 36136,
        36163, 36165, 36201, 36226, 36234, 36275, 36281, 36290, 36315, 36319, 36332,
        36344, 36361, 36368, 36386-7, 36412, 36419, 36428, 36489, 36500, 36515
        36602, 36617, 36643, 36676, 36683, 36703, 36717, 36722, 36724, 36726, 36734,
        36736, 36757, 36775, 36807, 36817, 36845-6, 36869, 36882, 36908, 36922,
        36967, 36986, 37010, 37041, 37059, 37082, 37093, 37121-2, 37125, 37171,
        37173-4, 37216, 37221, 37240, 37259, 37265-6, 37268, 37289-90, 37296, B37299,
        37302, 37323, 37360, 37368, 37384, 37389-90, 37412, 37434, 37440, 37456,
        37462, 37488-9, 37499-501, 37510, 37541, 37552, 37555, 37560, 37564, 37588,
        37637, 37710, 37721, B37727, 37751, 37796, 37814, 37845, 37847, 37867,
        37890, 37898-9, 37935, 37982, 37988, 38040, 38044, 38084-5, 38097, 38127,
        38148, 38174, 38210, 38231, 38262, 38293-4, 38299-300, 38392, 38407, 38431,
        38448, 38464, 38478, 38498, 38501, 38512, 38520, 38539, 38563-4, 38567-8,
        38578-9, 38590, 38602-3, 38636, 38666, 38702, 38704-5, 38727, 38743-4,
        38754-5, 38772, 38776, 38779, 38800, 38801, 38813-4, 38823, 38827, 38848,
        38854, 38859, 38883, 38897, 38949, 38979, 38995, 39002-4, 39015, 39032,
        39035, 39037, 39043, 39053, 39066-7, 39072, 39138, 39143, 39157, 39223,
        39244, 39249, 39262, 39271-2, 39342, 39354, 39380, B39394, 39398, 39464,
        39512, 39562, 39565, 39569, 39635, 39637, 39656, 39664-5, 39692, 39752,
        39755, 39765, 39794, 39831, 39895, 39897, 39904, B39951, 39968, 39976,
        39979, 40020, 40027, 40045, 40085, 40103, 40115, 40117, 40126, 40138, 40145,
        40170, 40219-23, 40225, 40235, 40238-42, 40248, 40271, 40273, 40275, 40280,
        40296, 40304, 40312, 40326, 40336, 40373

Biopotentials   see Chloroplast and chromatophore biopotentials

Biosphere production   see Ecosystem production ...

Blinks effect   see Emerson effect, Blinks effect

Books on photosynthesis   see General aspects ...

Boundary layer of air   see Resistance, leaf boundary layer

Bundle sheaths   see Carbon metabolism types ...; Carbon fixation pathways, compari-
        son in mesophyll and bundle sheath cells

C

$^{13}C/^{12}C$ ratio, $\delta^{13}C$
        23073; 25437, 25446-7, 25599, 25787, 25878, 25926, 26261, 26368, 27769,
        28383, 28406-7; 28567, 28748, 29273-4, 29488, 30332, 30745, 30822, 30952,
                                                                      (continued)

                        (continued)

                                                                        (continued)

CAM (continued)
> 32555, 32638, 32840, 32894, 33062, 33440-1, 33663-5, 33696, 33777, 33809,
> 33857-8, 33960, 34013, 34122, 34196, 34298, 34352-3, B34354, 34368-9, 34525,
> 34585, 34649, 34684, 34950, 34952, 35087, 35097, 35186, 35192, 35326, 35442,
> 35625, 35645, 35950, 36093, 36146, 36432, 36442-5, 36480
> 36647, 36663, 36672, 36701, 36713, 36916-8, 36939, 36997, 37022, 37028,
> 37177, 37249, 37274, 37287, 37317-8, 37398-9, 37564, 37628, 37631, 37858,
> 38088, 38177, 38185, 38235-7, 38253, 38341, 38400, B38477, 38640-3, 38784-5,
> 38788-9, 38907-8, 38961, 38996-7, 39132, 39216-7, 39324, 39382, 39431, 39479,
> 39500, 39603, 39687-8, 39724-6, 39830-1, 39926, 40010, 40180, 40212-4, 40236-
> -8, 40358, 40360

Canopy, CO_2 profiles
> 21549, 21581, 21633, 21801, 21803, 22593, 22597, 22645, 22797, 22984, 23578,
> 23700, 23814, 24004, 24015, 24278, 24296, 24689, 24947; 25195, 25238,
> 25431, 25571, 26348, 26531, 26783, 26870, 27166, 27630, 27658, 27728, 27886,
> 27981, 28195, 28197, 28256; 28521, 29064, 29289, 29372, 30084, 30438, 31272,
> 31710, 31760, 31920, B32150; 33245, 33632, 33703, 33871, 35151, 36059,
> 36066, 36212, 36467-8; 36636, 37696, 37992, 38100, 38302, 38442, 40288,
> B40331

Canopy density, thickness
> 21521, 21524, 21542, 21826, 21950, 22007, 22184, 22396, 22428-30, 22447,
> 22844, 22910, 22985, 23067, 23221, 23233, 23313, 23426, 23738, 23896, 23945,
> 24080, 24115, 24229, 24291, 24441, 24511, 24595, 24849, 24887,·25156
> 25169, 25172, 25238, 25412, 25520, 25523, 25607, 25615, 25659, 25781, 25792,
> 26020, 26027-8, 26031, 26182, 26202, 26373, 26501, 26569, 26747, 26804,
> 26876, 26904-6, 26944-5, 26947, 26954, 26964, 27014, 27199, 27209, 27234,
> 27336, 27643, 27647, 27656, 27735, 27740, 27920, 27983, 28022, 28169, 28212,
> 28265, 28350
> 28503, 28552, 28689, 28820, 29007, 29045, 29190, 29368, 29492, 29496, 29566·,
> 29574, 29633, 30011, 30029, 30146, 30376, 30409, 30438, 30674, 30689, 30874,
> 30976, 31004, 31113-5, 31180, 31535, 31584, 31777, 31822, 31870, 31934, 32504,
> 32545
> 32569, 32662-3, 32737-8, 32758, 32761, 32939, 33495, 33541, 33572, 33871,
> 33894, 33962-3, 33998, 34037, 34048, 34Q74, 34091-2, 34110, 34241-2, 34279,
> 34416, 34503, 34541, 34707, 34711, 34940, 35001, 35004, 35025, 35039,
> 35082-3,   35141, 35303, 35346, 35351, 35500, 35508, 35707-8, 35871, 36013,
> 36068, 36325, 36368
> 36599, 37198, 37298, 37343, 37389, 37412, 37541, 37605, 37623, 37689, 37710,
> 37772, 37796, 37909, 38040, 38071, 38219, 38257, 38346, 38407, 38583, 38636,
> 38760, 38839, 38883, 39129, 39138, 39555, 39740, 39748, 39752, 39764-5,
> 39897, 39976, 40304

Canopy horizontal structure
> 21633, 21761, 22854, 23039, 23412, 23856, 24093, 24215, 24244, 24474; 25546,
> 26047, 26086, 26202, 26844, 26879; 29676, 30555; 33695, 35079, 35082-3;
> 36962, 37383, 37825, 37908, 37995, 38000, B38211, 38572, 38605, 38800, 39682,
> B39711

Canopy, leaf age
> 21928, 22065, 22097, 22606, 24278, 25086; 25967, 27362, 27888, 28070, 28093;
> 28516, 28527, 30555, 31695; 35904; 40275

Canopy, leaf angles
> 21523, 21633, 21761, 22885, 22910, 23059, 23117, 23346, 23814, 24215, 24244,
> 24441, 24474, 24758, 24798, 24880, 24887
> 25195, 25264, 25660, 26027, 26080, 26086, 26188, 26844, 26877, 27630, 27728,
> 28051, 28092, 28195, 28231, 28265
> 28865, 29289, B29468, 29566, 29574, 29676, B29707, 30084, 30553, 30555,
> 30700, 30945, 31113-5, B31140, 31198, 31265, 31392, 31517, 31761, 31879,
> 31884, B32150, 32152

(continued)

                                                                        (continued)

Canopy, water vapour profiles
    21549, 21801, 22505, 22597, 23070, 23592, 23695, 23945, 24004, 24057, 24215,
    24947; 25195, 25571, 25661, 26515, 26531, 26848, 26947, 27630, 28011, 28197;
    28622, 29064, 29289, 29372, B29707, 30205, 31710, 31762, 31966-7; 36066;
    38100, B38211, 38520

Carbon-14   see Carbon isotopes ...

Carbon balance, plant
    22543, 22765, 22768, 22778, 22832, 24841, 24976; 25207, 25319, 25411, 25664,
    25666, 25674, 25753, 25771, 26073, 26383-5, 26591, 26883-4, 27011, 27097,
    27212, 27895, 28302, 28347; 28624, 28648, 29053, 29102, 29373, 29443, 29923,
    29954, 30476, 30524, 30745, 30947, 30976, 31592, 31606, 31916, 32339-41,
    32372, 32415, 32427; 32612, 32691, 32969, 33262, 33296, 34005, 34622, 34810,
    35129, 35161, 35224, 35559, 35721, 35808, 36044, 36281, 36445; 36817, 37049,
    37121, 37289, 37636, 37802, 37886, 37888, 37924, 38132, 38431, B39951, 40145,
    40291

Carbon dioxide   see CO_2

Carbon fixation in isolated chloroplasts and its products
    21673, 21756, 22530, 23310, 25148; 25671, 25870, 26314, 26772, 26900, 27284,
    27399; 30036, 30370, 31938, 32178-9; 32768, 34173, 36331, 36413; 37841,
    37850, 37876, 38113, 38408, 38412, 39619-20, 39974, 40206

Carbon fixation pathways, comparison in mesophyll and bundle sheath cells
    21814, B21833, 21874, 21942, 22397-9, 22635, 22698, 22833-4, 22838, 23877,
    24161-2, 24170, 24309, 24555, 24850; 25711, 25912, 26315-7, 26444, 26574, 26786,
    27007, 27364, 27597, 27621, 28493;   29534, 29709, 29830, 29865, 30232, 30274,
    30333, 30462, 30498, 30538, 30783-4, 30859, 30887, 30951, 31417, 31448-9,
    31452, 31466, 31611, 31998, 32223, 32462; 32826, 32987, 33080, 33869, 34013,
    34045, 34254, 34256, 34259, 34549, 34619-20, 34800, 34907, 35266, 35431,
    35434, 35468, 35471, 35704, B35862; 36719, 37084, 37120, 37391, 37542, 37627,
    37817, 37826, 37858, B38139, 38341, B38477, 38786, 39093, 39201, 39263,
    39265-6, 39275, 39284, 39460, 40345

Carbon fixation pathways enzymes, methods   see Enzymes of carbon fixation pathways...
    Ribulose 1,5-bisphosphate carboxylase ...; Phosphoenolpyruvate carboxylase...;
    Malic enzyme ...

Carbon fixation pathways, general aspects   see General aspects on carbon fixation ...

Carbon fixation pathways in photosynthetic bacteria   see Photosynthetic bacteria,
    carbon fixation pathways

Carbon fixation pathways, intermediary types of carbon fixation
    25138; 25443, 25445; 29326, 29817, 30273, 31039, 31637, 31697; 32654, 32913,
    34548, 34684, 35286, 35471, 36426; 36693-4, 37021, 38793, 39264-5, 39460,
    40238

Carbon fixation pathways, intermediates
    21673, 21785, 21885, 22098-9, 22119, 22252, 22397-8, 22530, 22552, 22721,
    22723-4, 22869, 22958, 22965, 23032, 23073, 23121, 23195-8, 23214, 23309-10,
    23370, 23586, 23715, 23726, 23771, 23841, 23926, 23938, 24008, 24059, 24066,
    24161, 24223, 24251, 24264, 24336, 24659, 24949
    25280, 25346, 25870, 26078, 26314-7, 26341, 26549, B26551, 26565-6, 26598,
    26721-2, 26770, 26864, 26900, 27247, 27255, 27359, 27461, 27524, 27621, 27797,
    28303-4, 28307
    28567, 28625, 28786, 29015, 29085, 29214, 29231, 29462, 29533-5, 29571, 29637,
    29669, 29728, 29892, 30036-7, 30098, 30106, 30144, 30197, 30233, 30276-7,
    30369, 30406, 30414, 30435, 30469, 30537, 30631, 30686, 30887, 30936, 30951,
    31164, 31235, 31248, 31266, 31315, 31320, 31338-9, 31343, 31353, 31417, 31420-
    -1, 31450, 31465, 31520, 31526, 31636, 32024, 32156, 32297, 32457, 32461
                                                              (continued)

Carbon isotopes ... (continued)
        38657, 38742, 38758, 38787, 38798, 38834, 38852, 38893, 38903, 38907, 38998,
        39064, 39087, 39137, 39367, 39414, 39428, 39585, 39668-9, 39717-8, 39741,
        39750, 39760-1, 39815, 39836, 39850, 39864, 39895, 39934, 39936, 39962,
        39974-5, 39993, B40015, 40080-1, 40145, 40159, 40168, 40267, B40331

Carbon metabolism types and algae productivity
        23294, 24644; 36592, 37453, 39553

Carbon metabolism types and biliproteins 23884

Carbon metabolism types and carbon fixation pathways
        22249, 22366, 23586-7, 23930, 24309, 24553, 24769, 24854; B25603, 25612,
        25768-9, 26006, 26155, 26316, 26428, 26786, 27116, 27198, 27292, 27419,
        27928, 28120, 28204, 28305; 29581, 30009, 30109, 30213, 31676, 32458; 32803,
        32840, 32867, 32987, 33699, 33705, 33960, 34199, 34980, 35425, 35645, 35881,
        35974, 36204, 36444-5; 37215, 37751, 38313, 38895, 39754, 40155, 40197

Carbon metabolism types and carotenoids
        21586, 21971, 24735; 25225, 25895; 33780, 34274, 35432; 37939

Carbon metabolism types and chlorophyll
        23884, 24735; 25925, 26786; 29446; 32574, 33305, 35432, 35478, 36574; 39803

Carbon metabolism types and chloroplast (chromatophore)
        22075, 22281, 23021, 23621, B23933, 24735; 25718, 25878, 27407, 27469, 27557,
        28119; 29581, 30433-4, 30557, 31778; 34155, 35961

Carbon metabolism types and ecosystem and plant productivity
        22123, 22489, 23448; B25603, 26979, 27539, 28005; 28873, 30077, 30917, 31159,
        31636; 32766, 33123, 33306, 33631, 35403, 36494; 37751, 38385

Carbon metabolism types and electron transport
        23884; B25603, 26473, 26786; 29383; 33197, 33433, 34848, 35432, 36150; 38569,
        38588, 38716

Carbon metabolism types and gas exchange
        21810, 21812-3, 21874, 22041, 22363, 22366, 23074, 23161, 23170, 23787,
        24100, 24735, 24861; B25603, 25760, 25768, 25967, 26091, 26240, 26475,
        26641, 26644-5, 26880, 27089, 27142, 27427, 27469, 28005; 28146; 28970,
        29050, 29091, 29166, 29489, 29985, 30124, 30270, B30511, 30754, 30822,
        30916, 31613, 31932, 32312, 32458; 32766, 32943, 33123, 33150, 33305-6,
        33440, 34452, 34625, 34836, 35766, 36059; 36693-4, 37021, 37298, 37516,
        37701, 37715, 37981, 38313, 38429, 38588, 39238, 39572, 39688, 40125

Carbon metabolism types and photorespiration
        22071; 26427, 26786, 27469, 27769; 33123, 33631, 34452, 35766; 37021,
        38429, 39688, 40114

Carbon metabolism types and resistances to CO_2 and water vapour transfer
        21813, 22075; 26091, 26240; 29232, 30270, 31836; 33167, 33305-6; 37021

Carbon metabolism types and respiration 28970, 32312; 36445

Carbonic anhydrase
        21529, 21580, 21671, 21943, 22024-5, 22027, 22876, 22906, 23075, 23231,
        23446, 23494, 24009-10, 24100, 24162, 24201, B24250, 24854, 24952, 25074,
        25106
        25329, 25503, 25650, 25657, 25670, 25699, 25768, 25875, 26023-4, 26118,
        26203, 26478, 26539, 26725, 26748, 26858, 27263, 27317-8, 27329, 27368,
        27605, 28176, 28205, 28220, 28229, 28304, 28368
        28613, 28901, 29503, 29709, 29776, 29990, 30630, 30991, 31107, 31136, 31471-
        -2, 31474, 31679, 31720

                                                                (continued)

Carbonic anhydrase (continued)
        32718, 33409, 33571, 33689, 34226, 34504, 34643-4, 34646, 35146, 35345,
        35485, 35822, 35860, 36424
        36812, 37229, 37503, 37755, 37902, 38391, 38758-9, 38949, 39154, 39283,
        39533, 39594, 40020, 40360

Carbonic anhydrase, methods
        25670, 26606, 27329; 30618

Carboxylation see Carbon fixation pathways

Carboxylation resistance see Resistance, carboxylation and excitation

Carotenes 21509, 21571, 21586, 21642, 21663, 21923, 22166, 22188, 22323, 22747,
        23115, 23169, 23259, 23384, 23399, 23568, 23807, 23883, 24069, 24145, 24413,
        24477, 24530, 24541, 24777, 24909, 25055-6
        25376, 25648, 25762, 25794, 25838, 25876, 26225-6, 26666-7, 26707, 27060,
        27066, 27110, 27165, 27237, 27405, 27480, 27527, 27586, 27863, 28094,
        28186, 28377
        28528, 28607,   28650-1, 28708, 28715, 28760, 28944-5, 28947-8, 28978, 29170,
        29209, 29234, 29241, 29324, 29422, 29456, 29480, 29573, 29614, 29698, 29897,
        30017-8, 30102, 30237-8, 30302, 30337, 30357, 30452, 30581, 30619, 30629,
        30730, 30760, 30802, 30851, 31076, 31189, 31349, 31405-6, 31503-4, 31559,
        31574, 31642, 31831, 31961, 32016, 32055, 32142, 32185, 32190, 32284, 32289,
        32349, 32441-2, 32481
        32595, 32741, 32991, 33030, 33173, 33178-9, 33264-5, 33284, 33351, 33472,
        33476, 33526, 33556, 33779, 33862, 34175, 34381, 34403, 34415, 34448, 34468,
        34477, 34533, 34581, 34595-6, 34607, 34670, 34707, 34791, 34845, 34869,
        34948, 35048, 35058, 35123, 35284, 35358, 35379, 35385, 35427, 35443, 35519,
        35768, 35786, 35844, 36004, 36051, 36154, 36193, 36232, 36405, 36446-8,
        36456, 36498, 36579, 36587
        36906, 37000, 37134, 37172, 37309, 37337, 37432-3, 37449, 37480, 37522,
        37574, 37613, 37647, 37684, 37686, 37732, 37739-40, 37878, B38021, 38102,
        38110, 38228, 38230, 38333, 38369-70, 38387, 38439, 38443, 38478-9, 38490,
        38561, 38581, 38687, 38769, 38774, 38819, 38842, 38849, 38904, 39039, 39170,
        39184, 39192, 39245-6, 39316, 39408, 39477, 39585, 39588, 39623, 39736,
        39840, 39862, 39866, 39901, 39948, 39986, 40072, 40087, 40164, 40174, 40245,
        40319

Carotenoids absorption spectra *in vitro*
        21855, 21891, 22323, 22459, 23248, 23340, 24709, 24800, 24978, 25016-7;
        25648, 25714, 25794, 26127, 26157, 26270, 26491, 27526, 27608, 27677, 27684,
        28094, 28377; 28569, 29172, 29434, 29897, 30137, 30207, 30582, 30730, 31189-
        -90, 31223, 31349, 31583, 31673, 31851, 32349, 32501; 32763, 32770, 32966,
        32991, 33476, 33952, 34086, 34214, 34318, 34381, 34594, 35854, 36155, 36498,
        36580, 36587; 37039, 37051, 37732, 37939, 38110, 38369, 38690, 38778, 38913,
        39170, 39437, 39866, 40245

Carotenoids absorption spectra *in vivo*
        21691, 21889, 22450-1, 22642, 23132, 23565, B24250, 24350, 24495-6, 24777,
        24829; 25600, 26225, 26808, 27193, 27608, 28027, 28029, 28060; 28770, 28917,
        29263-4, 29525, 29573, 29666, 29879, 29913, 30079, 30172, 30314, 30366,
        30475, 30487, 30974, 31368, 32017-8, 32103, 32288, 32500-1; 32924, 33203,
        33484, 33746-7, 34215, 34318, 34594, 34868, 36155; 36953, 36976, 37203,
        37337, 37447, 38687, 38690, 38850, 39170, 39437, 39448, 39827

Carotenoids and production of algae and water reservoirs
        23133, 23269, 23683-4; 25218, 26214, 27266, 27940, 28162, 28317, 28464-5;
        28790, 29802-3, 30356-7, 31047; 33535, B34348; 37417, 37466, 37540, 39209-
        -10, 40153

Carotenoids and production of higher plants
        21760, 23375, 34530, 24942; 28282; 30396, 32185-6

Carotenoids biosynthesis, precursors
        21678, 21904, 22188, 22290, 22322, 23277, 23324-5, 23375, 23398, 23400, 23675,
        23697, 23710, 23781, 24069, 24166, 24203, 24340, 24348, 24413, 24459, 24597,
        24642, 25056, B25064
        25416, 25553-4, 25816, 25852, 25876, 26069, 26153, B26185, 26226, 26328,
        26820, 27184, 27304, 27444, 27472, 27586, 27645, 27677, 27755, 27784, 27805,
        27972, 28217, 28356, 28379
        28752, 28944, 28948, 28978, 29241, B29468, 29640, 29698, 29760, 29904, 29947,
        30238, 30592, 30597, 30923, 31211, 31223, 31329, 31349-50, 31504, 31680,
        31779, 32215, 32229
        32595, 32731, 32741, 32763, 32875, 33284, 33519, 33796, 33862, 33952, 34163,
        34441, 34468, 34583, 34603-4, 34608, 34791, 35066, 35071, 35519, 35702,
        35763, 35768, 35854, 35952, 36231, 36262, 36409, 36411
        36611, 36885, 36894, 37000, 37062, 37109, 37155, 37231, 37248, 37294-5,
        37388, 37574, 37683-5, 37739-40, 37961, B38021, 38102, 38228, 38242, 38370,
        38439, 38479-80, 38504, 38877, 38904, 39170, 39269, 39316, 39426, 39429,
        39553, 39585, 39970, 40173, 40245

Carotenoids chemical structure
        22134, 23510, 23710, 24152, 24981; 25553, 25555, 25586, 25815, B26185,
        26186, 26678, 26915, 27192-3, 27885, 28060, 28466; 28767, 29241, 29804,
        29924, 30102, 30592, 33283, 33862, 34603-4, 34687, 35702, 35786; 36660,
        37000, 37172, 37231, 37433, 37683-4, 37864, 37939, 38004, B38021, 38476,
        B38477, 38778, 38904, 38913, 39007, 39901, 40174, 40363

Carotenoids complexes *in vitro*   31740; 32575; 37306

Carotenoids complexes *in vivo*
        21768, 23000, 24479, 24631; 25746, B26185, 26326, 26573, 26682, 26818, 27952,
        28111, 28466; 28770, 29505, 29876, 30314, 30366, 30672, 30789, 31385, 31625,
        31812, 31901; 33203-4, 33322, 34493, 34687, 35768, 36272, 36484; 37155,
        37309, 37447, 37707, 37717, 39285, 39901

Carotenoids degradation
        21594, 22068, 22914, 23044, 23050, 23429, 23710, 24208-9, 24583, 24731,
        24829; 25248, 25432, 25714, 25716, 26491, 26634, 26716, 26740, 27071, 27159,
        27387, 27527, 27711, 27850, 27855, 27864, 27885, 28239-40; 28944, 29241,
        29370, 30073, 30760, 31188, 31384, 31490, 31503, 31559, 31810, 31851, 32007,
        32037, 32067; 32731, 32998, 33476, 33610, 33796, 34380-1, 34941, 35443,
        35949; 36718, 36747, 37000, 37095, 37428, 37481, 37977, 38505, 38778, 39044,
        39092, 39245-6, 39447, 39485, 39840, 40174

Carotenoids determination see also Pigments determination, sampling and extraction

Carotenoids determination, column chromatography
        21622, 21904, 22167-8, 24230, 24409, 24709, 24717; 25794, 25815, 26157,
        B26185, 27645, 28094, 28712, 29209, 29434, 31189-90, 31961, 32304; 32966,
        33263, 33303, 33538, 33829, 34114, 34179, 34381, 34425, 34603, 36022, 36405;
        36588, 36981, 37039, 37939, B38021, 38396, 39184, 40054, 40329

Carotenoids determination, electrophoresis and other methods
        25815, B26185; 33009, 36448; 39170

Carotenoids determination, paper chromatography, thin-layer chromatography
        21622, 22166-8, 22323, 22616, 23026, 23384, 23781, 23953, 24129, 24163, 24230,
        24521, 24777, 24800, 24829, 24978, 25016-7, 25055; 25442, 25794, 25815,
        25876, B26185, 26290, 26312, 26820, 26922, 26956, 27066, 27197, 27511, 28191,
        28217; 28569, 28666, 28767, 29048, 29209, 29324, 29434, 29947, 30137, 30237,
        30269, 30582, 30629, 31189-90, 31430, 31574, 31673, 32016, 32093, 32304,
        32442; 32770, 33057, 33182, 33263, 33457, 33538, 34068, 34169, 34215, 34425,
        34603, 34855, 35823, 36022, 36180, 36579-80; 37039, 37051, 37225, 37939,
        B38021, 38370, 38521, 38561, 38675, 39116, 39170, 39477, 39866

Carotenoids determination, spectral methods
22166-8, 22860, 23026, 24230, 24829; 25794, 25815, B26185, 26311-2, 26556,
27511; 28569, 28607, 28767, 29048, 29481, 30018, 32304; 32770, 33538, 34169,
34212, 34425, 34624, 34855, 35015, 35522, 35652, 36405, 36579; 36981, 37225,
37529, 37864, B38021, B38477, 39116, 39484, 40335

Carotenoids energetic states *in vitro*
23533, 24573; 25808, 25976, 26082, 26832, 26913, 27015; 29234, 31475; 33580

Carotenoids energetic states *in vivo*
22020, 22101; 25380, 26808, 27114, 27173, 27868, 27952; 28698, 28824, 29161,
29997, 30157, 30366, 30525, 30780, 30915, 31219, 31475, 31490, 32251; 32947,
33139, 33203, 33337, 34435, 34797, 35745; 36990-1, 37156, 37547, 38332-3,
38621, 38638, 38687, 39907

Carotenoids, enzymes of synthetic and degradation processes 22188; 25553, 26153;
33827, 36499

Carotenoids fluorescence *in vitro* 21986, 22989, 23153; 35987

Carotenoids fluorescence *in vivo* 22101; 29014, 30366; 33203; 37447, B38139

Carotenoids in flowers 32740, 33457; 37732, 38504-5

Carotenoids in model systems
21614, 21855, 23340, 23510, 23533; 25393, 25691, 25808, 25975-6, 26832,
26913, 27015, 27045; 29172, 29234-5, 29662, 29695, 29925, 31475, 31740,
31742; 32575, 32991, 33580, 34115, 34435; 36952, 36976, 37134, 38322,
39285, 39915, 39947-9, 40004, 40032, 40363

Carotenoids in mutants see Mutants, carotenoids in

Carotenoids in photosynthesis mechanism
21807, 21892, 21954, 21986, 22020, 22101, 22134, 22154, 22289, 22322, 22450,
22580-1, 22619, 22644, 22711, 22774, 22904, 23169, 23292-3, 23421, 23946,
24133, 24136, 24189, 24209, 24316, 24318, 24451, 24488-91, 25057, B25064,
25093
25202, 25244, 25248, 25302, 25335, 25344, 25353, 25533, 25645, 25691, 25746,
25761, 25969, B26185, B26220, 26225, 26808, 26872, 26962, 27015, 27114,
27127, 27173, 27340, 27512, 27608, 27711, 27864, 27868, 27874, 27946, 27952,
28027, 28029
28551, 28640, 28671, 28698, 28708, 28725, 28778, 28816, 28824, 28876, 28892,
29078, 29148-9, 29161, 29172, 29263-4, 29359, B29468, 29516, 29527-8, 29695,
29836, 29876, 29879, 29948, 29991, 29996-7, 30157, 30172, 30181, 30194, 30198,
30337-8, 30366, 30395, 30475, 30523, 30525, 30780, 30788-9, 30877, 30915,
30972, 30974, 31009, 31061, 31157, 31179, 31219, 31475, 31485, 31489-90, 31503-
-4, 31573, 31642, 31740-2, 31811, 31971, 32007, 32014, 32017-9, 32045, 32071,
32114, 32132, 32251, 32376, 32379, 32489, 32500
32685, 32733-4, 32744, 32800, 32947, 33139, 33186, 33190, 33203-4,
33252-3, 33266, 33337, 33400, 33402, 33484, 33610,
33746-7, 33779, 33838, 34020, 34115, 34179, 34204, 34215, 34344, 34400,
34435, 34607, 34686-7, 34773, 34789, 34797, 35058, 35386, 35653, 35699,
35702, 35728, 35739, 35745, 35763, 35915, 36055, 36193, 36272, 36484, 36503
36660, 36704, B36769, 36772, 36796-8, 36952-3, 36990-1, 37117, 37181, 37275,
37447, 37547, 37610, 37613, 37705, 37707, 37739, 37785, 37954, 38025, 38239,
38322, 38332-3, 38354, 38458-9, 38621, 38638, 38647, 39119-20, 39285, 39436,
39448, 39554, 39588, 39625, 39676, 39731, 39901, 39907, 39915, 39986, 40247,
40302

Carotenoids in physiology of photosynthesis
23752, 24019, 24068, 24315, 24348, 24451; 25383, 27511, 27513, 27925; 30198;
33629, 36051, 36561

Carotenoids in seeds and fruits
    21571, 22323, 22914, 23277, 23324-5, 23575, 23878, 23890, 24069, 24071,
    24092, 24129, 24230, 24751; 25225, 25553, 25779, 26186, 26225, 27510, 27526,
    27644, 28186, 28401; 28978, 29241, 29324, 29414, 29622, 29702, 31361, 31832-
    -4, 31961, 32229; 33010, 33057, 34163, 34212, 34403, 34581, 34624, 34879,
    35251, 35379; 37172, 37428, 37734, 38476, 38774, 38849, 39114, 40329

Carotenoids precursors see Carotenoids biosynthesis, precursors

Cell counting methods see Chloroplast and cell counting methods

Chamber, assimilation see Assimilation chamber

Chemiosmotic hypothesis, proton transport in chloroplast
    21531, 21607-8, 21631-2, 21687, 21745, 21752, 21766, 21843-4, 21879, 21890,
    21912, 21997, 22017, 22103, 22146, 22179, 22215, 22233, 22324, 22425-6,
    22456, 22491, 22528, 22536, 22583, 22586, 22612-3, 22625, 22657, 22704,
    22706, 22721-2, 22773-4, 22894, 22900, 22904, 22912-3, 22946, 22969-72,
    22996, 23183-4, 23192, 23208, 23362, 23434, 23449, 23591, 23615, 23647,
    23656, 23662, 23726, 23842, 23847, 23891, 23921, B24024, 24028, 24042, 24081,
    24099, 24169, B24176, 24221, 24311, 24316, 24355, 24367, 24374, 24405, 24416-
    -7, 24497, 24525, 24545, 24671, 24790, 24856, 24870, 24938, 24973, 25006,
    25030, 25043, 25057, 25099, 25101, 25143, 25152
    25240, 25313-5, 25317, 25401, 25461, 25466, 25522, 25591, 25719-20, 25789,
    25804, 25980, 26056, 26103-5, 26193-4, 26198, 26210, B26220, 26259, 26268,
    26301, 26332, 26342-3, 26388, 26411-2, 26500, 26523, 26574, 26718, 26720,
    26767, 26850, 26980, 27006, 27075, 27092-3, 27262, 27335, 27353-5, 27454,
    27456, 27491, 27680, 27772, 27783, 27917, 27936, 27945, 28035, 28163-4,
    28227, 28292, 28327, 28410, 28438-9, 28447-8
    28523-4, 28562, 28581, 28633-6, 28640, 28642, 28676, 28679, 28692, 28744,
    28747, 28754, 28798, 28839, 28906-7, 29114, 29513, 29525, 29540-1, 29583,
    29596-7, 29694, 29720, 29748-9, 29847, 29870, 30001, 30004, 30039, 30114-5,
    30126, 30158, 30178-80, 30214, 30219, 30340, 30402, 30408-9, 30430, 30436,
    30590, 30617, 30644, 30772-3, 30830, 30884, 30909-10, 30965-6, 30974, 31061,
    31082, 31124, 31149, 31179, 31287, 31402, 31458-9, 31461, 31485-6, 31488,
    31524, 31549, 31560, 31562, 31598, 31617, 31683-4, 31689, 31693, 31715-7,
    31746, 31813, 31877, 31955, 32040, 32068, 32130, 32135, 32157-8, 32198, 32228,
    32245, 32309, 32320-1, 32330, 32332, 32376, 32476, 32480, 32487, 32489, 32498
    32596, 32631, 32636, 32657, 32669, 32697, 32703, 32732, 32744, 32792, 32807,
    32834, 32935-6, 33142, 33160-1, 33207, 33251-3, 33350, 33400, 33402, 33448,
    33452, 33553, 33560, 33606, 33687, 33691, 33743, 33754, 33762, 33778, 33819,
    33850, 33864, 33882, 33919, 33932, 33945, 33899, 33994-5, 34025-6, 34032-3,
    34039, 34093, 34127, 34186, 34202-3, 34205, 34282, 34300, 34355, 34431,
    34605, 34723, 34762, 34769, 34774, 34803-5, 34841, 34868, 34975, 35012,
    35054, 35068, 35111, 35188, 35209, 35356, 35392-3, 35402, 35487, 35538,
    35650, 35716, 35728, 35853, 35926, 36131, 36265, 36282, 36382, 36398, 36404,
    36449, 36484, 36490, 36496, 36502-3
    36677, 36699, 36755, B36768, 36796, 36841, 36847, 36862, 37117, 37167, 37218,
    37245, 37285, 37445, 37553, 37615, 37629, 37665, 37704, 37763, 37805-6,
    37822, 37841, 37844, 37850, 37877, 37891-2, 37907, 37913, 37919, 37944,
    38029-30, 38042, 38055, 38094, 38109, 38169, 38172, 38202, 38241, 38250,
    38339, B38477, 38531, 38730, 38750-1, 38820, 38882, 38896, 38924, 38969,
    39023, 39048, 39118, B39220, 39221, 39266, 39293-5,   39301, 39498,
    39521, 39554, 39584, 39608, 39676, 39777, 39842, 39859, 39913, 39978, 40055,
    40124, 40164, 40226, 40246-7, 40302, 40338, 40366

Chlorobium chlorophyll see Chlorophylls, *Chlorobium*

Chlorophyll absorption spectra *in vitro*
    21726, 21775, 21796, 21891, 21918, 22072, 22161, 22207, 22235, 22242, 22278,
    22378, 22445, 22458, 22534, 22633, 22642, 22959, 23044, 23089, 23097, 23340,
    23342, 23344-5, 23348-9, 23353, 23361, 23388-9, 23433, 23450, 23472, 23667,
    23812, 23857, 23936, 23966, 24065, 24180, B24250, 24273, 24327, 24357, 24400,
    24486, 24733, 24749, 24784, 24800, 24937, 24991, 25016-7, B25064, 25141
                                                              (continued)

Chlorophyll absorption spectra *in vitro* (continued)
        25204, 25517, 25563, 25567, 25614, 25714, 25748, 25974, 26041, 26043, 26053,
        B26220, 26270, 26508, 26609, 26712-3, 26718, 26738, 26863, 26871, 26897,
        26943, 27000, 27082, 27114, 27178, 27303, 27309, 27381, 27394, 27608, 27684-
        -5, 27794-5, 28215, 28222, 28251, 28324, 28377, 28481
        28570, 28699, 29026, 29083, 29390, 29538, 29543-4, 29662, 29774, 29925, 29943,
        29978, 30092, 30101, 30117, 30137, 30173, 30179, 30248, 30368, 30418, 30426,
        30514, 30582, 30650, 30670, 30729-30, 30795, 30848, 30934, 31254, 31271,
        31446, 31622, 31713, 31787, 31794, 31881, 31892, 32114, 32132, 32188, 32356,
        B32410, 32472, 32531
        32677, 32744, 32754, 32790, 32999, 33240-1, 33274, 33303, 33308, 33575, 33609,
        33782, 33816, 33830, 33863, 33865, 33952, 33988, 33996-7, 34001, 34014, 34056,
        34086, 34214, 34335, 34381, 34564, 34662, 34722, 34770, 34795, 34860, 34877,
        34981, 35050, 35156, 35167, 35193, 35300, 35449, 35613, 35617, 35668, 35683,
        35741, 35757, 35763, 35833, 35890, 36033, 36209, 36228, 36260, 36470-1,
        36520, 36564
        36654, 36834, 37007, 37009, 37069, 37254, 37256, 37264, 37307, 37460, 37529,
        37537, 37715, 37869, 37879, 37969-71, 38005, 38010, 38073, 38107, 38110,
        38153, 38187-8, 38220, 38322, 38365, 38470, B38477, 38502, 38537, 38571,
        38681, 38690-1, 38766, 38780-1, 38855, 39100, 39111, 39344, 39406, 39437,
        39443, 39520, 39522, 39580-1, 39806, 39835, 39862, 39915, 39943, 40135,
        40149, 40284, 40347

Chlorophyll absorption spectra *in vivo*
        21507, 21532, 21606, 21678, 21685, 21691, 21714, 21733-4, 21743, 21834,
        21889, 21898, 21906-7, 21918, 21970, 22020, 22058, 22079, 22087, 22134,
        22138, 22182, 22222, 22247-8, 22275, 22326, 22333-4, 22393, 22395, 22450-1,
        22458, 22461, 22483, 22509, 22558, 22580, 22595, 22600-1, 22605, 22615,
        22623, 22642, 22664, 22780, 22799, 22814, 22862, 22902, 22973, 23000, 23022,
        23029, 23043, 23097-8, 23110, 23132, 23139, 23205, 23208, 23278, 23344-5,
        23384, 23402, 23488, 23515, 23527, 23565, 23667, 23820, 23823, 23826-7,
        23839, 23874, 23876, 23879-80, 23884, 23892, 23916, B23933, 23965-7, 24041,
        24158, 24190, 24227, 24283, 24327, 24345, 24385, 24398, 24434, 24466, 24495-
        -6, 24631, 24727-9, 24747, 24756-7, 24796, 24818, 24908-9, 24916, 24930,
        25006, 25040, 25107, 25132-3
        25166, 25176, 25200, 25263, 25268, 25295, 25310, 25379, 25433, 25440, 25535,
        25557-8, 25570, 25668, 25688, 25703, 25741, 25746, 25749, 25757, 25782,
        25791, 25837-8, 25867, 25869, 25884, 25888, 25899, 25960, 26019, 26168,
        B 26220, 26326, 26406, 26454, 26483, 26545, 26556, 26573, 26658, 26710,
        26718, 26727, 26757-8, 26796, 26822, 26829-30, 26873, 26935, 26937, 26942,
        26985, 27050, 27071-2, 27114, 27163, 27222, 27279, 27340-1, 27490, 27512-3,
        27608, 27615, 27675, 27726, 27795, 27804, 27817, 27848, 27863, 27870, 27946,
        28008, 28029, 28036, 28114, 28122, 28127, 28200, 28217, 28226, 28258, 28260,
        28276, 28281, 28467, 28476
        28520, 28526, 28550, 28598, 28631, 28653, 28690, 28692, 28737, 28752, 28792,
        28817, 28836, 28917, 28932-3, 28938, 28958, 28964, 28967-8, 28976, 29030-1,
        29073-4, 29121, 29194, 29198-9, 29240, 29263-6, 29278, 29349, 29378, 29431,
        29474, 29476, 29505, 29525, 29558, 29573, 29580, 29662, 29681, 29743, 29788,
        29818, 29878-9, 29896, 29906, 29913, 29939, 29978, 30079-80, 30172, 30193-4,
        30227, 30248, 30295, 30330-1, 30366, 30475, 30487, 30552, 30570, 30616, 30669-
        -70, 30742, 30849, 30862, 30934, 30975, 31011, 31033, 31054, 31073, 31101,
        31110, 31138, 31186, 31206, 31280, 31299, 31361, 31429, 31444, 31480, 31503,
        31505, 31534, 31575, 31594, 31643, 31707, 31713, 31726, 31735, 31785, 31787-8,
        31805, 31846, 31892, 31900, 31929, 31980, 31984, 32005, 32021, 32081, 32084,
        32099-100, 32103, 32112, 32168, 32186, 32282, 32285, 32288, 32351, 32361,
        32408-9, B32410, 32472, 32481, 32522
        32563, 32597, 32619, 32626, 32676, 32849, 32889-90, 32978, 33024, 33026,
        33031, 33035, 33048, 33064, 33138, 33186, 33189, 33204, 33237, 33298, 33322,
        33374, 33381, 33390, 33392-3, 33403, 33429, 33468, 33552, 33569, 33579,
        33618, 33702, 33706, 33746-7, 33782, 33788, 33874, 33956, 33958, 34007,
        34014, 34024, 34111, 34174, 34244, B34248, 34305, 34327, 34357, 34371, 34431,
        34490-1, 34582, 34594, 34627-8, 34761, 34845, 34957, 35000, 35015, 35066,

(continued)

                                                                        (continued)

Chlorophyll and its products determination, spectral methods (continued)
        28346, 28348, 28565, 28962, 29026, 29293, 29561, 29743, 29919, 30142, 30164-5,
        30169, 30348, 30440, 30755, 30835, 31225, 31257, 31577, 31622, 31682, 31881,
        32083, 32116, 32405
        32626,32748,32940,32988,33018,33131,33577,33706,33733,33880,33927,34056,
        34069, 34082, 34086, 34107-8, 34144, 34169, 34335, 34594, B34706, 34722,
        34914, 35156, 35206, 35435, 35520-2, 35556, 35612, 35652, 35756, 35839,
        36082, 36211, 36293, 36350
        36588, 36819-20, 36905-6, 37085, 37157, 37225, 37423, B37528, 37635, 37800,
        37879-80, 38010, B38037, 38255, 38446, 38470,B38477, 39018, 39116, 39222,
        39484, B39693, 39698, 39757, 39789, 40184, 40283, 40305, 40315-6

Chlorophyll and production of algae and water reservoirs
        21543, 21681-2, 21726, 21742, 21776, 21805, 21953, B22052, 22121, 22287,
        22353, 22357, 22392, 22417, 22453-4, 22464, 22488, 22493, 22517, 22531,
        22538, 22551, 22578, 22592, 22646, 22742, 22756, 22764, 22807, 22858, 22898,
        22943, 22976, 23016-7, 23087, 23096, 23098, 23107, 23133, 23188-9, 23269-70,
        23275, 23278, 23284-5, 23301, 23314-6, B23407, 23441, 23542, 23569, 23590,
        23593, 23595, 23663, 23684, 23702, 23759, 23809, 23824, 23844, 23947, 23963,
        24051, 24058, 24060-2, 24090, 24118, 24140, 24194, 24200, 24202, 24308,
        24349, 24392-3, 24456, 24475, 24562, 24570-2, 24579, 24581-2, 24586, 24616-8,
        24680, 24686-7, 24699, 24708, 24732, 24741, 24763-4, 24786, 24803, 24857,
        24895-6, 24927, 24953, 24957, 24964, 24992, 25003, 25010, 25053, 25129-30,
        25157
        25163, 25165, 25210, 25218, 25234, 25270, 25390-1, 25426, 25509, 25541,
        25580, 25588, 25598, 25633, 25647, 25651, 25680, 25739, 25743, 25760, 25763,
        25841, 25849-50, 25894, 25952, 25960, 25994, 26002, 26012, 26089, 26130,
        26163, 26212, 26214, 26242, 26270-1, 26358, 26387, 26407, 26510, 26516,
        26518, 26529, 26540, 26543, 26582, 26655-6, 26823, 26838, 26849, 26953,
        27177, 27206, 27227, 27250, 27266, 27281, 27298, 27376, 27392-3, 27400,
        27459, 27567, 27613, 27642, 27648, 27650, 27662, 27732, 27766, 27913, 27940,
        27956-8, 27960-1, 27986, 27988, 28004, 28017, 28025-6, 28047, 28054, 28076,
        28088, 28118, 28126, 28128, 28140, 28162, 28174, 28190, 28226, 28317, 28354,
        28464-5
        28553, 28565, 28673, 28721, 28773, 28776, 28790, 28797, 28861, 28880, 28955,
        28977, 28982, 28984, 29011, 29041, 29077, 29127, 29151, 29171, 29185, 29226,
        29287, 29292-3, 29311, 29327, 29336, 29345, 29355, 29375, 29396, 29428, 29448,
        29494, 29588, 29653, 29655, 29674, 29678, 29716, 29802-3, 29809, 29842, 29916,
        29919, 30045-6, 30052, 30075, 30121, 30137, 30147-8, 30169, 30196, 30225-6,
        30231, 30247, 30281, 30305, 30315, 30339, 30356-7, 30381, 30389, 30407, 30450,
        30482, 30485, 30504, 30572, 30605-6, 30613, 30676, 30724-5, 30756, 30863-5,
        30873, 30906, 30987, 31031, 31047, 31215, 31311-2, 31316-7, 31382, 31397,
        31430, 31531, 31714, 31745, 31909, 31912, 31965, 32031-2, 32043-4, 32089,
        32107-8, 32137-8, 32166, 32273-4, 32348, 32356, 32368, 32380, 32404, 32423,
        32493, 32518, 32548
        32564-5, 32599, 32606, 32643, 32645, 32815, 32844, 32910, 32941, 32961,
        32998-9, 33018, 33044-5, 33070-1, 33083, 33238-9, 33274, 33297, 33333, 33340,
        33364-7, 33422, 33514, 33531, 33594, 33634-5, 33656, 33676, 33782-3, 33847,
        33851, 33866, 33896, 33923, 33927, 33965, 33977, 34002, 34030-1, 34034,
        34058, 34069, 34099, 34139, 34177, 34184, 34191-2, 34197, 34216, 34222,
        34240, B34248, 34272-3, 34387, 34394-5, 34489, 34514-5, 34537-8, 34567,
        34592, 34733, 34763, 34768, 34793, 34811, 34870, 34878, 34884, 34985, 34991,
        34994, 35015, 35020, 35027, 35063, 35070, 35127, 35196, 35287, 35360, 35407-
        -8, 35436, 35462, 35493, 35506, 35523, 35533, 35552-3, 35584, 35587, 35597,
        35635, 35654, 35690, 35734, 35761, 35799, 35841, 35852, 35880, 35896, 35908,
        35916, 35929, 35941-2, 36061, 36079, 36089, 36100, 36153, 36155, 36208,
        36285-7, 36295, 36334, 36369, 36390, 36420, 36512, 36553
        36592, 36657, 36758, 36835, 36868, 36873, 36886, 37008, 37029, 37058, 37070,
        37096, 37106, 37152, 37204-5, 37238, 37246, 37258, 37278-9, 37328-9, 37355,
        37369, 37417, 37437, 37442, 37450, 37466, 37483-4, 37540, 37573, 37606,
        37635, 37809, 37818, 37859, 37868, 37889, 37943, 38060, 38070, 38076,
        38078, 38086, 38251, 38274, 38307, 38337, 38359, 38378, 38402, 38409,
        38417, 38454, 38514, 38659, 38686, 38692-3, 38726, 38807, 38944, 38985,
        39060, 39083, 39108, 39122, 39148, 39209-10, 39255, 39327, 39346, 39370,
                                                                        (continued)

Chlorophyll and production of algae ... (continued)
    39530, 39559, 39633-4, 39696-7, 39722, B39759, 39781, 39789, 39836, 39844,
    39871, 39896, 39908, 39924, 39937, 39942, 39962, 39996, 40089, 40153,
    40163, 40172, 40195, 40232, 40234, 40362, 40364

Chlorophyll and production of higher plants
    21699, 21760, 21862, 22012, 22435, 22775, 23091, B23407, 23410, 23570, 23831,
    24040, 24512, 24530, 24942, 25013, 25015, 25123; 25422, 25644, 27386, 28282;
    29067, 29233, 29419, 30396, 30581, 31510, 31638, 31839, 32185-6, 32259,
    32300; 32892, 33110, 33647, 34438, 34932, 35226, 35454, 36230; 36669, 38215,
    39981, 39984

Chlorophyll biosynthesis, precursors
    21506, 21528, 21532, 21534, 21558, 21575, 21642, 21664, 21678, 21701, 21703,
    21710, 21741, 21831, 21872, 21918, 21922, 22001, 22011, 22067, 22094, 22181,
    22265, 22290, 22307, 22317, 22322, 22335, 22337, 22341-2, 22391, 22438-40,
    22467, 22469, 22472-3, 22506, 22600-2, 22619, 22670, 22691, 22735, 22789,
    22799-800, 22813-6, 22827, 22888, 22902, 22911, 22990, 23035, 23095, 23101,
    23108-9, 23277, 23291, 23312, 23398, 23400, 23436, 23451, 23527-8, 23531,
    23564, 23581, 23614, 23619, 23639, 23642, 23674, 23697, 23704, B23767,
    23768, 23781, 23812, 23823, 23829, 23831, 23874-5, 23879-83, 23913, 23915,
    23924, 23951, 23960, 23968, 24019, 24109, 24117, 24158, 24168, 24177-80,
    24209, 24305, 24330, 24340, 24348, 24364, 24381, 24398, 24400, 24402, 24413,
    24459-60, 24463-5, 24561, 24590, 24597, 24642, 24653-4, 24696, 24727, 24756,
    24792, 24800, 24912, 24916, 24924, 24962, 24986-7, 24991, 24995-6, 25000,
    B25064, 25121, 25132-3, 25144
    25166, 25200, 25204, 25237, 25268-9, 25293, 25295, 25334, 25344-5, 25440,
    25462, 25498, 25504, 25517, 25564, 25567, 25596, 25616, 25689, 25700, 25747-
    -8, 25852-3, 25899-900, 25904, 25918, 25933, 25936-8, 25988, 26053, 26057,
    26069-70, 26094, 26097, 26099, 26113, 26139, 26143, 26153, B26185, 26191-2,
    26205, 26215, 26248, 26276, 26278, 26310, 26340, 26351, 26377, 26399, 26415,
    26437, 26481, 26545, 26558, 26562, 26594, 26603-4, 26630, 26667, 26681,
    26706, 26737, 26768, 26842, 26890, 26926, 26936-7, 26949, 26980, 27043,
    27076-8, 27081, 27083, 27260, 27303-4, 27314-5, 27329, 27380, 27397, 27407,
    27444, 27452, 27505, 27564-5, 27637, 27659, 27677-8, 27685, 27703, 27718,
    27727, 27755, 27775, 27784, 27795-6, 27805, 27822, 27831-2, 27837, 27851-4,
    27927, 27972, 27997, 28039, 28156, 28166, 28170, 28217-8, 28277, 28319,
    28362, 28373, 28379, 28476
    28499, 28519, 28525-6, 28532, 28585-6, 28597, 28628, 28637, 28639, 28656-7,
    28667-8, 28674, 28709, 28719, 28745, 28752, 28762, 28793, 28811, 28836, 28910,
    28912, 28956, 29012, 29030, 29043, 29058-9, 29150, 29220, 29259, 29329, 29341,
    29347-9, 29361, 29392, 29403, 29433, B29468, 29476, 29523, 29546, 29573,
    29593-4, 29610-1, 29623, 29640, 29708, 29760, 29773, 29833-4, 29859, 29903-5,
    29939, 29947, 30006, 30027-8, 30176, 30200-2, 30215-6, 30229, 30240, 30252,
    30311, 30318-20, 30330-1, 30334, 30424-5, 30514, 30537, 30540-1, 30560, 30597,
    30602-3, 30610, 30612, 30660, 30673, 30740-1, 30790-2, 30795, 30827, 30837,
    30848-50, 30867, 30893-4, 30923, 30992, 31042, 31054, 31096, 31101, 31116,
    31139, 31192, 31211, 31223, 31285, 31329, 31427, 31444-5, 31538, 31542-3,
    31581, 31615, 31622-3, 31682, 31713, 31715, 31719, 31735, 31779, 31795, 31829,
    31846, 31855, 31880, 31982, 32001, 32063, 32084, 32086, 32202, 32232, 32253,
    32291, 32298-9, 32302, 32351, 32360-1, 32381-2, 32397-8, 32414, 32421, 32451,
    32488, 32511, 32531, 32549
    32566, 32580, 32590, 32593, 32595, 32597, 32651, 32661, 32670, 32704, 32731,
    32763, 32771, 32776, 32778-9, 32783, 32790, 32799, 32805, 32820, 32854,
    32864, 32875-6, 32889-90, 32931, 32974-5, 32985, 33012, 33026, 33031, 33047,
    33094-5, 33108, 33202, 33205, 33298, 33301-2, 33332, 33389-90, 33392-3,
    33429, 33455, 33482, 33494, 33519, 33577, 33579, 33596, 33603, 33609, 33630,
    33639, 33648, 33654, 33684, 33692, 33727, 33781, 33796, 33874-6, 33909,
    33952-5, 34042, 34045, 34087, 34094, 34140, 34174, 34194, 34220, 34235-7,
    34245, 34296, 34327, 34329, 34340-2, 34390, 34397, 34420, 34441, 34468,
    34522, 34533, 34608, 34627, 34691, 34699, 34703, 34739, 34746, 34770-1,
    34790, 34795-6, 34837-8, 34846, 34862, 34867, 34869, 34872, 34920-2, 34934-5,
    34948, 35015, 35032, 35034-5, 35066, 35071-2, 35132-3, 35137, 35143-5,
                                                        (continued)

Chlorophyll biosynthesis... (continued)

Chlorophyll chemical structure

Chlorophyll complexes *in vitro*

Chlorophyll complexes *in vivo*
                                                                    (continued)

Chlorophyll complexes *in vivo* (continued)
        25166, 25174, 25184, 25186, 25200, 25228, 25241, 25246, 25263, 25268-9,
        25297, 25311, 25323, 25379, 25440, 25456, 25462, 25488, 25564, 25570, 25606,
        25616, 25650, 25668, 25703, 25746, 25773, 25791, 25820, 25822, 25837-8,
        25881, 25888-9, 25899, 25957, 26048, 26053, 26063, 26099, 26101, 26143,
        26160, 26168, 26172, B26185, B26220, 26252-3, 26293, 26326-7, 26352-3,
        26362, 26414, 26457, 26545, 26556, 26573, 26594, 26604, 26608-9, 26632,
        26672, 26674, 26682-3, 26700, 26709, 26718, 26757-8, 26761, 26774, 26818,
        26822, 26890, 26927, 26930, 26935-8, 26977, 26985, 26995, 27043, 27078,
        27128, 27137, 27222, 27341-2, 27407, 27489, 27512-3, 27615, 27619, 27668,
        27683, 27690, 27717, 27771, 27795, 27817, 27823, 27852, 27863-4, 27867,
        27943, 27952, 28021, 28028, 28036, 28045, 28055, 28111, 28122, 28127, 28222,
        28232, 28234, 28238, 28258, 28260, 28279, 28281, 28345, 28376, 28410, 28416,
        28467
        28499, 28519, 28528-9, 28535, 28574-7, 28586-7, 28590, 28595-6, 28637-8, 28644,
        28657, 28691, 28700, 28713, 28752, 28755-6, 28836, 28878, 28886, 28892, 28905,
        28958, 28964-8, 28989, 28991, 29001, 29026, 29030-1, 29107, 29118, 29173,
        29194-5, 29197-200, 29250, 29263, 29275, 29329-30, 29339, 29348-9, 29361,
        29406, 29410, 29431-2, B29468, 29483, 29519, 29522, 29525, 29580, 29609, 29662,
        29681-2, 29738, 29754, 29756-7, 29773, 29777, 29785, 29787-8, 29818, 29875-6,
        29896, 29903, 29905, 29907, 29922, 29938-9, 29991, 29997, 30027-8, 30051,
        30067, 30135, 30179, 30181, 30188, 30202, 30248, 30252-3, 30321-2, 30330-1,
        30335, 30366, 30425-6, 30472, 30481, 30488, 30497, 30552, 30570, 30612, 30669-
        -70, 30672, 30687-8, 30694, 30740, 30742, 30828, 30850, 30857, 30870, 30893,
        30975, 31033, 31046, 31054, 31097, 31101, 31104, 31108, 31129, 31138, 31143,
        31186, 31298, 31330, 31354, 31364, 31369-71, 31375, 31385, 31391, 31426, 31429-
        -30, 31480-1, 31486, 31503, 31511, 31538, 31543, 31581, 31621, 31625, 31666,
        31735, 31756, 31785-6, 31788, 31812, 31846, 31892-3, 31900, 31929, 31981-2,
        31992, 32007, 32045, 32071, 32081, 32099-101, 32112-3, 32124, 32130, 32168,
        32214, 32228, 32246, 32266, 32282, 32285-6, 32298, 32307, 32360, 32393, 32432,
        32441, 32452, 32522
        32563, 32566,   32572, 32588, 32594,   32597, 32600,   32650-1, 32661, 32675-6,
        32704, 32805, 32849, 32889-90, 32925-6, 32992-4, 33024-6, 33031, 33035, 33037,
        33064, 33100, 33108, 33118, 33147, 33185-7, 33204-5, 33259, 33298, 33311,
        33322, 33337, 33353, 33381, 33389, 33392, 33403, 33405, 33468-9, 33515-6,
        33577, 33579, 33617-8, 33667, 33709, 33778-9, 33782-3, 33811, 33814, 33842,
        33874, 33880, 33954, 33956-8, 33967, 33983, 34007, 34013, 34111, 34205,
        34237, B34248, 34322, 34327, 34350, 34367, 34390, 34441, 34482-3, 34491,
        34493, 34529, 34569, 34582, 34610, 34628, 34670, 34688, 34707, 34766-7,
        34845, 34910, 34934, 34990, 35026, 35042, 35058, 35066, 35114, 35137, 35166,
        35181, 35274, 35376, 35380, 35386, 35398, 35502, 35571, 35665-6, 35668,
        35719, 35725, 35739, 35757-8, 35777, 35835, 35844, 35897, 35899, 35988,
        36030, 36039, 36062, 36095, 36118-9, 36177, 36233, 36235, 36244-5, 36258,
        36272, 36355, 36377, 36399-400, 36409, 36411, 36447, 36480, 36539, 36546,
        36585
        36611, 36631, 36651, 36660, 36668, 36690, 36702, 36712, 36747, 36771, 36854-
        -6, 36895, 36924, 36943-4, 36979, 26990, 36996, 37019, 37055-6, 37064-5,
        37076, 37078, 37138, 37155-6, 37161, 37181, 37183, 37262, 37275, 37310,
        37315-6, 37321, 37336, 37350, 37358, 37366-7, 37385, 37410, 37446-7, 37485,
        B37528, 37534, 37554, 37556, 37598, 37607, 37610, 37640-1, 37646, 37674,
        37692, 37707, 37717, 37749-50, 37767, 37872, 37877, 37954-5, 37957-8, 37964,
        37998, 38007, 38091, 38095-6, 38119, 38153-5, 38164, 38178, 38193, 38233,
        38245, 38254, 38334, 38373, 38380, B38415, 38447, 38471, 38480, 38502,
        38530, 38548, 38555-7, 38571, 38618, 38648, 38723-4, 38730, 38832-3, 38862,
        38878, 38896, 38900, 38918, 38941, 38963, 38972, 39000, 39114, 39149, 39168,
        39180, B39191, 39231, 39258, 39282, 39285, 39292, 39310, 39318, 39379,
        39444-6, 39452, 39517, 39577, 39588, 39597-8, 39606, 39610, 39612-3, 39624,
        39646, 39745-6, 39762-3, 39801, 39806, 39822, 39901-2, 39909, 40016, 40048,
        40069, 40102, 40122, 40131, 40136, 40234, 40246-7, 40359

Chlorophyll degradation
        21796, 21857, 21886, 21889, 21896, 21961, 21969, 22068, 22102, 22235, 22300,
        22312, 22323, 22342, 22377, 22390, 22408, 22468, 22499, 22534, 22581, 22633,
                                                                        (continued)

Chlorophyll degradation (continued)

Chlorophyll delayed light emission, luminescence *in vitro*

Chlorophyll delayed light emission, luminescence *in vivo*
                                                                    (continued)

Chlorophyll delayed light emission, luminescence *in vivo* (continued)
        32620, 32657, 32672, 32703, 32728, 32744, 32791, 32954, 33144, 33482, 33484,
        33552, 33617-8, 33749, 33794, 33991, 34032, 34105-6, 34207, 34277, 34348,
        34418, 34420, 34526, 34601, B34706, 34728, 34759, 34778-90, 34848, 35093,
        35245, 35269, 35556-7, 35756, 35839, 35858, 35945, 36045, B36081, 36276,
        36278, 36469, 36586
        36699, 36739, 36745, 36905, 36968-9, 37129, 37137, 37280, 37469, 37482,
        376623, 37668, 37705, 37723, 37767, 37861, 37996, 38138, 38254, 38460, 38651,
        B38706, 38773, 38989, 39349, 39373, 40076, 40346

Chlorophyll determination see Chlorophyll and its products determination ...

Chlorophyll energetic states *in vitro* (*cf.* also Chlorophyll in model systems)
        21721, 22046, 22174, 22384-6, 22534, 22642, 22871, 22887, 23130-1, 23359,
        23364, 23378, 23433, 23533, 24197, 24327, 24573, 24622, 24783, 25141
        25219-20, 25394, 25735-6, 25800, 26082, B26220, 26459, 26556, 26670, 26673,
        26679, 26832, 27015, 27780-1, 27844, 28021, 28036, 28161, 28214, 28273,
        28290
        28563, 28627, 28954, 29080, 29125, 29369, 29662, 29774, 29978, 30248, 30388,
        30641, 30793, 30997, 31074, 31618, 31713, 32123, 32148, 32208, 32531
        32642, 32790, 33185, 33241, 33414, 33580, 33763, 33824, 33988, 34205, 34336-7
        34543, 34565, 34662, 34679, 34697, 34909, 35050, 35316, 35741, 35820, 36073,
        36356, 36470
        36675, 36737, 36911, 37104, 37160, 37206, 37324, 37460, 37537, 37869, 38051,
        38072, 38320, 38362, 38855, 39111, 39520, 39851, 39941

Chlorophyll energetic states *in vivo*
        21607, 21816, 21856, 22020, 22022, 22057, 22088, 22125, 22134, 22182, 22193,
        22275-6, 22288, 22299, 22443, 22458, 22558, 22580-1, 22632, 22642, 23046,
        23130, 23206, 23434, 23437, 23520, 23631, 23874, 23954, 23966, 24158, B24250,
        24306, 24327, 24480, 24496, 24811, 24813
        25201, 25244-5, 25355, 25544, 25631, 25703, 25733-4, 25751, 25846, 25881,
        26060, 26394-6, 26410, 26556, 26610, 26672, 26675-6, 26680, 26720, 26764,
        26942-3, 27100, 27114-5, 27128, 27173, 27282, 27341, 27378-9, 27866-7, 27952,
        28131, 28136, 28155, 28234, 28332, 28409
        28515, 28699, 28724, 28824, 28876, 28884-6, 28934, 28997-8, 29026, 29074,
        29124, 29134, 29161, 29310, 29358, 29439, 29445, 29453, 29497, 29505, 29521,
        29619, 29621, 29666, 29679, 29681, 29710, 29754-5, 29779, 29818, 29878, 29896
        29936, 29946, 29969, 29971-2, 29978, 30159, 30179, 30183, 30248, 30359, 30366
        30385, 30424, 30479, 30539, 30625, 30742, 30780, 30794, 30813, 30939, 30970,
        31063, 31074, 31108, 31219, 31299, 31369, 31444, 31571, 31638, 31642, 31709,
        31980, 32021, 32123-4, 32162, 32251, 32357-8, B32410
        32588, 32620, 32635, 32658, 32672, 32675, 32754, 32880, 32884, 32890, 32925-
        -6, 32992, 33024, 33035, 33063, 33120, 33139, 33141, 33185, 33187-8, 33258,
        33337, 33356, 33579, 33640, 33649, 33735, 33746-7, 33788, 33814, 33880, 33884
        33964, 34013, 34023-4, 34277, 34338, 34349, 34361, 34418, 34566, 34572, 34584
        34610, B34706, 34756, 34759, 34797, 34832, 34860, 34893, 35085, 35107, 35218-
        -20, 35274, 35311, 35366, 35576, 35658, 35738, 35745, 35820, 35847, 35945,
        36046, 36176, 36239, 36247, 36469, 36518, 36546, 36583
        36627, B36768, 36770, 36911, 36932, 36955, 36971, 36990-1, 36995, 37032,
        37064-5, 37156, 37161, 37180, 37254, 37314, 37358, 37370, 37373, 37385,
        37426, 37494, 37545-7, 37598, 37661, 37705, 37743, 37767, 37773, 37811,
        37869, 37914, 37996, 38090, 38164, 38244-5, 38254, 38351, 38361, B38477,
        38595, 38621, 38653, 38687, B38706, 38714, 39026, 39091, 39278, 39303,
        39452, 39497, 39516-7, 39592, 39613, 39745, 39909, 40011, 40048, 40076,
        40084, 40246, 40259

Chlorophyll energetics model see Model ...

Chlorophyll, enzymes of synthesis and degradation (other than chlorophyllase)
        21701, 21703, 21710, 22093-4, 22275-6, 22307, 22341, 22439-40, 22670, 22672,
        23035, 23095, 23312, 24364, 24443 25345, 25462, 25881, 25937, 26153, 26205,
        26248, 26558, 26681, 26750, 26949, 27509, 27775, 27822, 27831; 28628, 29059,
                                                                              (continued)

Chlorophyll, enzymes ... (continued)
        29708, 30740, 31005, 31054, 31330; 32778, 33651, 33732, 33783, 33827, 33874-
        -5, 34194, 34235-6, 34340, 34522, 34846, 35204, 35582, 35629, 35809, 35923,
        35992, 36402; 36759, 37357, 37523, 37572, 37703, 37936, B38118, 38130, 38225,
        39429, 39597, 39848

Chlorophyll fluorescence *in vitro*
        21597, 21636, 21721, 21917-8, 22160, 22235, 22278, 22480, 22483, 22534,
        22664, 23012, 23130, 23153, 23185, 23205, 23327, 23348, 23359, 23433, 23831,
        23970, B24250, 24327, 24399-400, 24421
        25316, 25517, 25567, 25973, 26041-2, B26220, 26713, 27082, 27114, 27781,
        28030, 28214-5, 28434, 28481
        28626, 29026, 29543-4, 29548, 29662, 29774, 29943, 30173, 30367, 30388, 30641,
        30657, 30997, 31374, 31577, 31992, 32224, 32241, 32381, 32531
        32784, 32790, 33407-8, 33414, 33580, 33830, 33941, 33987-8, 34280, 34662,
        34679, 34740, 34850, 35041, 35562, 35638, 35918, 36033, 36046, 36073, 36211,
        36228, 36470
        36737, 36819, 37009, 37103, 37160, 37307, 37374, 37376, 37537, 37596, 38072,
        38108, 38281, B38477, 38502, 38536, 38681, 38780-1, 38855, 39095, 39179,
        39406, 39520, 39522, 39851, 39915, 40260, 40347

Chlorophyll fluorescence *in vivo*
        21506, 21567, 21597, 21600, 21607-8, 21664, 21679, 21690, 21706, 21718,
        21720, 21733, 21744, 21749, 21767, 21774, 21790, 21810, 21825, 21856, 21898-
        -9, 21901, 21918, 21924, 21962-5, 21993, 22020, 22044, 22062, 22078-9,
        22088, 22090-1, 22182, 22214, 22264-5, 22275-7, 22291, 22298-300, 22325,
        22328-9, 22332, 22359, 22405, 22426, 22458, 22462-3, 22483, 22514, 22527,
        22558, 22580-1, 22605, 22615, 22629, 22632, 22642, 22683, 22721, 22748,
        22766, 22794, 22815-6, 22857, 22933-4, 22936, 22955-7, 22973, 23017, 23021,
        23023-4, 23031, 23042-3, 23102-3, 23118, 23120, 23132, 23140-1, 23153, 23184-
        -5, 23187, 23192, 23205, 23208-9, 23240-1, 23323, 23327, 23394-6, 23402,
        23434, 23441, 23446, 23586, 23616-8, 23622, 23639, 23647, 23656, 23726,
        23728-30, B23767, 23768, 23794, 23830, 23843, 23916, B23933, 23934, 23954-7,
        23970-1, 24041, 24103, 24133, 24158, 24179, B24250, 24307, 24327, 24330,
        24356, 24371-3, 24401-3, 24417, 24421, 24451, 24478, 24480, 24496, 24575,
        24577, 24631, 24653, 24721, 24728, 24735, 24737, 24756, 24792, 24825, 24864,
        24870, 24874, 24910-1, 24917, 24930, 25002, 25012, 25050, 25076, 25081-3,
        25107, 25113-4, 25151-2
        25166, 25201, 25211, 25240, 25245-6, 25263, 25293, 25295, 25315, 25339,
        25344, 25353, 25356, 25378-9, 25385-6, 25396, 25456, 25476, 25444, 25570,
        25606, 25616, 25630, 25636, 25645, 25733, 25746, 25819-20, 25840, 25846,
        25861, 25868, 25892, 25899, 25994, 26048, 26063, 26100-1, 26143, 26168,
        26175, 26196, 26224, 26293-4, 26298, 26326, 26396, 26406, 26411-2, 26468,
        26480, 26523, 26538, 26559, 26582, 26588, 26608, 26615, 26632-3, 26675-6,
        26687, 26703-4, 26710, 26718, 26767, 26792, 26794, 26796-7, 26813, 26829-30,
        26899, 26935-8, 26955, 27064, 27067, 27084, 27111, 27114-5, 27128, 27140,
        27173, 27273, 27319, 27332, 27345, 27378-9, 27391, 27505, 27512, 27619,
        27675, 27678, 27717, 27746-7, 27762-5, 27780, 27809, 27817, 27861, 27863,
        27868, 27870, 27879, 27901, 27917, 27927, 27946, 27952, 27954-5, 28028,
        28055, 28104, 28111, 28122, 28234, 28236, 28260, 28276, 28418, 28449,
        28491
        28518-9, 28550-1, 28585-6, 28588-9, 28591, 28595, 28636, 28657, 28676, 28678,
        28680-3, 28698, 28724-6, 28737, 28752, 28771, 28839, 28859, 28867, 28878,
        28886, 28892, 28903, 28934-6, 28938, 28956, 28963-9, 28994, 28998-9001, 29012,
        29014, 29017, 29025-7, 29030, 29074, 29096, 29134, 29194-5, 29244, 29268,
        29276, 29282, 29341, 29348-9, 29436, 29449, 29525-6, 29528-9, 29534, 29546-7,
        29592, 29595, 29599-600, 29602, 29619-21, 29635, 29661, 29681-2, 29712-3,
        29755-7, 29807, 29818, 29831, 29836, 29842, 29872, 29878, 29906, 29928, 29943,
        29946, 29955, 29991, 29997, 30000-1, 30012, 30108, 30154-6, 30158-60, 30176,
        30183-4, 30193-4, 30225, 30228-9, 30248, 30256, 30299, 30302, 30323, 30326-8,
        30330-1, 30347, 30360, 30366, 30429, 30439, 30467, 30495, 30525, 30539, 30570,
        30587, 30590-1, 30596, 30600, 30616, 30644; 30708-10, 30718-20, 30736, 30759,
        30791-2, 30794, 30833-5, 30838-9, 30848, 30850, 30859, 30875, 30877, 30892,
                                                                    (continued)

(continued)

Chlorophyll in model systems (continued)
        32575, 32601, 32642, 32730, 32904, 32991, 33308, 33382, 33564, 33580, 33616,
        33618, 33667, 33683, 33743, 33779, 33885, 33941, 33964, 34001, 34032, 34115,
        34187, 34211, 34277, 34313, 34336, 34419-20, 34679, 34756, 34774, 34909-10,
        34973, 35000, 35041, 35050, 35316-7, 35366, 35422, 35741-2, 35779, 35832-4,
        36040, 36130-1, 36228-9, 36260, 36352, 36470
        36622, 36675, B36769, 36911, 36952, 36972, 37104, 37134, 37136, 37160, 37219,
        37254, 37306-7, 37324, 37459-60, 37570, 37594-5, 37597, 37743, 37879, 37910,
        38004-6, 38051-2,   38072-3, 38107, 38152-3, 38187-8, 38309, 38311,
        38320-2, 38362, 38536-7, 38600-1, B38706, 38765, 38780-1, 38855, 38876,
        39285; 39406, 39491, 39495, 39520, 39522, 39592, 39666, 39778, 39861-2,
        39915, 39941, 40004, 40031-2, 40093, 40135

Chlorophyll in mutants see Mutants, chlorophyll in

Chlorophyll in photosynthesis mechanism
        21643, 21646, 21662, 21664, 21690, 21807, 21816, 21843, 21892, 21899, B21914,
        21954, 21963, 21991, 22020, 22046, 22106, 22125, 22134, 22193, 22198, 22221,
        22243, 22288-9, 22299, 22390, 22443, 22450, 22458, 22558, 22568, 22580,
        22583, 22596, 22640, 22748, 22933, 22970, 22972, 23000, 23046, 23101-2, 23118,
        23141, 23200, 23215, 23292, 23327, 23347, 23402, B23407, 23434, 23437, 23558,
        23566, 23574, 23584, 23720, 23761, 23825-6, 23857, 23876, 23916, 23934,
        23955, 23957, 23966, 24133, 24154, 24191, 24227, B24250, 24280, 24306-7,
        24327, 24374, 24548, 24575, 24622, 24811-3, 24876, 24914, 24931, 25051, 25057,
        B25064
        25166, 25185-6, 25202, 25241, 25244-5, 25263, 25323, 25344, 25348, 25353,
        25355, 25379, 25441, 25490, 25533, 25570, 25606, 25616, 25631, 25691, 25733,
        25751, 25761, 25773, 25834, 25846, 25879-81, 25888, 25918, 25957, 26043,
        26160, B26185, B26220, 26223-4, 26257-8, 26298, 26352, 26394-5, 26398,
        26413, 26456, 26479, 26523, 26538, 26556, 26561, 26573, 26609-10, 26614,
        26672, 26676, 26680, 26718, 26720, 26734, 26761, 26765, 26872, 26914, 26938,
        26943, 26974, 26990, 27015, 27064, 27072, 27083-4, 27100, 27114-5, 27127-8,
        27137, 27173, 27222, 27254, 27273, 27282, 27309, 27340, 27512, 27518-9,
        27528, 27608, 27615, 27619, 27668, 27764, 27864, 27867-8, 27878, 27952,
        28021, 28029, 28036, 28127, 28131, 28155, 28234, 28258, 28260, 28290, 28332,
        28345, 28409-10
        28515, 28519, 28535, 28547-8, 28550, 28575-6, 28588, 28590, 28595-6, 28644,
        28657, B28677, 28691, 28698, 28700, 28720, 28724-5, 28729, 28737, 28751,
        28822, 28824, 28834, 28859, 28877, 28884-5, 28892, 28903, 28934, 28969, 28989,
        28991, 28997, 28999, 29001, 29010, 29012, 29025, 29027, 29124, 29134, 29149,
        29161, 29174, 29199, 29244, 29263-4, 29275, 29278, 29310, 29339, 29358,
        29406, 29431-2, 29440, 29445, 29452-4, B29468, 29518-21, 29525, 29528, 29580,
        29614, 29620-1, 29666, 29679, 29714, 29738, 29754, 29756-7, 29784-5, 29818,
        29836, 29876, 29879, 29896, 29922, 29971-2, 29978, 29997, 30085, 30135, 30138,
        30155, 30158, 30172, 30179, 30181, 30183, 30189, 30198, 30248, 30256, 30320,
        30322-3, 30326-8, 30347, 30359, 30366, 30418, 30422-3, 30436, 30442, 30479,
        30523, 30525, 30539-40, 30594, 30625-6, 30669, 30708, 30718, 30742, 30759,
        30779-80, 30782, 30795-6, 30812-3, 30831, 30870, 30877-8, 30881, 30915, 30926,
        30939, 30970, 31009, 31027, 31030, 31033, 31046, 31060, 31104, 31108, 31138,
        31143, 31185, 31194, 31219, 31288, 31297, 31354, 31362, 31364, 31369, 31371,
        31390-1, 31429, 31444, 31487-9, 31505, 31507, 31556, 31575, 31588, 31642,
        31666, 31713, 31740-2, 31788, 31794, 31805, 31893, 31906, 31955, 31971, 31979,
        31981-2, 32007, 32049, 32099, 32101, 32112-3, 32123, 32127, 32130, 32132,
        32148, 32162, 32168-9, 32216, 32223, 32238-9, 32246, 32249, 32280-1, 32357-8,
        32379, 32393, B32410, 32432, 32441, 32444
        32587, 32619-20, 32650, 32658, 32661, 32667, 32672, 32675, 32714, 32733,
        32744, 32792, 32880, 32884, 32890, 32925-6, 32934, 32992, 32994, 33024,
        33037, 33063, 33100, 33120, 33139, 33141, 33186, 33188, 33190, 33204, 33258,
        33311, 33316, 33337, 33381, 33389-90, 33403, 33486, 33515-6, 33563-4, 33605,
        33609, 33626, 33708-9, 33745-8, 33750, 33778-9, 33788, 33814, 33864, 33885,
        33927, 33956, 33964, 33982, 34013, 34020, 34042, 34115, 34186, 34204-5, 34237,
        34277, 34313, 34338, 34344, 34349-50, 34357, 34367, 34419, 34441, 34461,
        34566, 34569, 34572, 34584, 34601, 34607, 34610, 34688, 34756, 34760, 34767,
        34775-6, 34795-6, 34843, 34845, 34860, 34893, 34965, 34968, 34990, 35058,
                                                              (continued)

Chlorophyll in photosynthesis mechanism (continued)
     35075, 35107-8, 35114, 35218, 35220, 35238, 35269, 35291, 35311, 35317,
     35366, 35386-7, 35422, 35428, 35432, 35526, 35665-6, 35670, 35719, 35725,
     35745, 35820, 35847, 35897, 35945, 35988, 36023-4, 36062, B36081, 36118,
     36193, 36246-7, 36256, 36272, 36276, 36311, 36352, 36355, 36377, 36411, 36430,
     36447, 36469, 36484, 36546, 36567, 36583
     36627, 36646, 36660, 36675, B36768-9, 36770-2, 36944, 36952, 36955, 36990-1,
     36995-6,   37055-6, 37064, 37078, 37117, 37165, 37167, 37180-1, 37206,
     37254-5, 37262, 37275, 37310, 37314-6, 37358, 37367, 37370, 37375, 37385-6,
     37446,   37485, 37494, 37524, 37534, 37545-7, 37565, 37584, 37593-4,
     37607, 37610, 37661, 37675, 37705-7, 37743, 37750, 37759-60, 37764, 37767,
     37785, 37835, 37877, 37954, 37957, 37964-5, 37996, 38096, 38111, 38150,
     38153-5, 38160, 38204, 38232-3, 38244-5, 38254, 38319-22, 38334, 38382,
     B38415, 38447, 38492, 38526, 38548, 38557, 38571, 38595, 38604, 38613, 38637,
     38654, 38669, 38698-9, 38706, 38723, 38730, 38833, 38887, 38957, 39000,
     39023, 39027, 39091, 39149, 39222, 39231, 39285, 39318, 39349, 39379, 39444,
     39451, 39462, 39502, 39516-7, 39577, 39588, 39592, 39613, 39615, 39627,
     39646, 39666, 39681, 39731, 39745, 39835, 39901-2, 39909-10, 40029, 40048,
     40069, 40076, 40084, 40122, 40140, 40182, 40246-7

Chlorophyll in physiology of photosynthesis
     21662, 21697, 21721, 21776, 21793, 21842, B21914, 22010, 22085-6, 22089,
     22184, 22192, 22413, 22453, 22623, 22815, 22848, 22851, 23242, 23258, 23278,
     23470, 23739, 23754, 23780, 23845, 23867, 23895-7, 24019, 24148, 14197,
     24280, 24321, 24323, 24348, 24378, 24401, 24715, 24804, 24821, 24882
     25184, 25259-60, 25293-4, 25383, 25404, 25406, 25441, 25616, 25810, 25853,
     25867, 26099, 26123, 26353, 26512, 26597, 26684, 26700, 26869, 27036, 27069,
     27129, 27317, 27350, 27460, 27463, 27508, 27511, 27513, 27567, 27925, 28008,
     28128, 28172-3, 28188, 28379, 28433
     28656, 28776, 28797, 28835-6, 28843, 28921, 28940, 29004, 29040, 29163, 29191,
     29564, 29593-4, 29826, 30057, 30131, 30198, 30263, 30281, 30320, 30575, 30613,
     30643, 30722, 30896, 30935, 31118-9, 31230, 31281, 31340, 31510, 31550, 31587,
     31627, 31801, 32121, 32187-9, 32230, 32289, 32300, 32311, 32417
     32574, 32666, 32769, 32899, 33034, 33070, 33076, 33118, 33379, 33499, 33630,
     33647, 33700, 33736, 33816, 33869, 33892, 34070, 34390, 34473, 34515, 34702,
     34729, 34782, 34870, 34932, 35062, 35191, 35270, 35339-40, 35475, 35609,
     35763, 35836, 35866, 35886, 36051, 36332, 36519, 36551
     36865, 36985, 37271, 37348, 37602-3, 37659, 37678, 37930, 38059, 38087,
     38117, B38139, 38194, 38215, 38326, 38426, 38493, 38515, 38644, 38657,
     38663, 38783, 38791, B38892, 39055, 39117, 39126, 39149, 39159, 39171,
     39252, 39460, 39627, 39808, 39895, 40009, 40098, 40156, 40201, 40239

Chlorophyll in seeds and fruits
     22323, 22468, 22911, 22914, 23277, 23319, 23324-5, 23395, 23406, 23829-30,
     23878, 23890, 24071, 24129, 24299, 24824; 25634, 26511, 27265, 27520,
     27526, 27561; 28613, 29414, 29702, 29920, 30441, 30514, 30537, 31361,
     31833; 32739, 33057, 33348, 34581, 34678, 35251, 35379, 35609, 35746,
     35810, 36293; 36644, 36920, 36924, 37428, 37775, 38106, 38803, 39114,
     39163, 39435, 39587, 39814, 40112

Chlorophyll luminescence see Chlorophyll, delayed light emission ...

Chlorophyll, methods see Chlorophyll and its products determination ...

Chlorophyll number see Chlorophyll in physiology of photosynthesis

Chlorophyll precursors see Chlorophyll biosynthesis ...

Chlorophyll unit see Photosynthetic (chlorophyll) unit

Chlorophyllase
     21683, 23599, 24361, 24643, 24727-9; 25462, 25938, 26097, 26107, 26404,
     26777, 27855, 27885, 28039, 28111, 28320; 30186, 31221, 31385, 31605, 31782,
                                                                (continued)

Chlorophyllase (continued)
    31852, 31855, 31991, 32077-8; 32865, 33562, 33693, 34639, 34691, 34693,
    34934-5, 34941, 35614, 35810, 35814, 36032, 36098; 36644, 37520, 38345,
    38440, 38775, 38871, 39597, 39805, 39849, 39880, 40130, 40335

Chlorophyllase and other enzymes of chlorophyll synthesis and degradation, methods
    23599; 31775; 36032, 36098; 37936, 38871, 39848

Chlorophylls a,b content and their ratio
    21509, 21520, 21534, 21536-7, 21564, 21575, 21595, 21606, 21608, 21613,
    21626, 21629, 21642, 21662, 21678, 21697, 21726, 21734, 21743, 21747, 21760,
    21767, 21772-3,   21825, 21842, 21872, 21880, 21894, 21923, 21925, 21948-
    -9, 21974, 22056, 22067, 22079, 22086, 22197, 22213, 22248, 22252, 22265,
    22290, 22323, 22326, 22338, 22350, 22394, 22397, 22413, 22435, 22466-7,
    22476, 22516, 22518, 22527, 22534, 22555-6, 22565, B22572, 22573, 22618-20,
    22624, 22673, 22694, 22793, 22800, 22805, 22815-6, 22818, 22841, 22851,
    22862, 22926, 22928, 22959, 22967, 22973, 22980-1, 22988, 23057, 23080,
    23083, 23086, 23092, 23149, 23193, 23200, 23226, 23248, 23259, B23262, 23263,
    B23264, 23269, 23272, 23288, 23350, 23356, 23383-4, 23404, B23407, 23417,
    23428-9, 23443, 23460, 23463-4, 23488-9, 23515, 23526, 23528, 23536, 23550,
    23560, 23562, 23567, 23570, 23583, 23586, 23590, 23638, 23642, 23687, 23718,
    23751-3,   B23767, 23771, 23775, 23779, 23794, 23804, 23810-1, 23820,
    23822, 23835, 23876, 23883, 23890, 23916, 23960, 23964, 23973, 24008, 24019,
    24040-1, 24068, 24075-8, 24084, 24086, 24111, 24122, 24127, 24148, 24163,
    24190, 24226, 24228, 24280, 24284, 24299, 24308, 24328, 24364, 34411, 24414,
    24421, 24457-9, 24462-4, 24479, 24482, 24505, 24523-4, 24530, 24540-1, 24545,
    24548, 24560, 24584, 24597, 24603, 24615, 24654, 24713, 24715, 24775, 24785,
    24818, 24824, 24879, 24892, 24912, 24915, 24923, 24929, 24942, 24984-5,
    25002, 25016-7, 25019, B25064, 25102, 25117, 25152
    25166, 25176, 25180, 25184-6, 25204, 25217, 25228, 25237, 25240, 25245-6,
    25261, 25287, 25293-4, 25297, 25335, 25364, 25368, 25379, 25423-4, 25445,
    25456, 25470, 25474, 25480, 25518, 25525, 25620, 25679, 25686, 25700, 25705,
    25716, 25795, 25819, 25838, 25862, 25867, 25872, 25887, 25896, 25898, 25905-
    -6, 25913, 25918, 25930, 25962, 25988, 25991-2, 26019, 26033, 26057, 26078,
    26099, 26101, 26125, 26131, 26139, 26148, 26172, 26181, B26185, 26219,
    B26220, 26246, 26270-1, 26284, 26327, 26353, 26361, 26392, 26400-1, 26404,
    26408, 26415, 26440, 26470, 26493, 26495, 26508, 26511, 26518, B26551, 26554-
    -5, 26573, 26575, 26597, 26614, 26617, 26634, 26666-7, 26680, 26683, 26700,
    26707, 26734, 26749, 26757, 26777, 26779, 26786, 26795, 26800, 26812, 26830,
    26855, 26872-3, 26910, 26917, 26925, 26927-8, 26930, 26934, 26977, 26985-6,
    26990, 26994, 27027, 27035, 27052, 27057, 27060-1, 27066, 27110, 27189,
    27216, 27222, 27237-9, 27270, 27284, 27288, 27309, 27348, 27373, 27380-1,
    27386, 27388, 27394, 27405, 27443-4, 27488-9, 27494, 27508, 27527, 27547,
    27552, 27559-60, 27568, 27581, 27599, 27649, 27671, 27683, 27689-90,
    27715, 27717, 27794-5, 27816, 27824, 27834-5, 27848, 27852, 27854, 27863,
    27874, 27884-5, 27924, 27926, 27947, 27978, 27980, 27996, 28055, 28057,
    28117, 28122, 28180, 28187, 28205, 28221-2, 28240, 28258, 28261, 28276,
    28290, 28297, 28317, 28357, 28361, 28422, 28467, 28477
    28518, 28529, 28532, 28550, 28568-9, 28584-7, 28589, 28595, 28607, 28625,
    28650-1, 28655-6, 28667, 28690-1, 28715, 28718, 28752, 28760, 28790-1, 28801,
    28827, 28836, 28880, 28896, 28962, 28990-1, 29004, 29039, 29067, 29104, 29150,
    29155, 29170, 29191, 29210, 29218, 29220, 29240, 29251, 29253, 29259, 29262,
    29268, 29323, 29368, 29384, 29386, 29396, 29410, 29422, 29438, 29443, 29448,
    29450, 29461, 29473, 29480, 29501, 29534, 29538, 29562, 29598, 29603, 29614,
    29661, 29669, 29677, 29682, 29687, 29704, 29706, 29711, 29717, 29753-4, 29760,
    B29763, 29765, 29777, 29803, 29807, 29833, 29835, 29876, 29898, 29939, 29943,
    29979, 29994, 30006, 30017, 30067, 30092, 30101, 30123, 30186, 30188, 30193,
    30200, 30207, 30230, 30253, 30274, 30298, 30302, 30334, 30348, 30363, 30421,
    30441, 30452, 30481, 30494, 30496, 30536, 30548, 30550, 30570, 30576,
    30579-81,   30594-7, 30611-2, 30623, 30659, 30683, 30687, 30721, 30759-60,
    30800, 30802, 30828, 30837, 30851, 30866, 30881, 30911-2, 30946, 30951, 30962,
    30993, 31012, 31063, 31109, 31116, 31119-20, 31128, 31144, 31155, 31221,
    31230, 31232, 31236, 31243, 31251, 31276, 31281, 31324, 31329-30, 31336,
                                                                    (continued)

Chlorophylls *a*, *b* ... (continued)
    31340, 31361, 31384-5, 31395, 31405-6, 31426, 31480, 31489, 31502-3, 31540,
    31551, 31559, 31575, 31579, 31587-8, 31621, 31625, 31673, 31735, 31737, 31784,
    31795-6, 31839, 31877-8, 31885, 31930, 31971, 31979, 31991-2, 32005, 32008-9,
    32055, 32061, 32065, 32086, 32093, 32101, 32113, 32116, 32133, 32142-3, 32175,
    32185-8, 32190, 32192, 32212, 32216, 32230-1, 32237, 32254, 32258-9, 32261,
    32266, 32284-6, 32289, 32298-300, 32379, 32385, 32401, 32421, 32424, 32432,
    32441, 32481, 32488, 32501, 32526
    32566, 32572, 32590-1, 32593-5, 32601, 32605, 32626, B32630, 32632, 32639,
    32643-7, 32660, 32665-7, 32691, 32706, 32716, 32731, 32739-40, 32742, 32749,
    32769-70, 32819, 32829, 32844, 32865, 32889, 32892, 32910, 32929, 32931,
    32940, 32981, 33002, 33012-4, 33026, 33030, 33064, 33079, 33084, 33122,
    33173, 33177, 33219, 33245, 33268-9, 33280, 33305, 33331, 33336, 33339,
    33348-9, 33351, 33379, 33387, 33410, 33423, 33435, 33472, 33481, 33489,
    33499, 33502, 33513, 33523, 33540, 33543, 33589, 33608, 33655, 33670, 33676,
    33733, 33736, 33761, 33790, 33795-6, 33805, 33816, 33825, 33835, 33842,
    33892, 33909, 33952, 33956, 33987, 33997, 34008, 34016, 34026, 34030, 34042,
    34045, 34098, 34111, 34137, 34153, 34175, 34198, 34214, 34219, 34312, 34318-
    -9, 34437-8, 34441, 34457, 34464, 34468, 34480, 34482-3, 34490-1, 34493,
    34502, 34549, 34567, 34572, 34581, 34588, 34594-6, 34606-7, 34620, 34630,
    34670, 34693, 34702, B34706, 34707, 34722, 34739, 34765-6, 34779, 34790,
    34845, 34869, 34894, 34897, 34917, 34922, 34924-5, 34934, 34937, 34940,
    34945, 34948, 34956, 34959, 34990, 35021, 35035, 35044-5, 35048-9, 35051,
    35059, 35074, 35081, 35092, 35096, 35123, 35132, 35158, 35161, 35167, 35181,
    35232, 35241, 35248, 35250-1, 35261, 35270, 35309, 35321, 35342, 35348,
    35363-4, 35379, 35385, 35408, 35417, 35427, 35432, 35434, 35437, 35440,
    35449, 35454, 35471, 35477, 35496, 35583, 35609, 35617, 35619, 35638, 35657-
    -8, 35763, 35835-6, 35844, B35862, 35867, 35921, 35956, 35970, 35988, 36004,
    36051, 36062, 36083-4, 36097, 36109, 36113, 36137, 36142, 36145-6, 36151,
    36154-5, 36177, 36180, B36191, 36193, 36195-6, 36213, 36234, 36258, 36270,
    36293, 36305, 36310, 36316, 36332, 36355, 36357, 36411, 36446-7, 36450,
    36456, 36480, 36495, 36503, 36519, 36534
    36598, 36617, 36628, 36638, 36668-9, 36688, 36701-2, 36712-3, 36717, 36752,
    36766, 36803-4, 36807, 36816, 36825, 36845, 36906, 36922, 36927, 36936,
    37014, 37051, 37062, 37077-8, 37093, 37101, 37112, 37130, 37138, 37157,
    37171, 37173, 37202, 37233, 37257, 37265, 37267, 37271, 37294-5, 37303,
    37317, 37321, 37337, 37341, 37346, 37362, 37401, 37449, 37479-80, 37502,
    37517, 37519-20, 37556, 37568, 37602-3, 37607, 37612-3, 37620-1, 37645-6,
    37668, 37686, 37717, 37732, 37735, 37740, 37747-8, 37756-7, 37770, 37784,
    37835, 37871, 37878, 37918, 37921, 37925, 37998, 38104, 38106, 38110,
    B38118, 38158, 38166, 38194, 38227-30, 38290, 38292, 38315, 38342, 38345,
    38352-3, 38363, 38365, 38376, 38380-2, 38387, 38426, 38439, 38443, 38463,
    38468, 38478-9, 38490, 38492, 38497, 38515, 38542, 38546, 38554, 38557-8,
    38561-2, 38578, 38685, 38691, 38723-4, 38731, 38745, 38769, 38775, 38803,
    38839, 38850, 38852, 38854, 38872, 38877, 38896-7, 38936, 38951, 38980,
    38984, 39037, 39052, 39065, 39073, 39076-7, 39083, 39114, 39126, 39133,
    39149, 39182, B39191, 39206-7, 39209-10, 39237, 39253, 39284, 39300, 39310,
    39325, 39346, 39380, 39399, 39408-10, 39429, 39435, 39437, 39460, 39477,
    39494, 39539, 39543, 39557, 39568, 39587, 39597-8, 39601, 39606, 39612,
    39625, 39701, 39710, 39741, 39747, 39763, 39805, 39808, 39820, 39835, 39862,
    39877, 39880-1, 39884, 39981, 39895, 39909, 39915, 39943, B39951, 39958,
    39968, 39977, 39995, 39997, 40014, B40015, 40069, 40072, 40124, 40128,
    40156, 40164, 40173, 40179, 40198, 40201, 40239, 40248, 40269, 40284, 40303,
    40342, 40352, 40355

Chlorophylls *c*, *d*
    21595, 21726, 21775, 22095, 22538, 22767, 22865, 22926-8, 22930, 23016,
    23269, 23393, 23590, 24233, 24308, 24451, 24882, 25055-6
    25442, 25648, 25651, 25849, 25895, 26025, 26116, 26127, B26185, B26220,
    26270-1, 26404, 26453, 26508, 26518, 26521, 27381, 27511, 27513, 27649,
    27689, 27921, 27940, 28162, 28317, 28377, 28434
    28698-9, 28790, 29396, 29677, 29803, 29991, 30137, 30582, 30592, 30616,
    30629, 31257, 31430, 32123, 32216, 32258, 32288, 32442, 32472
                                                        (continued)

                                                                    (continued)

Chloroplast and chromatophore chemical composition (continued)
28554, 28574, 28650, 28709, 28723, 28733, 28804, 28812, 28836, 28840, 28910,
28923, 28974, 29210, 29218, 29299-300, 29393, 29431, 29477, 29571, 29627,
29698, 29741, 29745, 29758, 29765, 29808, 29813, 29830, 29932, 29938, 29948,
29998, 30036-7, 30068, 30089, 30123, 30145, 30149-50, 30334, 30341-2, 30428,
30472, 30534, 30549, 30551, 30562, 30646, 30731, 30801, 30816, 30852, 30854,
30900-1, 30933, 30958, 31082, 31104, 31109, 31120, 31137, 31294, 31323, 31380,
31424, 31434-5, 31538, 31557, 31590, 31593, 31616, 31639, 31644, 31662, 31815,
31892, 31937, 32079, 32086, 32132, 32149, 32291, 32336, 32389
32671, 32889, 32893, 32902, 32957, 33030, 33058, 33098, 33117, 33202, 33227,
33237, 33247, 33373, 33387, 33419, 33481, 33517, 33598, B33628, 33659, 33686,
33769, 33774, 33779, 33792, 33838, 33900, 33938, 33940, 34159, 34180, 34232-4,
34297, 34307, 34586, 34704-5, 34743, 34850, 34852, 34890, 34892, 34927, 34929,
34931, 34956, 34975, 34986, 35067, 35071, 35114, 35130, 35190, 35241, 35296,
35361, 35400, 35419, 35499, 35548, 35751, 35755, 35859, B35862, 35987, 36005,
36128, 36262, 36302, 36461, 36478, 36488, 36573
36611, 36666, B36768, 36884, 36888, 36894-5, 36898, B37037, 37048, 37119,
37158, 37203, 37322, 37335, 37337, 37340-1, 37411, 37425, 37468, 37526,
37558, 37716, 37747, 37921, 38003, 38146, 38460, B38477, 38479, 38545-6,
38573, 38598, 38604, 38748, 38837, 38840, 38864, 38884-5, 38889, 38896,
38910, 39040, 39044, 39146, 39227, 39248, 39365, 39432, 39450, 39508,
39700, 39772, 39887, 39915, 40017, 40058, 40123, 40189, 40295

Chloroplast and chromatophore dimensions
21663, 21851-2, 21863, 21871-2, 22117, 22148, 22290, 22467, 22496, 22565,
B22572, 22573, 22624, 22870, 22924-5, 23034, 23092, 23350, 23392, 23549,
B23767, B23933, 24075-6, 24088, 24196, 24218, 24253, 24315, 24340, 24347,
24817, 24847, 25008, B25064
25530, 25770, 25878, 26077, 26095, 26674, 26715, 26818, 26842, 26926, 26928,
27249, 27316, 27443, 27488-9, 27499, 27715, 28180-1, 28288
28649, 28684, 28883, 28909, 29110, 29112, 29121, 29260, 29470, 29510, 29704,
29871, 30020, 30199, 30298, 30329, 30667, 30753, 30826, 30879, 30896-7, 31106,
31530, 31568-9, 31644, 31801, 32085, 32125-6, 32222, 32420, 32551-2
32895, 32920, 32929-30, 33030, 33405, 33410, 33493, 33518, 33736-7, 34282,
34401, 34472, 34705, 34710,   34924-5,   34956, 35051, 35065, 35128, 35400,
35585, 35749, 35776, 35842, B35862, 36190, B36191
36733, 36980, 37119, 37158, 37237, 37281, 37568-9, 37693-4, 38026-7, 38221,
38540, 38546, 38624, 38854, 39145, 39374, 39449, 39548, B39611, 39632,
39765, 39773, 39881, 39972, 39977, 40177, 40353

Chloroplast and chromatophore distribution in cells see Chloroplast and chromatophore number and distribution

Chloroplast and chromatophore fragments
21572, 21578, 21599-600, 21606, 21613, 21632, 21685, 21690, 21743, 21745,
21825, 21888, 21922, 21983, 21996, 22079, 22164, 22243-4, 22289, 22326,
22345, 22349-50, 22374-5, 22406, 22462, 22483, 22501, 22514, 22539, 22596,
22605, 22626, 22667, 22705, 22708, 22748, 22773-4, 22780-1, 22856, 22906,
22913, 22922-3, 22980-1, 23056-7, 23093, 23110, 23120, 23122, 23127, 23139-
-41, 23169, 23187, 23242, 23310, 23345, 23373, 23421, 23473, 23515-6, 23586,
23642, 23655, 23724, 23730, 23742, 23852, 23860, 23876, 23916, B23933,
23934-5, 24008, 24099, 24116, 24165, 24190, 24226, 24298, 24385, 24395,
24415, 24423, 24443, 24455, 24457-9, 24462, 24478-80, 24487, 24493, 24495-6,
24591, 24603, 24728, 24876-7, 24879, 24909, 24923, 24954, 25002, 25006,
25019, 25030, 25057, 25083, 25099, 25113-4
25180, 25192, 25196, 25211, 25217, 25228, 25245-6, 25263, 25287, 25323,
25347-8, 25470, 25504, 25703, 25745, 25837, 25884, 25927, 25968, 26022,
26063, 26101, 26172-4, 26176, 26219, B26220, 26441, 26463, 26468, 26561,
26602, 26615, 26677, 26694, 26710, 26793, 26829-30, 26910, 26917, 27075,
27104, 27155, 27312, 27399, 27486, 27581, 27603, 27675-6, 27681, 27713,
27784, 27789, 27816, 27829, 27831, 27851-2, 27863, 27870, 27934, 27968,
27978, 28111, 28184, 28236, 28276, 28332, 28452

(continued)

Chloroplast and chromatophore fragments (continued)
28514, 28549-50, 28554, 28562, 28574, 28582, 28584, 28586-7, 28590, 28698,
28701, 28720, 28751, 28771, 28784, 28878, 28886, 28997, 29037, 29161, 29210,
29244, 29384, 29406, 29438, 29441-2, 29449, 29502, 29546-7, 29580, 29600,
29602, 29672, 29682, 29687, 29698, 29738, 29768, 29872, 29932, 29949, 29955,
30013, 30050, 30123, 30135, 30181, 30188, 30207, 30245, 30254-5, 30321,
30326-7, 30334, 30359-60, 30425, 30539, 30559, 30717, 30780, 30789, 31011,
31030, 31078-80, 31137, 31161, 31206, 31295, 31298, 31323, 31375, 31391,
31426, 31434, 31479-80, 31534, 31549, 31564, 31588, 31598, 31612, 31621,
31625, 31629-32, 31642-3, 31707, 31725-6, 31735, 31796, 31857, 31930, 31955,
31992, 32005, 32007, 32100, 32114, 32178, 32266, 32392-3
32594, B32630, 32659-61, 32674, 32682, 32701, 32720, 32834, 32839, 32932,
32994, 33016, 33024, 33064, 33143, 33201, 33341, 33487, 33498, 33578, 33712,
33719-20, 33755, 33788-9, 33806, 33880, 33882, 33900, 33930-1, 33958, 34007,
34042, 34102, 34141, 34166, 34204, 34224, 34244, 34277, 34286, 34317, 34349-
-50, 34357, 34371, 34431, 34490, 34628, 34703, 34725-6, 34776, 34845, 34889,
35009, 35058, 35066, 35076, 35114, 35236, 35328, 35548, 35669-70, 35697,
35739, 35745, 35825, 35993, 36003, 36055, 36095, 36119, 36130, 36245, 36310,
36355, 36377, 36405, 36430, 36503, 36507
36625, 36660, 36702, 36740-1, 36773, 36786-7, 36828, 36836, 36854, 36894,
36953, 36976-7, 37019, 37045-6, 37117, 37137, 37170, 37223, 37251-2, B37528,
37539, 37547, 37667-8, 37670, 37730, 37749-50, 37840, 37878, 37917, 37958,
37961, 37967, 37979, 38007, 38024, 38039, 38096, 38129, 38136, B38139,
38165, 38183, 38193, 38233, 38254-5, 38291, 38334, 38373, 38479, 38510,
38548, 38555, 38571, 38617-20, 38638, 38669, 38773, 38775, 38835, 38898,
38905, 38925, 38927, 38972, 38989, 39001, 39198, 39247, 39292, 39295, 39318,
39384-5, 39444-5, 39453, 39577, 39597, 39612, 39615-7, 39625, 39763, 39774,
39797, 39841, 39880, 40011, 40029, 40031, 40107, 40122

Chloroplast and chromatophore number and distribution
21663, 21747, 21863, 21871-2, 21974, 22082, 22313, B22572, 22573, 22624,
22978, 22988, 23034, 23071, 23092, 23350, 23689, B23767, B23933, 24075-6,
24086, 24100, 24196, 24237, 24353, 24467, 24499, 24817-9, 24847, 24985,
25031
25528, 25530, 25655, 25781, 25906, 26674, 26853, 26926, 26928, 26994, 27443,
27488, 27497, 27499, 27657, 28288, 28357
28649, 28655, 28684, 28883, 28909, 28925, 29110, 29112, 29128, 29360, B29468,
29470, 29704, 29871, 30199, 30298, 30329, 30643, 30660, 30896-7, 31039,
31155, 31204, 31604, 31768, 31781, 31801, 32061, 32222
32717, 32895, 32929-30, 33098, 33158, 33410, 33417, 33513, 33602, 33736-7,
33761, 34319, 34472, 34751, 34894, 34924-6, 34948, 35045, 35051, 35776,
35813, 35842, B35862, 36190, B36191, 36534
36879, 36927, 37568-9, 37693-4, 38033, 38221, 38540, 38614, 38747,
38854, 39040, 39062, 39145, 39539, B39611, 39765, 39773, 39881, 39971-2,
39977, 40009, 40177

Chloroplast and chromatophore replication, ontogeny
21528, 21653, 21747, 21850, 21871-2, 22011, 22068, 22083, 22147, 22180,
22193, 22237, 22277, 22307, 22312-3, 22323, 22338, 22361-2, 22401, 22423,
22469, 22492, 22602, 22692, 22800, 22816, 22945, 23094-5, 23109, 23129,
23279, 23319, 23353, 23369, 23482-3, 23527, 23642, 23674, 23676-7, 23689,
23766, B23767, 23835, 23882, 23903, 23919, 23969, 23987, 24086, 24157,
24184, 24209, 24237, 24340, 24346-8, 24382-3, 24506, 24545-6, 24590, 24735,
24792-3, 24867, 24958-9, 24996-7, 25000, 25008, 25075, 25147
25287-8, 25373, 25377, 25500, 25528, 25530, 25551, 25616, 25621, 25655,
25689, 25745, 25837, 35853, 25904, 26077, 26142, 26201, 26279-80, 26340,
26447-8, 26462, 26467, 26649, 26842, 26890, 27054, 27081, 27157, 27167,
27217, 27249, 27316, 27351, 27407, 27443, 27482, 27497-9, 27577, 27614,
27657, 27717, 27818, 27884, 27922, 27967, 27985, 28149, 28217, 28363-4,
28430, 28436
28585, 28590, 28651, 28736, 28805, 28836, 28883, 28896, 28909, 28947, 29055-6,
29128, 29150, 29260-1, 29318, 29329, 29360, 29437, 29465, 29517, 29813, 29922,
30003, 30012, 30020, 30027, 30054, 30082, 30216, 30242, 30287, 30312, 30467,
(continued)

Chloroplast and chromatophore replication, ontogeny (continued)
        30557, 30560, 30668, 30673, 30690, 30747, 30753, 30790, 30814, 30826, 30893-4,
        30900-1, 30992, 31103-4, 31133, 31222, 31318, 31443, 31468, 31504, 31644,
        31680, 31719, 31743-4, 31757, 31778-9, 31807, 31830-1, 31854, 31876, 31878,
        31880, 31973, 31979, 32029, 32061, 32084, 32302, 32365-6, 32381, 32388, 32397,
        32399-400, 32421, 32452, 32473
        B32592, 32595, 32661, 32872, 32876, 32883, 32927, 32930, 33087-9, 33098,
        33108, 33202, 33205, 33227, 33319, 33373, 33380, 33389, 33480-1, 33502-3,
        33589, 33592, 33598, 33603, 33625, 33827, 33849-50, 33889, 33959, 33980,
        34029, 34237, 34320, 34331, 34397, 34401, 34457, 34540, 34579, 34606, 34685,
        34746, 34771, 34920, 34922, 34928, 34930, 35011, 35067, 35071, 35128, 35241,
        35289-90, 35384, 35497, 35515, 35519, 35590, 35602, 35625, 35689, 35731,
        35763, 35814, B35862, 35922, 35959-60, 36056, 36101, 36177, 36249, 36341,
        36351, 36370, 36383, 36459, 36487-8, 36563
        36590, 36611, 36639, 36690, 36696, 36739, 36894-5, 36980, 36998, B37037,
        37077, 37100, 37119, 37127, 37228, 37237, 37322, 37336, 37363-4, 37457,
        37539, 37716, 37725, 37925, 37960, 37978, 38146, 38183, 38214, 38260,
        38273, 38301, 38342, 38449, 38466, B38477, 38479-80, 38504-5, 38517, 38547,
        38682-4, 38774, 38874, 38884, 38922, 38977, 39045, 39062, 39082, 39151,
        39169, 39197, 39260, 39273, 39300, 39316, 39357, 39432, 39506, 39539, B39551,
        39582, 39742-3, 39798, 39814, 39969, 40051, 40063-4, B40105, 40164, 40177-8,
        40231, 40314, 40323, 40337

Chloroplast and chromatophore volume changes
        22017, 22232, 22526, 22536, 22557, 22560, 22820-1, 22891, 23726, 23847,
        24078, 24136, 24187, 24867, 25160
        25621, 26276, 26435, 26467, 26496, 27018, 27028, 27351, 27355, 27454, 27497,
        27556, 27576, 27796, 28449
        28981, 29305, 29352, 29458, 29596, 29932, 30036, 30110-1, 30199, 30889, 30910,
        30984, 31207, 31210, 31530, 31568-9, 31814, 32445
        32795, 32931, 32934, 33098, B33628, 33933, 34128, 34316, 35065, 35094, 35555,
        35579-80, 35603-5, 36047
        37336, B37831, 39054, 39374-5, 39392-3, 39632, 40290, 40321, 40323, 40354

Chloroplast envelope see Chloroplast outer membrane ...

Chloroplast immobilization see Photosystems stabilization ...

Chloroplast, isolated, carbon fixation in see Carbon fixation in isolated chloro-
        plasts ...

Chloroplast, isolated, gas exchange by
        21717, 21818, 22140, 22452, 22536, 22666, 22721, 22833-4, 22980-1, 23183,
        23195, 23492, 23715, 23841, 24501, 24631
        25372, 25712, 25833, 26008, 26206, 26333-4, 26565-6, 26703, 27453, 27583-4,
        27618, 27641, 27655, 27665, 27906-8, 27943, 28000, 28112, 28198, 28227,
        28263, 28307, 28334, 28396, 28449, 28457
        28556, 28562, 28592, 28599, 28653, 28665, 28789, 28886, 29076, 29081, 29141,
        29382, 29533, 29615, 29891-2, 29942, 30291, 30430, 30518, 30526, 30561,
        30604, 30609, 30765, 30769, 30777, 30853, 30977, 30986, 31058, 31292, 31320,
        31322, 31451, 31453, 31483, 31488, 31520, 31523-6, 31626, 31873, 31897, 31939,
        31944-5, 32049, 32079, 32177, 32223, 32432, 32502
        32607-9, 32627, 33387, 33424, 33426-7, 33459, 33585, 33869, 33929, 33933,
        33944, 33982, 33986, 33994, 34044, 34062-3, 34141, 34180, 34208, 34317,
        34423, 34657, 34799, 34891, 34925, 34956, 35059, 35329, 35375, 35402, 35428,
        35447, 35470, 35540, 35822, 35848, 35918, 35993, 36508, 36541, 36555
        36661-2, 36817, 36884, 36975, 37038, 37282, 37393, 37509, 37542, 37875,
        37928-32, 38015, 38027, 38029, 38056, 38115-6, 38142, 38249-50, 38377,
        38433, 38452, 38487, 38560, 38623, 38731, 38736, 38896, 39225, 39247, 39301,
        39334, 39336, 39351-2, 39363, 39474, 39627, 39675, 39694, 39767, 39838,
        39970

Chloroplast isolation
        21565, 21716, 22111, 22304, 22370, 22463, 22590, 22626, 22836, 22891, 22996,
        23357, 23415, 23705, B23933, 24149, 24292, 24823, 24948
                                                                    (continued)

Chloroplast isolation (continued)
      25180, 25188, 25301, 25757, 25770, 26153, 26206, 26302, 26554, 26790, 26894,
      27263, 27279, 27394, 27562, 27583-4, 27615, 27968, 27985, 28253, 28304, 28331,
      28371, 28451
      28514, 28538, 28575, 28896, 29141, 29376, 29464, 29477, 29571, 29687, 29960,
      30190, 30210, 30216, 30295, 30301, 30551, 30558-9, 30998-9, 31323, 31325,
      31426, 31726, 32079, 32247, 32335, 32390, 32395, 32513
      32632, 32933, 33198, 33207, 33387, 33424, 33426-7, 33450, 33755, 33844,
      33995, 34044, 34180, 34557, 34956, 35066, 35089-90, 35333, 35479, 35540,
      35764, 36005, 36147, 36486, 36504
      B36768, 36822, 37186, 37293, 37335-6, 37393, 37468-9, 37758, 37800, B38037,
      38056, 38089, 38178, 38408, 38452, B38477, 38487, 38736, 38838, 39278, 39333,
      39335, 39337, 39359, 39504, 39514, 39548, 39838, 39860, 40164

Chloroplast, localization of electron transport chain in thylakoid see Electron
      transport chain localization in thylakoid

Chloroplast movements
      21676, 21905-7, 22702-3, 24166, 24365-6, 24770, 24890, 25160; 25558-60,
      26088, 26321-4, 26346, 26435, 26951, 27047, 27051, 27719, 27727, 27756-7;
      28684, 28825, 28871, 28950, 29479, 29553, 29761, 29815, 31084, 31289, 31541,
      31847, 32126; 32873, 32895, 32920, 33615, 34396, 35119, 35585, 35714, 35747,
      36153, 36323, 36326, 36453; 37001, 37006, 37348, 37379, B37831, 37832-3,
      38445, 39050, 39356, 39486-90, 39527-9, 39825, 39969, 40251, 40369, 40371

Chloroplast outer membrane, chloroplast envelope
      21546, 21612-3, 21651, 21718, 21755, 21844, 21871, 22033, 22114, 22223,
      22281, 22338, 22536, 22694, 22732, 22891, 22967, 23209, 23353, 23507, 23755,
      23924, 24049, 24370, 24828, 24997, 25137
      25500, 25745, 25770, 25833, 25979, 26036, 26090, 26126, 26234, 26276, 26279,
      26334-5, 26342, 26447, 26566, 26767, 26816-7, 26893, 27186, 27241, 27304,
      27468-9, 27562, 27640, 28249, 28364, 28456
      28650, 28953, 29047, 29076, 29111, 29654, 29728, 29814, 29884, 29888, 30036,
      30190, 30690, 30937, 31035, 31093, 31325, 31366, 31387, 31680, 31811, 31924,
      32193
      33117, 33122, 33146, 33373, 33425, 33659, 33674, 33839, 33849, 33933, 33938,
      34249, 34525, 34563, 34704, 34975, 34986, 34989, 34996, 35385, 35474, B35862,
      36045
      36611, B36768, 36794, 36898, 37135, 37154, 37203, 37235, 37335-6, 37841,
      37844, 37849, 37863, 37932, 37978, 38092-3, 38101, 38301, 38540, 38614,
      38700, 38812, 39135, 39623, 40021, 40078

Chloroplast proteins (and other photosynthetic proteins), methods
      21565, 21600, 21685, 21860, 22780, 23281, 23321, 23482, 23498, 23821, 24379;
      25199, 25228, 25890, 26158, 26432, 26455, 27214; 28575, 29474, 30694, 31029,
      31080, 31476, 31580, 31590, 31643, 31812, 31815, 31892, 31900; 32890, 33058,
      33383, 34816, 34908, 34968, 35382, 36147, 36377, 36504; 36667, 37153, 37408,
      37468, 37534, B38477, 39631

Chloroplast ultrastructure (cf. also Chloroplast outer membrane; Stroma of chloro-
      plast; Thylakoid, granum)
      21546, 21582, 21595, 21601, 21606-8, 21610, 21651, 21662, 21664, 21700,
      21756, 21767, 21778, 21791, 21825, 21848, 21850-2, 21871, 21873, 21899,
      21978, 21984, 22009, 22033, 22117, 22147, 22175-6, 22193, 22200, 22223-4,
      22233, 22240-1, 22277, 22312-3, 22322, 22328, 22336, 22355, 22393-5, 22415,
      22422-4, 22449, 22467, 22473, 22492, 22538, 22642, 22647, 22698, 22740,
      22743, 22783, 22793, 22800, 22815-7, 22890, 22926, 22945, 22975, 23031,
      23093, 23108-9, 23129, 23280, 23310, 23319, 23322-3, 23325, 23350, 23357,
      23366, 23369, 23392, 23402, 23450, 23528, 23530, 23575, 23631, 23642, 23647,
      23670, 23674, 23680-1, 23689, 23705, 23755, B23767, 23769, 23773, 23820,
      23822, 23849, 23883, 23903-5, 23919, 23924, 23969, 23987, 24030, 24067,
(continued)

Compensation irradiance
        21633, 22535, 23443, 24336, 24758, 25032; 25168, 25549, 26300, 26370, 26661,
        26863, 27198, 27431, 27613, 28077, 28194, 28229, 28462; 28687, 28856, 29053,
        29155, 29191, 29242, 29337, 29563, 29675, 30099, 31067, 31515, 31885, 32181;
        33496, 33565, 33736, 33759, 34061, 34445, 34792, 35616, 35762, 36051, 36259;
        36726, 37269, 37895, 38020, 38065, 39063, 39688, 39874, B40105, 40201,
        40250, 40293

Compensation point, CO_2 see CO_2 compensation concentration

Compensation point, light see Compensation irradiance

Competition in ecosystem
        21667, 21845, 21862, B22358, 22447, 22907, 23338, 24614; 25613, 26947, 27336,
        27740; 28689, 29334, 29456, 30029, 30034, 30463, 31004, 32328; 33226, B33343,
        34541, 35102, 35508; 36962, B37727, 38082, 38764, 38990, 39074, 39785

Conductance for transfer of gases see Resistance ...

"Contribution" of individual organs see Biomass distribution and redistribution;
        Photosynthate translocation ...

Correlations within plant 21786, 22015; 25881; 30409, 30680, 31933-4, 32436

Cosmic radiation see Ionizing radiation ...

Counting methods, chloroplast and cell see Chloroplast and cell counting methods

Coupling factor 1 see ATPase ...

Cover, vegetative see Canopy ...; Ecosystem ...

Crassulacean Acid Metabolism see CAM

Cultivar differences, carbon fixation pathways
        22050, 22975, 24523; 26118, 26777, 27053, 27384, 28385; 29633, 30098, 30474;
        33633, 33658, 35183-4; 36736, 37557, B38139, 38612, 38872, 39741, 39927

Cultivar differences, carotenoids
        21571, 22178, 22696, 22860, 23149, 23193, 23526, 23810-1, 24375, 24892;
        25479, 26187, 26777, 27110, 27925; 29414, 29762, 30396, 30481, 30802, 31144,
        31384-5, 32261, 32284; 32722, 34214, 34403, 34425, 34493, 34879, 35363-4,
        36262; 37039, 37428, 37449, 37620, 37686, 37732, 37735, 38769, 38849, 39114,
        39540, 39758, 40328

Cultivar differences, chlorophyll
        21513, 21949, 22012, 22085, 22178, 22300, 22696, 22851, 22911, 23149, 23193,
        23288, 23526, 23570, 23810-1, 24019, 24108, 24285, 24523, 24714-5, 24892
        25310, 25364, 25404, 25479, 26050, 26187, 26552, 26757, 26777, 27405, 27460,
        27520, 27690, 27925, 28200, 28282
        28799, 29004, 29238, 29251, 29410, 29414, 29457,
        29470, 29711, 29762, 29979, 30031, 30103, 30131, 30396, 30441, 30481, 30712,
        30734, 30802, 31144, 31213, 31281, 31384-5, 31550, 31581, 32057, 32261, 32284,
        32300, 32307
        32605, 32647, 32729, 32881, 33002, 33034, 33513, 33761, 33815, 34153, 34214,
        34242, 34463-4, 34493, 34782, 34815, 34839, 35184, 35280, 35339, 35363-4,
        35389, 35437, 35454, 35512, 36213, 36262, 36528
        36613, 36920, 37039, 37157, 37428, 37449, 37557, 37612, 37620, 37686, 37732,
        37735, 38166, 38345, 38353, 38591, 38657, 38725, 38769, 38872, 38984, 39054,
        39065, 39114, 39147, 39207, 39459, 39738-9, 39741, 39758, 40328, 40343

Cultivar differences, chloroplast
        24892, 25087; 25277, 27647, 27883; 28586, 28656, 28668, 28840, 29470, 30933;
        32893, 33761, 35296, 35482, 35875, 36047, 36262; 36733, 36923, 37557, 38991,
        39054, 40016, 40078

Cultivar differences, ecosystem and plant productivity
        21513, 21570, 21949-50, 22212, 22260, 22725, 22755, 22879, 23117, 23544,
        23734-5, 23763, 23777, 23779, 23795, 23862, 24080, 24142, 24252, 24363,
        24375, 24609, 24646-7, 24798, 25059, 25125
        25232, 25264, 25357, 25364, 25405, 25412, 25576, 26050, 26134, 26265, 26466,
        26778, 27210, 27384-6, 27389, 27778, 27819, 27982, 28041, 28262, 28300
        28552, 28620, 28624, 28785, 28941, 29060, 29213, 29410, 29427, 29475, 29492,
        29633, 29992, 30031, 30113, 30162, 30221, 30461, 30555, 30648, 30663, 30677,
        30712, 30734, 30802, 31013, 31021, 31180, 31213, 31273, 31281, 31359, 31367,
        31463, 31591, 31933-4, 32057, 32059, 32062, 32171, 32243, 32261, 32431, 32434,
        32505, 32530, 32545
        32567, 32737, 32965, 32973, 33059, 33066, 33092, 33101, 33222, 33513, 33698,
        33846, 33870, 33894, 34074, 34091-2, 34152, 34309, 34403, 34503, 34666,
        34781-2, 34834, 35184, 35202-3, 35389, 35394, 35454, 35512-4, 35534, 35541,
        35549, 35795, 35977, 36053, 36077, 36178, 36213, 36381, 36528, 36548
        36643, 36691, 36721, 36736, 36777, 36857, 36881, 36994, 37198, 37250, 37296,
        37313, 37440, 37489, 37555, 37612, 37829, 37847, 37974, 38346, 38512, 38535,
        38572, 38769, 38873, 38883, 39013-5, 39033-4, 39053, 39065, 39141, 39157,
        39207, 39272, 39280, 39569, 39656, 39727, 39741, 39852, 39904, 39967, 40044,
        40079, 40191, 40271, 40304, 40326

Cultivar differences, electron transport chain
        22298, 22409, 23672, 23944, 24286, 24821, 25087, 25134; 25392, 26001, 26766,
        27161; 29711, 30103, 30964, 31489; 32604, 32786, 33761, 34214, 36300-1,
        36556; 37615, 38166, 38316, 38769, 38994, 39404, 40097, 40099, 40342-3

Cultivar differences, gas exchange
        21570, 21677, 21967, 22042, 22085, 22212, 22298, 22678, 22725-6, 22796,
        22851, 22952, 22966, 23172, 23763, 23779, 23795-6, 23922, 24142, 24252,
        24523, 24715, 24787
        25261, 25264, 25312, 25328, 25404, 25415, 25615, 25708, 25828, 25967, 26118,
        26392, 26647, 26777, 27210, 27384, 27389, 27460, 27778, 27830, 27896, 27951,
        28018-9, 28092, 28282, 28300, 28375
        28629, 28632, 28785, 29004-5, 29251, 29470, 29711, 29874, 30047, 30103, 30131,
        30221, 30257, 30663, 30705, 30712, 30743, 31281, 31550, 31596, 31978, 32057,
        32163, 32252, 32300, 32545
        32689, 32751, 32810, 32965, 32973, 33002, 33110, 33112, 33604, 33698, 33761,
        33894, 33984, 34153-4, 34289, 34463, 34665, 34782-3, 34835, 35184, 35210,
        35222, 35295-7, 35339, 35622, B35862, 35920, 36010, 36064, 36213, 36222,
        36417
        36613, 37067, 37296, 37313, 37555, 37614, 37649, 38074, 38166, 38346, 38353,
        38360, 38523, 38532, 38574-5, 38657, 38710, 38753, 38868, 38873, 38984,
        39033, 39065, 39087, 39131, 39147, 39304, 39407, 39420, 39423, 39459, 39599,
        39741, 39815, 39852, 39874, 39980-1, 40079, 40103, 40313

Cultivar differences, photorespiration
        23796; 27460, 27778, 27951, 28092; 28785; 32786, 32965, 33633, 34154, 34524,
        35339, 36417; 36736, 37067, 37451, 37649, 39033, 39065, 39599

Cultivar differences, resistances to CO_2 and water vapour transfer
        21529, 24449, 24649; 25264, 25364, 26118; 28785; 30162, 30221, 30931, 31596,
        32252, 32485; 32810, 32973, 33984, 34109, 35222, 35340, 35563, 35907, 36417;
        36736, 37418, 37440, 37649, 38075, 38166, 38346, 38532, 39005, 39036, 39147,
        39420, 39423, 39599, 39656, 40313

Cultivar differences, respiration
        22952, 23796, 25137; 27389, 28092, 28300; 29992, 30804, 31281, 32163; 32965,
        36178, 36213, 36299; 36736, 37067, 37649, 38166, 38523, 39874

Cultivation of algae and photosynthetic bacteria   see Algae and photosynthetic bac-
        teria, cultivation; Algae mass cultures productivity

Cuticular CO_2 and O_2 exchange 23873; 26826; 33355

Cuticular resistance   see   Resistance, cuticular

Cytochromes
        21562, 21567-8, 21611, 21613, 21638, 21669, 21689, 21773, 21807, 21810,
        21825, 21839, 21923-4, 21938, 21965, 21970, 21993, 21996, 22021, 22063,
        22141, 22145-6, 22269, 22322, 22325, 22478, 22499, 22501, 22525, 22580-1,
        22588, 22605, 22640, 22648, 22667, 22683, 22698, 22759, 22774, 22794, 22809-
        -12, 22935, 22972, 22983, 23057, 23123-4, 23126, 23148, 23173, 23200, 23253,
        23261, 23276, 23292, 23335, 23421, 23515, 23519, 23548, 23574, 23580, 23611,
        23647, 23704, 23724, B23767, 23843, 23847, 24027, 24095-7, 24163, 24189,
        24221, 24241, B24250, 24281, 24327, 24329, 24340, 24403, 24418, 24466, 24487,
        24631, 24635, 24690-1, 24723, 24808, 24879, 24908-9, 25006, 25030, B25064,
        25068, 25070, 25076
        25183, 25266, 25385-7, 25393, 25427, 25470, 25511, 25565, 25606, 25669,
        25701, 25704, 25757, 25855, 25890, 25893, 25964, 25988, 26084, 26172, 26178,
        B26220, 26334, 26382, 26423-5, 26442-3, 26456, 26486, 26608, 26660, 26694,
        26794, 26796, 26807, 26826, 26872, 26990, 27006, 27086, 27116, 27254, 27275,
        27304, 27309, 27340, 27345, 27447-9, 27518-9, 27535, 27537, 27615, 27617,
        27659, 27668, 27708, 27809, 27841-2, 27863, 27868, 28045, 28135-6, 28184,
        28189, 28232, 28237, 28253, 28333, 28376, 28423-4, 28469
        28525, 28580, 28590, 28599, 28641, 28643, 28668, 28670-1, 28720, 28746, 28751,
        28778, 28812, 28814, 28816, 28836, 28845, 28851, 28867, 28893-5, 28904, 28910,
        28967-8, 29074, 29078, 29180, 29182-3, 29192, 29238, 29245-6, 29263, 29358-9,
        29525-6, 29528-9, 29602, 29640, 29677, 29687, 29785-6, 29807, 29818, 29831,
        29840, 29928, 29942, 29959-60, 30012, 30014, 30050, 30118, 30123, 30134,
        30145, 30208-9, 30245-6, 30260-1, 30303, 30336-7, 30340, 30386, 30398, 30428,
        30443, 30491-3, 30594, 30780, 30839, 30845, 30859, 30878, 30983, 31000, 31007,
        31009, 31025, 31027, 31035, 31053-4, 31109, 31137, 31288, 31365, 31368, 31389,
        31455, 31462, 31478, 31511, 31522, 31534, 31556, 31626, 31665, 31736, 31785-6,
        31907, 31960, 32005, 32039, 32045, 32072, 32114, 32127, 32168-9, 32250, 32266,
        32301, 32315, 32393, 32407, B32410, 32417, 32419, 32421, 32453, 32464-5, 32481
        32537
        32602, 32614, 32675, 32680, 32701-3, 32711, 32733, 32756-7, 32764, 32799,
        32801, 32842, 32860-1, 32889, 32898, 32906-9, 32955, 32992, 33032, 33036,
        33114, 33119, 33142, 33189-90, 33251, 33253, 33294, 33307, 33309, 33400,
        33402-3, 33406, 33502, 33551-2, 33568-9, 33597-8, 33688, 33708, 33710, 33901,
        33919, 33964, 33983, 34004, 34042, 34138, 34205, 34224, 34277, 34321, 34357,
        34377-8, 34393, 34483, 34566, 34717, 34727, 34762, 34789, 34848, 34858,
        34891, 34974, 34990, 35058, 35066, 35104, 35177, 35236, 35245, 35322, 35420,
        35427, 35479, 35526, 35531, 35566, 35573, 35596, 35736, 35739, 35811-2,
        35818, 35863, 35897, 35915, 35917, 35969, 36118, 36129, 36342, 36355, 36377,
        36385, 36389, 36397-8, 36409, 36411, 36476, 36484-5, 36505, 36547, 36573
        36611-2, 36621, 36649-50, 36710, 36772, 36798, 36911, 36935, 37042, 37115,
        37117, 37129, 37151, 37211, 37218, 37252-3, 37326, 37370, 37445-6, 37455,
        37475, B37493, 37526, 37558, 37632, 37651, 37793, 37821-2, 37842, 37875,
        37885, 37901, 37914, 37954-7, 37964, 37968, 38039, 38111, 38159, 38240,
        38263, 38342, 38381, B38415, 38613, 38619-20, 38630, 38656, 38665, 38683,
        38833, 38905, 38957, 38969-72, 39012, 39048, 39118, 39120, 39296, 39315,
        39356-8, 39379-80, 39447-8, 39491, 39674-7, 39679, 40055, B40105, 40182,
        40201, 40251, 40294, 40301

Cytochromes, methods
        22021, 23253, 23276, 24635, 24685; 25988, 26660, 27965; 29840, 30492; 32702,
        32764, 32907, 34004, 34224, 35811-2, 36505, 36573; 36621, 37151, 37651,
        37821, 37885, 39358

D

Dark CO$_2$ fixation
        21783, 22537, 22556, 23013, 23099, 23121, 23210, 23300, 23333, 23379, 23610,
        23786, 23926, 24083, 24999, 25018

                                                                        (continued)

Dark CO_2 fixation (continued)
     25446, 25516, 25528, 25780, 25784, 25863, 26004, B26551, 26563, 26645, 27107,
     27427, 27542, 28000-1, 28064, 28077, 28390
     29040, 29092-3, 29115, 29151, 29274, 29363, 29495, 29849, 30435, 30695,
     30822, 30887, 30896, 31038, 31057, 31421, 31442, 31676, 32032, 32440
     32633, 32742, 32912, 33665, 34369, 34385, 34427, 34429, 34496, 34585, 35029,
     35087, 35097, 35281, 35293, 35527, 35631, 35678, 36044, 36068, 36241, 36512,
     36549
     36701, 36880, 36939, 37185, 37209, 37212-3, 37791, 37994, 38114, 38181,
     38237, 38327, 38374-5, 38521, 38757, 38768, 38997, 39153, 39366, 39536,
     39578, 39619, 39726, 40210, 40261

Data recording and processing 25549, 25728, 26635; 33039, 36000; 37068

Decapitation see Defoliation, decapitation ...

Defoliation, decapitation, ear and root removal, effect on biliproteins 33535

Defoliation, decapitation, ear and root removal, effect on carbon fixation pathways
     27384; 30088; 32717, 34937; 37093

Defoliation, decapitation, ear and root removal, effect on carotenoids
     22581; 27239

Defoliation, decapitation, ear and root removal, effect on chlorophyll
     22006, 23529, 23756, 24127; 26775, 27239, 27384; 29104, 29523, 30253, 31627,
     31737; 32863, 33301, 34508, 36213; 37093, 37490, 38131

Defoliation, decapitation, ear and root removal, effect on chloroplast 28649; 32717

Defoliation, decapitation, ear and root removal, effect on ecosystem and plant pro-
          ductivity
     21522, 21634, 21786, 22096, 22352, 22663, 23499, 23774, 23776, 23895, 24127,
     24275, 24326, 24511, 24598, 24744, 24820, 24970
     25189, 25336, 25818, 25832, 25917, 26213, 26452, 26465, 26688, 26775, 26945,
     27199, 27538, 27664, 27720, 27991, 28050, 28392
     28074, 28873, 29046, 29104, 29381, 29475, 29559, 29771, 29794, 29895, 30513,
     30554, 31305, 31399, 31606, 31748, 31763, 31792, 31926, 31934, 32431
     32567, 32713, 33019, 33079, 33181, 33193, 33226, 33291, 33832, 33859, 33936,
     33950, 34037, 34176, 34541, 34936-7, 35056, 35073, 35455, 35549, 35712,
     35888, 35966, 36112, 36213, 36275, 36406, 36458
     36623, 36722, 36777, 36788, 37093, 37164, 37169, 37290, 37297, 37490, 38043,
     38210, 38636, 38645, 38818, 38822, 39034, 39070, 39140, 39565, 39682-4,
     39755, 39852, 39858, 40199

Defoliation, decapitation, ear and root removal, effect on electron transport chain
     25189; 31737

Defoliation, decapitation, ear and root removal, effect on gas exchange
     21538, 21634, 21786, 22006, 22210, 22952, 23218-9, 23430, 23756, 23962,
     24275, 24598, 24820; 25328, 25765, 25832, 26121, 26775, 27384, 27501, 27594,
     28091; 28605, 28649, 29046, 29104, 29133, 30088, 30253, 30516, 31606, 31627,
     32201; 32689, 32713, 32717, 33079, 33291, 33832, 33859, 33879, 34132, 34176,
     34936-7, 35283, 35673, 36077, 36213, 36406, 36457, 36530; 36742, 37093-4,
     37290, 37709, 38210, 38943, 38968, 39546, 39755

Defoliation, decapitation, ear and root removal, effect on photorespiration 33831

Defoliation, decapitation, ear and root removal, effect on resistances to CO_2 and
          water vapour exchange
     23218, 23430; 25832, 26775; 31627, 32194; 32713, 34384, 35283, 36457; 36742

Defoliation, decapitation, ear and root removal, effect on respiration
     22952, 23962, 24275; 29104, 30516; 33859, 34508, 36077, 36213, 36406; 36742,
     37290, 39755

Defoliation, methods 24275

Desiccation of tissue see Water saturation deficit

Deuterium oxide, tritium oxide
     21641, 24943; 25271; 30649; 35452; 36878, 36927, 37208-10, 37741, 38066, 39383,
     40012

Development, leaf, plant see Leaf and plant development and ageing

Dew see Precipitation, dew ...

Dew measurement see Aerodynamic methods ...

Dew point see Humidity of air ...

Dichroisms determination (methods and results)
     21715, 21898, 21900, 21917, 22113-4, 22394, 22593, 22596, 22979, 23055,
     23707, 23909, 23998, 24154, 24327, 24385, 24784, 24810
     25263, 25337, 25351, 25746, 25837-8, 26061, 26158, 26374, 26401, 26942, 26992,
     27088, 27128, 27193, 27340-1, 27758, 27952, 28258-9, 28414
     28892, 28932-3, 28943, 28960, 28968, 29099, 29227, 29544, 29580, 29666, 29738,
     29799, 29872, 29924-5, 29929, 29999, 30181, 30366, 30514, 30632, 30678, 30989,
     31011, 31076, 31186, 31445-6, 31622, 31642, 31713, 31805, 31901, 32254, 32282
     32563, 32790, 32890, 33064, 33133-4, 33203-4, 33498, 33574; 33772, 33798,
     33910, 33993, 34085, 34204, 34662, 34957, 34977, 35166, 35486, 35697, 35846,
     36041, 36119, 36160, 36229, 36374, 36400, 36504, 36564
     36589, 36924, 36943, 36973, 36990, 37166, 37181, 37448, 37598, 37823, 37864,
     B38021, 38096, 38162, 38245, 38648, 38766, 38817, 38913, 39229-31, 39326,
     39362, 39406, 39452, 39613, 39769-70, 39978, 40311, 40321-2

Differentiation of tissues see also Leaf and plant development and ageing; Ontogeny...

Differentiation of tissues and carotenoids 21872

Differentiation of tissues and chlorophyll 21872, 21925, 23048

Differentiation of tissues and chloroplast 21872, 23490; 40121

Differentiation of tissues and electron transport chain 21925

Differentiation of tissues and gas exchange 21593, 21696; 36870

Diffusion, diffusion coefficient see CO_2 transfer, theory

Diffusion (diffusive) conductance see Resistance ...

Diffusion (diffusive) resistance see Resistance ...

Diurnal changes in algae productivity
     21972, 21990, 22292, 22493, 22578, 22654, 22858, 22866, 22943, 23759, 24144,
     24420, 24634, 25112; 26051, 26568, 27392, 27470, 27611, 27838, 27959-60,
     27970, 27987, 28076, 28129; 29151, 29226, 29426, 29511, 29582, 29825, 30151,
     30676, 30906, 31911, 32091, 32137; 32910, 33586, 33852, 34099, 34359, 34395,
     34793, 35299, 35872, 35995, 36334; 37058, 37204, 37279, 37442, 37540, 37606,
     37635, 37945, 37985, 38086, 38277, 38473, 38693, 39060, 39714, 39759, 39924

Diurnal changes in biliproteins 35178-9

Diurnal changes in carbon fixation pathways
     22038, 22303, 24126, 24656; 25620, 25787, 25954, 26299, 27052, 27147, 27183,
     27541, 28065; 29204, 29816, 30332, 30630, 30665, 30695, 30928, 31062, 32139,
     32364; 33076, 33739, 33777, 33857-8, 34298, 34368, 34684; 36672, 36917,
     36997, 37022, 37697, 38177, 38253, 38612, 38640-3, 38788-9, 39132, 39217,
     39431, 39725, 40180, 40213

Diurnal changes in carotenoids 22303, 23471; 28188; 31142; 32910, 35179, 35348,
36153; 36628

Diurnal changes in chlorophyll
22303, 23441, 24194, 24285; 25557-8, 25679, 26242, 27622, 28076; 29623,
29674, 30381, 31142, 31362-3, 32241; 32910, 33076, 35178-9, 35348, 36153;
36628, 37645, 38274, 38426, 38515, 39826, 39944

Diurnal changes in chloroplast
21905, 22744, 24867; 25559-60, 27497; 34396, 34578, 35845, 36045; 36923,
37001, 38467, 38573, 39825, 40314

Diurnal changes in ecosystem and plant productivity
21549, B22052, 22108, 22116, 22144, 22999, 23592, 23802, 23910, 23945,
23983, 24143, 24214, 24244, 24259, 24438, 24441, 24567, 24641, 24656, 24689,
24887, 24907
25203, 25264, 25527, 25662, 25666, 25683, 26027, 26080, 26337, 26372, 26783,
26954, 27052, 27087, 27139, 27220, 27270, 27630, 27656, 27658, 27812, 28087,
28158, 28283, 28350, 28397, 28408
29086, 29289, 30010, 30061, 30084, 30438, 31168, 31198, 31517, 31762, 32182
32693, 32721, 32796, 33167, 33632, 33741, 34006, 34467, 35262, 36066, 36212,
36225
36601, 36636, 37241, 37591, 37772, 37992, 38442, B38739, 39088, 39342-3,
39549, 39563, 39760, 40079, 40216, 40288

Diurnal changes in electron transport chain
23626-7, B24176, 24652, 24866-7; 30630, 32242; 37851, 38027, 38276, 38515,
38956, 40024

Diurnal changes in gas exchange
21517, 21529, 21802-3, 21841, 22014, B22052, 22053, 22152-3, 22165, 22231,
22284, 22302-3, 22380, 22413, 22494, 22686, 22858, 22875, 22948, 22984,
22999, 23007, 23018, 23040, 23105, 23171, 23257, 23295, 23298-9, 23341,
23371, 23376, 23467, 23493, 23524, 23556, 23596, 23626-7, 23648-9, 23666,
23693, 23712, 23786, 23980, 23988, 24055, 24063, 24126, 24214, 24289, 24302,
24341, 24377-8, 24427, 24438, 24440, 24507, 24532, 24588, 24601, 24614,
24619, 24675, 24689, 24715, 24758-9, 24826, 24866-8, 24934, 24988, 24990,
25112
25371, 25527, 25559, 25571, 25596, 25666, 26027, 26046, 26075, 26085, 26121,
26166, 26265, 26303, 26368, 26403, 26506, 26515, 26580, 26583-5, 26591,
26674, 26690, 26801, 26811, 26905, 26909, 26912, 27095, 27107, 27139, 27180,
27267, 27271, 27359, 27574, 27638, 27750, 27768, 27773, 27782, 27888, B27915,
28014, 28066-7, 28076, 28085, 28087, 28152, 28193, 28230, 28342
28527, 28614, 28732, 28785, 28844, 28971, 29016, 29066, 29102, 29131, 29206,
29249, 29281, 29306, 29333, 29372-3, B29468, 29605, 29676, 29732, 29816,
29817, 29820, 29931, 30084, 30249, 30381, 30476, B30511, 30515-6, 30630,
30695, 30706, 30808, 30822, 30898-9, 30904-5, 30927, 30930, 30943, 30946-7,
31062, 31068, 31142, 31258, 31261, 31278, 31321, 31362-3, 31515, 31585, 31613-
-4, 31701, 31925, 31994, 32023, 32057, 32139, 32242, B32318, 32364, B32515
32633, 32694, 32814,   32871, 32877, 32969, 33002,
33052, 33076, 33208, 33231, 33262, 33335, 33358-9, 33664, 33845, 33857-8,
33915, 33984, 34011, 34157, 34170, 34227, 34239, 34289, 34352, 34368, 34376,
34453, 34467, 34471, 34497, 34511, 34578, 34644, 34650, 34677, 34684, 34689,
34755, 34886, 34899, 34962, 34966, 35092, 35096, 35126, 35153, 35172-3, 35179,
35224, 35262, 35449, 35476, 35494, 35508, 35616, 35622, 35624, 35673, 35789,
35791, 35912, 35950, 36050, 36067-8, 36132-3, 36152-3, 36183, 36199, 36207,
36225-6, 36358, 36417, 36424, 36550-1
36672, 36683, 36686, 36809, 36910, 36916, 36923, 37049, 37162, 37241,
37287, 37327, 37381, 37389, 37397, 37434, 37495, 37540, 37656, 37851,
37874, 38027, 38077, 38083, 38085, 38112, 38132, 38140-1, 38161, 38235-6,
38253, 38274, 38286, 38400, 38426, 38435, 38515-6, 38573, 38640, 38710,
B38722, 38745, 38935, 38956, 38997, 39025, 39088, 39185, 39342, 39347,
39431, 39500, 39639, 39649, 39697, 39725, 39784, 39826, 39867, 40023, 40158,
40212, 40233, 40238, 40291, 40314, 40357

Diurnal changes in photorespiration 26299; 36417

Diurnal changes in resistances to CO_2 and water vapour transfer
        21550, 21929, 22314-5, 22505, 22662, 22761-2, 22875, 22962, 23167, 23295,
        23299, 23679, 23693, 23945, 24215, 24302, 24378, 24675, 24826
        25283, 25327, 25371, 25406, 25430, 25454, 25653, 25742, 25843, 25947,
        25999, 26015, 26237, 26265, 26303, 26612, 26739, 26909, 27139, 27269,
        27271, 27356, 27360, 27574, 27636, 27749, 27912, 28065, 28067, 28089,
        28125, 28257, 28350, 28462
        28500, 28732, 28810, 29020, 29162, 29179, 29204, 29232, 29373, 29394-5, 29551,
        29816, 30212, 30249-50, 30392, B30511, 30547, 30583, 30630, 30738,,30818,
        30905, 30930, 31068, 31268, 31321, 31512, 31516, 31536, 31613, 31762, 31888,
        31994, 32051, 32110, 32120, 32139, 32364, 32369
        32693, 33167, 33358-9, 33398, 33451, 33857-8, 33984-5, 34109-10, 34135,
        34453, 34511, 34541, 34690, 34954, 35096, 35192, 35355, 35476, 35508, 35529,
        35536-7, 35963, 36050, 36120, 36207, 36225, 36358, 36417
        36683, 36916, 37177, 37300, 37413-6, 37434, 37441, 37473, 38016, 38075,
        38236, 38401, 38435, 38455, 38534, 38737, 38848, 39090, 39328, 39511, 39561,
        39649, 39659, 39661, 39725, 39813, 40079, 40169, 40213, 40318

Diurnal changes in respiration
        22999, 23105, 23257, 23371, 23467, 23712, 24015, 24055, 24427, 24532, 24758,
        24934, 24988; 28527, 29065, 29820, 30476, 30516, 30904, 30947, 31077, 31362,
        32023, 32339, 32364; 32969, 33262, 33890-1, 33915, 34170, 34368, 35172,
        35789, 36199, 36424; 36672, 36939, 38141, 38880, 39088, 39342, 40233, 40291

Drought and carbon fixation pathways 26549; 33660, 36443; 36880, B38238, 38253

Drought and carotenoids 24527; 36936, 37582, 37735

Drought and chlorophyll
        22290, B23262, 24482; 25992, 26183, 26758; 29410; 35906, 36213, 36343, 36394;
        36936, 36960, 37171, 37377, 37582, 37735, 38345, 39312, 39524

Drought and chloroplast 22290, 23028; 26126; 28529, 30879; 33660, 35661

Drought and ecosystem and plant productivity
        21787, 22379-80, 24378, 24629, 24649, 24798, 25123; 25970, 26685, 27325,
        27658, 27710, 28241, B28482; 29103, 29410, 30565, 30979, 31015, 31868,
        32010, 32200; 33022, 34416, 34541, 35273, 35534, 35594, 35632, 35713, 36016,
        36343, 36394; 36637, 36880, 37027, 37082, 37171, 37384, 37462, 37681, 37838,
        38083, 38134, 38246, 38455, 38743, 38754, 39053, 39059, 39549, 39688,
        B39903, 40250

Drought and electron transport 24808, 24928; 29537; 36880, 37013

Drought and gas exchange
        23218, 23300, 23465, 23696, 23775, 24595; 25431, 25859, 25949-50, 25992,
        26494, 26685, 26690, 27107, 27656; 29415, 30448, 31258, 31536; 32698,
        32986, 33440, 33660, 34295, 34368, 34632, 34677, 35594, 35632, 35636-7,
        35713, 36213, 36343; 36880, 37013, 37171, 37742, 38075, B38238, 38253,
        38851, 39058-9, 39524, 39599, 39688

Drought and photorespiration 37516, 39599, 39688

Drought and resistances to CO_2 and water vapour transfer
        21550, 22964; 29467, 31536; 34651, 35594; 39005, 39599, 40169

Drought and respiration 29537, 30448; 33660, 35636, 36213; 36880

Dry-matter production, gravimetric determination
        22455, 23194, 23418, 24604; 26812, 27436, 28058, 28475; 28632, 30427, 30663,
        30748; 36543; 37814, 40333

E

Ear removal   see Defoliation, decapitation, ear and root removal ...

Ecosystem production, primary productivity (terrestrial) (*cf*. also Biomass...)
        21525, 21602, 21633, 21660, 21674, 21704, 21794, 21802, 21808, 21832,
        B21833, 21845, 21853, 21862, 21876, 21878, 21880, 21882, 21902, 21944,
        21950, 21973, 21977, 22007-8, 22014-5, 22030, 22034, B22052, 22053, 22118,
        22123, 22165; 22171, 22222, 22273, 22282, 22315, 22331, 22343, 22367-9,
        22379-80, 22414, 22428-31, 22437, 22503, 22532, 22543, 22570, 22593, 22631,
        22639, 22716, 22750, 22768, 22797, B22822, 22823, 22829, 22832, 22849,
        22878-9, 22885, 22893, 22896, 22910, 22919, 22937, 23007, 23015, 23027,
        23038-9, 23063-4, 23067, 23091, 23104, 23135-8, 23145, 23216, 23220, 23223,
        23232, 23245, 23303-4, 23339, 23405, 23409-12, B23413, 23448, 23466, 23497,
        23499, 23578, 23589, 23620, 23645, 23650, 23658, 23733, 23736, 23740-1,
        23777, 23783, 23795, 23814, 23818, 23828, 23871, 23893-4, 23907, 23911,
        23923, 23925, 23943, 23983, 23991, 24006-7, 24014, 24019-20, 24038-9, 24044,
        24080, 24105, 24110, 24115, 24125, B24132, 24146, 24150, 24181, 24185-6,
        24207, 24210, 24219, 24267, 24276, 24303, 24314, 24331, 24378, 24424-5,
        24428, 24438, 24504, 24508-9, 24511, 24595, 24609, 24633, 24646-7, 24666,
        24744, 24752, 24758, 24765, 24773, 24780, 24798, 24808, 24940, 24969, 25003,
        25013-5, 25022-3, 25039, 25047, 25066, 25103, 25115, 25126, 25136-7, 25149-50
        25198, 25229, 25238-9, 25253-4, 25264, 25274-5, 25299, 25303, 25312, 25331,
        25336, 25357, 25412, 25435, 25472-3, 25475, 25480-1, 25518, 25523, 25538,
        25576, 25581, 25593-4, 25608, 25615, 25623, 25626, B25637, 25674, 25737,
        25753, 25801, 25811, 25818, 25827, 25857, 25886, 25910, 25921, 25946, 25970,
        25978, 25982, 25989, 25997, 26017, 26029, 26031, 26050, B26054, 26055,
        26068, 26076, 26086-7, 26092, 26096, 26106, 26112, 26124, 26132, 26145,
        26151, 26182, 26189, 26202, 26204, 26256, 26263, 26267, 26284-6, 26331,
        26373, 26419, 26450, 26452, 26461, 26490, 26509, 26530, 26534, 26547,
        26599, 26648, 26685, 26688-9, 26719, 26724, 26735, 26742, 26778, 26784,
        26787, 26803-4, B26846, 26847, 26881, 26883, 26895, 26904, 26906, 26919-20,
        26927, 26945, 26952, 26954, 26964, 27014, 27190, 27209, 27228, 27242, 27293-
        -5, 27333, 27336, 27343, 27371, 27385, 27409, 27413, 27415, 27422, 27428,
        27457, 27466, 27507, 27538, 27563, 27604, 27656, 27673, 27707, 27710,
        27728, 27734-5, 27740, 27748, 27760, 27778-9, 27793, 27830, 27893-4, B27915,
        27920, 27930,   27981-3,   28013, 28020, 28031, 28041, 28087, 28093,
        28098-100, 28109-10, 28113, 28143, 28158, 28160, 28169, 28171, 28175,
        28192, 28197, 28212, 28228, 28262, 28265, 28282, 28294, 28300, 28322, 28343,
        28352, 28381, 28397, 28408, 28431, 28459-60, 28470, 28474, 28480
        28503, 28516, 28521, 28533, 28544, 28552, 28571, 28624, 28632, 28648, 28689,
        28703-4, 28711, 28788, 28800, 28809, 28813, 28857, 28864, 28899, 28937,
        28941, 28972, 29007, 29009, 29053, 29060, 29067, 29086, 29116, 29123, 29129-
        -30, 29158, 29163, 29166, 29168, 29201, 29213, 29221-2, 29236, 29254, 29290,
        29334, 29362, 29373, 29377, 29380, 29413, 29419, 29448, B29468, 29492, 29496,
        29501, 29506, 29515, 29524, 29550, 29559, 29565-6, 29575-6, 29633, 29656,
        29685, 29722, 29789-90, 29797, 29806, 29811, 29883, 29887, 29967, 29975-6,
        29984, 30029, 30044, 30049, 30070-1, 30096, 30113, 30120, 30162, 30195,
        30290, 30376, 30387, 30400, B30405, 30412, 30427, 30438, B30477, 30483,
        30489, 30505-6, 30524, 30532-3, 30554, 30565, 30573, 30674, 30689, 30691,
        30734, 30861, 30874, 30916, 30929, 30945, 30947, 30953, 30967, 30979, 30994,
        31001, 31021, 31032, 31131, 31156, 31180, 31201, 31208, 31249-50, 31259,
        31265, 31272-3, 31304, 31306-7, 31332, 31367, 31463, 31496, 31500, 31586,
        31595, 31608, 31647, 31670, 31748, 31753, 31777, 31822, 31826, 31840-2, 31848,
        31868, 31875, 31916, 31922, 31926, 31935, 31952, 31996, 32010, 32012, 32033,
        32046, 32052, 32059, 32133, 32151, 32186-7, 32210-1, 32243, 32296, 32322,
        32325, 32394, 32403, 32415, 32431, 32434, 32436, 32504-5, 32545
        32567, 32574, 32629, 32634, 32640, 32648, 32652, 32662, 32686, 32699, 32737,
        32761, 32783, B32852, 32868, 32882, 32888, 32918, 32950, 32973, 32980,
        33002, 33023, 33043, 33054-5, 33073-4, 33085, 33097, 33123, 33128, 33166,
        33175, 33184, 33196, 33209, 33222, 33255, 33260, 33293, 33331, 33394, 33412,
        33488, 33495, 33528, 33541, 33547, 33567, 33620, 33647, 33668, 33730, 33756,
                                                                        (continued)

Ecotypes, geographical types, and resistances to CO_2 and water vapour transfer
22284; 26116, 26532; 31861, 31864; 33435, 35355; 37347, 38285

Ecotypes, geographical types, and respiration
22950; 26116, 27337, 27638; 29409, 30271

Efficiency, photochemical (*cf.* also Irradiance and gas exchange, analysis of light
curves)
28783, 29333, 29675, 29790, 29975, 30047, 30722, 30916, 31359, 31470, 32175;
33433, 33833, 34436, 34452, 34495, 34646-7, 34942, 35003, 35561, 35564,
36037, 36195-6; 36602, 37702, 38839, 39033, 39179, 39420, 39882, 39960, 40083,
40154, 40250, 40293

Electron paramagnetic resonance see EPR, NMR

Electron spin resonance see EPR, NMR

Electron transport chain activity
21646, 21697, 21744, 21825, 21915, 22106, 22142, 22151, 22164, 22215, 22243,
22427, 22514, 22526, 22567, 22576, 22605, 22683, 22706, 22720, 22800, 22821,
22882, 22973, 23043, 23148, 23395, 23794, 23851, 23920, 24225, 24496, 24664,
24684, 24762, 24791, 24909, 25002, 25102
25217, 25321, 25750, 25892, 26366, 26554, 26762, 27857, 28225, 28285
28511, 28589, 28695, 28990, 30415, 30630, 30765, 31051, 31338, 31521,
31738, 31918-9, 32097, 32161, 32315
32664, 32667, 32674, B32723, 32726, 32769, 32795, 32860, 32866, 32899,
32908, 32934, 33994-5, 34042, 34286, 34468, 34643, B34706, 34766-7, 34984,
35716
36803, 36892, 36931, 36993, 37170, 37469, 37843, 37922, 37937, 38053, 38275,
38394, B38477, 38515, 38723, 39306, 39330, 39427, 39612, 39730, 40039,
B40105, 40200

Electron transport chain activity, methods
21744, 23876; 35889, 36082; 36892, 37980, 39625

Electron transport chain components see Cytochromes; Ferredoxin ...; Ferredoxin-
NADP reductase; NADP ...; O_2 evolution ...; Photosystems; Plastocyanin;
Quinones

Electron transport chain components and carbon fixation pathways 35302; 37311

Electron transport chain components and chloroplast 39845, 40311

Electron transport chain components and photorespiration 23981-2

Electron transport chain, general aspects see General aspects on carbon fixation ...

Electron transport chain localization in thylakoid
21535, 21575, 21646, 21778, B21979, 22017, 22100, 22346, 22354, 22528, 22580,
22644, 22657, 22704-6, 22969-71, 23093, 23631, B23767, 24095, 24097, 24136,
24327, 24395, 24471, 24491, 24789-90, 25006, 25030
25182, 25228, 25241, 25244, 25331, 25339, 25421, 25626, 25762, 25789, 25884,
25907, 26193, 26259, 26268, 26343, 26423, 26445, 26538, 26609, 26767, 26859,
26942, 27137, 27140, 27305-6, 27399, 27617, 27904, 27966, 27979, 28137,
28163-4, 28237, 28303, 28334, 28409-10
28574, 28581, 28587, 28590, 28744, 28751, 29001, 29114, 29182-3, 29309-10,
29352, 29438, 29738, 29837, 29870, 29932, 29958, 30000, 30125-6, 30156,
30178-9, 30245, 30361, 30411, 30525, 30771, 30788, 30958, 30969, 31027,
31030, 31097, 31172, 31186, 31287, 31335, 31410-1, 31413-4, 31426, 31483,
31485-6, 31549, 31566, 31610, 31642, 31665, 31718, 31740, 31742, 31810,
31883, 31901, 32077, 32157-8, 32160-1, 32204, 32250, 32260, 32266, 32282,
32285, 32386, 32443-4, 32471, 32487, 32550

(continued)

Electron transport chain localization in thylakoid (continued)
    32651, 32697, 32732, 32744, 32890, 32905, 33056, 33089, 33253, 33259, 33380,
    33400, 33402, 33406, 33880, 33919, 33989, 34168, 34185, 34202-5, 34277,
    34433, 34860, 34893, 34910, 35013, 35076, 35386, 35393, 35487, 35538, 35753,
    35988, 36130-1, 36173-4, 36265
    36622, 36706, 36856, 36892, 36944, 36990, 37047, 37167, 37181, 37193, 37211,
    37218, 37275, 37339, 37705, 37739, 37800, 37913, 38049, 38054, 38425,
    B38477, 38730, 38733, 38795, 38820, 38861, 38905, 38972, 39231, 39285,
    39646, 39686, 39745-6, 39901, 39913, 39915, 40246-7

Electron transport chain model   see Model ...

Electron transport chain, serological analysis
    21613, 21743, 21777-8, 21792, 22217, 22574, 23253, 23577, 23624, 23798,
    24095, 24097, 24130, 24135, 24182, 24351, 24471, 24860, 25030; 25891, 26711,
    27138, 27745, 28281, 28315; 28778, 28851-2, 28886, 29044, 29145, 29374,
    29474, 29626, 30042, 30361-2, 31414, 31590, 31643, 31665, 31754, 32069;
    32905-6, 34066, 34372, 35419-20, 35697-9; 36756, 40096, 40311

Emerson effect, Blinks effect
    21638, 22580, 23724, 23989, 24724, 24932; 25244, 25362, B26220, 26861,
    27950, 28286, 28376, 28395-6; 29340, 30892, 32533-4; 33982, 34568-9, 35307,
    36150; 36937, 37561, 37610, 37787, 37966, 38349

Energy balance, leaf   B22358, 22875; 25724; 33062, 34664

Energy content in biomass
    22173, 22258, 22297, 22570, 22765, 22778, 22918-9, 23137, 23408, 23630,
    24153, 25025; 25210, 25358, 25367, 25581, 25807, 25817, 26140, 26153, 27728,
    28283, 28470; 29193, 29791, 30459, B30477, 30533, 30943, 31035, 32324;
    33113, 33331, 33412, 34000, 34150, 34444, 34577, 34612, 34626, 35056, 35338,
    35561, 36053; 36656, 36973, 37031, 37054, 37795, 38257, 38314, 38567, 38799-
    -800, B38892, 39002-4, 39013, 39262, 39752, 40227, 40336

Energy content in biomass, methods
    23408, B24024; 25413; 33134

Energy utilization, plant and ecosystem
    21629, 21832, 21973, B21979, 22036, 22123, 22185, 22297, 22302, 22543,
    22750, 22755, 22765, 22829, 23003, 23106, 23145, 23147, 23423, 23814, 23862,
    23925, 24229, 24419, 24438, 24647, 24780, 24782, 24863, 24969, 25023, 25025
    25331, 25357, 25772, 25903, 25946, 25978, 26189, 26267, 26426, 26719, 26869,
    26881, 27128, 27180, 27333, 27635, 27688, 27779, 27786, 27894, B27915,
    28192, 28247, 28470
    28544, 28837-8, 28858, 29018, 29057, 29193, 29332, 29373, 29530, 29797, 29812,
    30113, 30375, 30378, 30380, 30505-6, 30512, 30524, 30750, 30917, 30921, 30946,
    30976, 31040, 31130-1, 31283, 31358-9, 31411, 31586, 31817, 31849, 32006,
    32175
    32634, 32874, 32888, 32897, 32914, 32916, 32960, 33103, 33123, 33412, 33541,
    33833, 34000, 34152, 34241, 34436, 34444, 34545, 34623, 34645, 34720, 34942,
    34959, 35002, 35062, 35080, 35105, 35129, 35158, 35258, 35488, 35597, 35759,
    35788, 35790, 35928, 36078, B36191, 36222, 36368, 36535
    36930, 36944, 36973, 37054, 37673, 37772, 37776, 37778, 37795, 37890, 37935,
    37966, 38028, 38190, 38257, 38272, 38294, B38528, 38563-4, 38755, 38839,
    B38892, 39015, 39053, 39512, 39565, 39727, 39831, 39836, 39883, 39932,
    40138, 40277, 40345

Enzymes and carbon fixation pathways   23504; 33807, 34018, 34065, 34559, 35294,
    36452; 39132, 39183, 39476, 40189

Enzymes and carotenoids   30073, 32037-8; 34544; 39039, 39170, 39840

Enzymes and chlorophyll   21943, 23140-1, 23951, 24305; 27027; 28619, 29477, 32009;
    32597, 33145, 33353, 34165-7, 34371, 35437; 36631, 37371, 39039

Enzymes and chloroplast (chromatophore)
21551, 21857, 22739-41, 23742-3, 24328, 24360, 24423, 24828, 24855; 25277, 25398, 26893, 27003, 27223, 27816; 28619, 28727, 29393, 29430, 29634-5, 29745, 29808, 29908, 30267-8, 31050, 31294, 31380, 31639; 32679-80, 33197, 33374, 34142, 34492, 35009, 35748, 36238, 36245, 36351, 36486; 36631, 36666, 36730, 36829, 36855, 37223, 37235, 37322, 37340-1, 37371, 37411, 37507, 37509, 38003, 38038, 38049, 38092, 38129, 38540, 38672, 38889, 38911, 38959, 39202, 39221, 39591, 39772, 39838, 39845, 39929, 40071, 40218, 40282

Enzymes and electron transport chain
21551, 23014, 23139, 23141, B23407, 23934, 24184, 24415, 24874; 25362, 25427, 26554, 27019, 27603, 27618, 27763, 27943-4, 28203; 29417, 30535, 30859, 31007, 31389, 31483, 31942; 32954, 33299, 33353, 34165, 34167, 34527, 34674, 34996, 35732, 36324, 36464; 36606, 36891, 36894, 36931, 37371, 37425, 37704, 37728, 37813, 38049, 38054, 38541, 39193, 39202, 39845, 39955, 40055, 40257

Enzymes and gas exchange
23504; 26067, 27382, 27618; 31730; 39637

Enzymes and photorespiration   22347, 22942, 23231, 23981-2; 33593

Enzymes and photosynthates   35353; 39637

Enzymes and respiration   21814, 23231, 24657, 24659, 24795

Enzymes of biliprotein synthesis and degradation   see   Biliproteins, enzymes ...

Enzymes of carbon fixation pathways other than RuBPC, PEPC and malic enzyme
21576, 21578-80, 21655, 21672-3, 21702, 21724, 21814, 21840, 21933, 21937, 21942, 22018, 22054, 22081, 22150, 22157, 22239, 22250, 22274, 22303, 22635, 22671, 22695, 22697, 22699, 22713, 22785-6, 22802, 22838, 22958, 22977, 22980-1, 23030-1, 23033, 23060, 23062, 23072, 23336-7, 23370, 23397, 23419, 23586, 23674, B23767, 23788, 23797, 23868, 23949, 23974, 24002, 24008, 24106, 24112, 24161, 24164-5, 24223, 24255, 24342, 24346, 24352, 24357, 24360, 24380, 24383, 24386, 24397, 24658, 24682, 24734, 24769, 24804, 24931, 24933, 24984, 25033, 25065-6, 25073, 25097, 25101, 25138
25211-3, 25221, 25226, 25278, 25325, 25338, 25431, 25516, 25528, 25583-4, 25671-2, 25730, 25768-9, 25786-7, 25833, 25867, 25944, 26003-4, 26059, 26165, 26188, 26203, 26249, 26280, 26314, 26317, 26525-6, 26597, 26622-4, 26685, 26755, 26786, 26806, 26858, 26866, 26872, 26921, 26940, 27007-8, 27053, 27091, 27147, 27183, 27263, 27297, 27315, 27329, 27364, 27382, 27422, 27431, 27473, 27509, 27542, 27569, 27621, 27639, 27755, 27759, 27772, 27793, 27906, 27909, 28040, 28151, 28183, 28220, 28285-6, 28303, 28342, 28355, 28357, 28403, 28419, 28430, 28490, 28492
28555, 28557-8, 28560, 28609, 28613, 28662-3, 28775, 28818, 28853-4, 28909, 28911, 28989, 29061-2, 29075, 29085, 29098, 29117, 29132, 29215, 29231, 29267, 29271, 29385, 29478, 29534, 29579, 29669, 29772, 29777, 29792, 29807-8, 29833, 29860-1, 29863, 29912, 29945, 29990, 30125, 30190-1, 30204, 30232, 30234-5, 30266-7, 30276, 30350, 30538, 30630, 30707, 30893-4, 30897, 30936, 30951-2, 31042, 31062, 31089, 31164, 31227, 31250-1, 31300, 31337, 31343, 31433, 31450--2, 31466, 31472, 31519, 31530, 31576, 31698-9, 31973, 31977, 32176, 32223, 32290-1, 32335, 32337-8, 32366, 32429, 32440, 32454-5
32628, 32664, 32767, 32866-7, 32931, 32958-9, 32968, 33004, 33102, 33148-9, 33344, 33450, 33518-9, 33533, 33612, 33651, 33660, 33739, 33809, 33827, 33849, 33913, 33943, 34062-4, 34199, 34256, 34259, 34292, 34429, 34473, 34525, 34549, 34560, 34573, 34575-6, 34586, 34631, 34653, 34675, 34692, 34746, 34749, 34800, 34922, 34932, 34974, 35051, 35192, 35248, 35266, 35378, 35402, 35411, 35431, 35442, 35467, 35471, 35593, 35736, 35755, 35881, 36111, 36192, 36234, 36274, 36328, 36366, 36463, 36486, 36491, 36533, 36577, 36603, 36647, 36663, 36665, 36730, 36800, 36849, 36901, 37034-5, 37120, 37170, 37197, 37273, 37317-8, 37360, B37458, 37486, 37697-8, 37827, 37841, 37858, 37930-1, 37934, 38178, 38297, 38338, 38341, 38419, 38566, 38606, 38748, 38770-1, 38835, 38845, 38895, 38951, 39062, 39064, 39153, 39217,
(continued)

Enzymes of carbon fixation pathways ... (continued)
        39218, 39235, 39237, 39264, 39266, 39416, 39438, 39460, 39467, 39476, 39505,
        39593, 39641-4, 39713, 39724, 39747, 39809, 39876, 39973, B40105, 40159,
        40203, 40210, 40218, 40256-7, 40261, 40270, 40272, 40282, 40360

Enzymes of carbon fixation pathways other than RuBPC, PEPC, malic enzyme, malate
        dehydrogenase, methods
        21579, 21937, 22081, 22699, 22713, 22802, 23868, 24002, 25033, 25065-6;
        25235, 25656, 26356, 26978, 28492; 28557, 28663, 29117, 29215, 29863, 30268;
        33102, 36491, 36577; B37458, 40282

Enzymes of carotenoid synthesis and degradation see Carotenoids, enzymes ...

Enzymes of chlorophyll synthesis and degradation see Chlorophyll, enzymes ... ;
        Chlorophyllase

Enzymes of electron transport chain, methods 21617; 31730; 34229-30; 39221

Enzymes of glycollate cycle, methods 32547; 32952, 33478; 39150

Enzymes of photorespiration see Photorespiration enzymes; Enzymes of glycollate
        cycle, methods

Epidermis see Leaf epidermis ...

EPR, NMR (methods and results)
        21583, 21617, 21643-6, 21816-9, 21970, 21981, 21983, 22060, 22088, 22090,
        22126-7, 22146, 22373, 22375-6, 22390, 22459, 22561, 22642, 22682, 22708,
        22887, 23055, 23057, 23076, 23084, 23089, 23139, 23192, 23215, 23253, 23437,
        23445, 23472, 23546, 23583, 23600-1, 23655, 23729-30, 23799, 23966, 24211,
        24304, 24327, 24337, 24344, 24453, 24638, 24640, 24651, 24760-2, 24783-4,
        24844, 24877, 24904, 24917, 24936-7, 25019
        25276, 25331, 25477, 25488, 25517, 25563, 25629, 25736, 25773, 25844,
        25893, 25963, 25968, 26176, B26220, 26394-5, 26457, 26559, 26561, 26609,
        26616, 26657, 26679-80, 26751, 26830, 26942-3, 26974, 27033, 27088, 27099-
        -100, 27192-3, 27277, 27282, 27308-9, 27326, 27452, 27528, 27706, 27868,
        27875, 28037, 28060, 28131, 28135-6, 28155, 28234, 28258, 28332, 28334-5,
        28376, 28420, 28437-8
        28570, 28572, 28667-8, 28720, 28751, 28769, 28823, 28832-3, 28935-6, 28946,
        28956, 29038, 29079-80, 29124, 29143, 29310, 29358, 29438-42, 29452-4, 29555,
        29689, 29715, 29740-1, 29768, 29779, 29784, 29796, 29799, 29830, 29880,
        29969-72, 29991, 30138, 30188, 30252, 30254-5, 30260-1, 30302, 30386, 30436,
        30444, 30657, 30717, 30812-3, 30926, 31030, 31033, 31048, 31063, 31097,
        31099, 31122, 31133, 31253, 31297, 31369-71, 31390, 31401, 31410, 31446,
        31511, 31565-6, 31605, 31635, 31638, 31649, 31713, 31785-6, 31809, 31857-8,
        31902, 31976, 32072, 32114, 32123, 32127, 32134, 32148, 32208, 32214, 32228,
        32301, 32370, 32468, 32474, 32476, 32481, 32495
        32580, 32674, 32724, 32857-9, 32885, 32898, 32962, 32972, 33067, 33277,
        33355-6, 33403, 33482, 33485, 33490, 33494, 33515, 33564, 33590-1, 33609,
        33640, 33667, 33716-20, 33751, 33778, 33800, 33821, 33824-6, 33865, 33930-1,
        34009-10, 34068, 34223, 34234, 34277, 34311, 34337-9, 34356-7, 34431, 34490,
        34584, 34603, 34725-7, 34753, 34779, 34911, 34965, 35050, 35086, 35108,
        35282, 35292, 35386, 35388, 35397, 35414-5, 35600, 35642, 35670, 35741,
        36095, 36129, 36139-40, 36164, 36220, 36276-7, 36311, 36352-3, 36357, 36376,
        36430, 36483, 36490, 36507
        36606, 36608, 36619, 36673, 36689, 36739, 36779, 36823, 36860-3, 36885,
        36919, 36971-2, 37032, 37050, 37079-81, 37161, 37193, 37200, 37255, 37420,
        37432-3, 37454, 37545-7, 37565, 37611, 37667-70, 37749, 37760, 37763, 37773,
        37840, 37882-4, 37900, 37948, 37971, 38048, 38052, 38063, 38111, 38153-4,
        38204, 38233, 38240, 38291, 38320, 38361, 38386, 38536, 38538, 38549,
        38571, 38593, 38600-1, 38615, 38667-9, 38913, 38927, 38957, 39091, 39203,
        39320, 39384-5, 39387-8, 39458, 39614, 39622, 39630, 39681, B39693, 39842,
        39887, 39909-10, 39920, 39975, 40021, 40031, 40140, 40183, 40192, 40279

Ethylene   see   Gases, organic ...

Evolution   see   Phylogeny ...

Excitation resistance   see   Resistance, carboxylation and excitation

Exhaust gases   see   Pollution of air ...

Exposure chamber   see   Assimilation chamber

Extension growth, leaf dimensions
        21533, 21853, 21878, 21882, 21928-9, 22007, 22012, 22015, 22097, 22113,
        22228, 22335, 22401, 22485, 22516, 22663, 22701, 22727, 22844, 22859, 22885,
        22999, 23223, 23423, 23431, 23465-6, 23493, 23500, 23521, 23587, 23604,
        23645, 23650, 23669, 23695, 23779, 23813, 23950, 23961, 23975, 23984, 24000,
        24076, 24085, 24123, 24231, B24271, 24444, 24628, 24673, 24718, 24744, 24770,
        25042, 25045-6, 25059
        25170, 25209, 25230, 25250, 25318-9, 25481, 25521, 25526, 25538, 25605,
        25666, 25725, 25754, B25775, 25785, 25811, 25826, 25829, 25943, 25955,
        26003, 26017, 26073, 26093, 26197, 26213, 26284, 26400, 26461, 26466, 26643,
        26798, 26904, 26906, 26927, 26931, 26954, 27011, 27108, 27142, 27172, 27190,
        27229, 27270, 27404, 27457, 27488-9, 27540, 27579, 27625, 27701, 27725,
        27899, 27922, 27981, 28041, 28134, 28207, 28336, 28400, 28454, 28456
        28688, 28986, 29007, 29042, 29072, 29103, 29136, 29204, 29236, 29351, 29559,
        29794, 29822, 29826, 29871, 29954, 29968, 29984, 30257, 30283, 30371, 30427,
        30460, 30480, 30699, 30711, 30739, 30764, 30844, 30861, 30916, 30979, 31004,
        31013, 31021, 31041, 31114, 31164, 31178, 31261, 31284, 31496, 31563, 31608,
        31706, 31761, 31827, 31840-1, 31853, 31870, 31887, 31925, 31947, 32052, 32170,
        32175, 32219, 32341, 32412
        32589, 32593, 32612, 32640, 32648, 32794, 32918-9, 32939, 32986, 33020,
        33066, 33085, 33111, 33164-5, 33255, 33288, 33342, 33416, 33495, B33510,
        33536, 33611, 33725, 33728, 33740, 33808, 33898, 33962, 34038, 34088,
        34265-6, 34275-6, 34279, 34326, 34334, 34479, 34536, 34960, 35017, 35033,
        35073, 35102, 35122, 35169, 35248, 35542, 35595, 35620, 35708, 35785, 35787,
        35849, 35870, 35883, 35904, 35912, 35931, 36036, 36070, 36097, 36116, 36132,
        36205, 36254, 36345, 36407, 36418, 36516
        36601, 36691, 36717, 36746, 36879, 37043, 37086, 37121-2, 37199, 37241, 37265,
        37265, 37268, 37290, 37349, 37416, 37439, 37473, 37501, 37564, 37589-92,
        37689, 37721, 37796, 37836, 37847, 37874, 37982, 38068. 38097, 38120, 38124,
        38127, 38148, 38217, 38221, 38428, 38500, 38577, 38590, 38705, 38743, 38754,
        38760, 38848, 38860, 38886, 38984, 39068, 39097, 39224, 39244, B39394,
        39419, 39442, 39561, 39567, 39690, 39765, B39951, 40020, 40027, 40117,
        40142, 40208, 40219, 40240, 40274, 40293, 40297

Extraction of pigments   see   Pigments ...

Exudation of photosynthates   see   Photosynthate translocation ...

F

Fatty acids   see   Lipids, fatty acids ...

Ferredoxin, ferredoxin-NADP reductase, methods
        21589, 22372, 23055, 23577, 24213, 24334, 24648, 24707, 25105; 26066, 26654,
        27310, 27628; 28543, 29271, 29718, 29872, 30042, 30304, 31100, 31783, 32069,
        32333; 32905-6, 32922, 32945, 33399, 33650, 34067, 34089, 34147, 34745,
        35516, 35816, 36063, 36102, 36374; 36947, 37131, 37272, 37419, B37458,
        37885, B38037, 38630, 39816, 40256

Ferredoxin, flavoproteins, rubredoxin
        21530, 21547, 21548, 21583-4, 21589, 21591, 21611, 21620, 21638, 21668,
        21737-8, 21743, 21865, 21935, 21937, B21979, 21980-1, 21983, 22051, 22126,
        22346-7, 22373-4, 22588, 22658-60, 22669, 22693, 22736, 22773, 22853, 22979,
        23057, 23124, 23150, 23272, 23517, 23525, 23563, 23577, 23600-1, 23708,
        23719-20, B23767, 23798-9, 23847, 24041, 24074, 24223, 24304, 24327, 24362,
        24380, 24418, 24615, 24648, 24651, 24663-4, 24706, 24720, 24723, 24784,
        24875, 24917, 24944-5, 25006, B25064, 25091, 25104-5, 25124, 25145
        25211, 25347, 25470, 25505, 25511, 25565, 25583, 25585, 25626-9, 25635,
        25939-40, 25942, 25963, 25968, 26173, B26220, 26254, 26266, 26305-6, 26310,
        26442, 26636-7, 26653-4, 26671, 26677, 26695, 26901, 26942, 27019-20, 27048,
        27076, 27088, 27230, 27310, 27516, 27519, 27668, 27772, 27870, 27963, 27979,
        28080, 28103, 28105, 28218-9, 28306-7, 28409
        28539, 28543, 28592-4, 28599, 28667-8, 28670, 28720, 28750, 28765, 28778,
        28816, 28832, 28851-2, 28859, 28868, 28888, 29021, 29037-8, 29070, 29143,
        29180, 29245-6, 29271, 29310, 29412, 29435,   29438-9,   29442, 29525,
        29531, 29570, 29686-9, 29796, 29798-9, 29851-4, 29880, 29998, 30034, 30042-3,
        30059, 30118, 30254, 30284, 30304, 30307-9, 30336, 30419, 30443, 30491,
        30493, 30509, 30716, 30746, 30774, 30777, 30853, 30963, 31025, 31100, 31151-3,
        31196, 31253, 31274, 31337, 31418, 31437, 31462, 31473, 31698-9, 31783, 31786,
        31844, 31874, 31918, 31976, 32039, 32069, 32072, 32333, 32454-5, 32464, 32495
        32571, 32602, 32608, 32628, 32656, 32673-4, 32732, 32853, 32905-6, 32945,
        32968, 33084-5, 33253, 33294, 33355-6, 33399, 33402, 33461-2, 33477, 33483,
        33487, 33550, 33612, 33650, 33710-3, 33901-5, 33910, 33930-1, 34066-8,
        34080-1, 34147, 34172, 34205, 34277, 34321, 34328, 34356, 34358, 34574, 34585,
        34726, 34780, 34974, 34982, 35009, 35066, 35075, 35207, 35333, 35456, 35516,
        35539, 35726, 35729, 36063, 36074, 36099, 36129, 36324, 36373-5, 36389,
        36463-4, 36514, 36517
        36606, 36612, 36665, 36707, 36710, 36823, 36848, 36919, 36948, 37033-5,
        37050, 37079-80, 37200, 37272-3, B37458, 37565, 37583, 37643-4, 37823, 37841,
        37885, 37942, 38029-30, 38041, B38139, 38142, 38199,
        38240, 38263, 38298, 38377, 38381, B38415, 38483, 38527, B38528, 38593,
        38733, 38811, 38911, 38921, 39193, 39213, 39252, 39326, 39331, 39384-5,
        39451,   39468, 39588-9, 39769-70, 39816, 39846-7, 39863, 39885, 40247,
        40256-7, 40301

Ferredoxin-NADP reductase, pteridines
        21638, 21743, 22027, 23184, 23261, 23272, B23767, 23798, 24213; 26066,
        B26220, 26310, 26677, 27020, 27603, 27628-9, 27963, 28105, 28218; 28778,
        28816, 28851-2, 28859, 29037, 29043, 29143, 29246, 29397, 29531, 29718,
        30303, 30491, 30535, 30777, 30983, 31003, 31152-3, 31874, 31964, 32072;
        32905-6, 32921-2, 32945-6, 33289, 33477, 33850, 33901, 33910, 34168, 34205,
        34321, 34745, 35036, 35420, 35816; 36611, 36929, 36947-8, 37034, 37272,
        37419, 37425, B37458, 38030, 38054, 38269, 38298, 38381, 38630, 38733,
        38868, 39250, 39588, 39817, 40256, 40301, 40334

Flashes of light   see   Irradiation, flash ...

Flavoproteins   see   Ferredoxin ...

Flooding and carbon fixation pathways   34278, 35049; 40083

Flooding and carotenoids   23682

Flooding and chlorophyll
        23682; 25992, 26749; 29136, 29338, 30857; 34271, 35049, 35437, 35906; 37349,
        38035

Flooding and ecosystem and plant productivity
        24609; 25435, 25566, 25784, 26499, 26729, 27240, 27248, 27625, B28482;
        28873, 29136, 30351-2, 30857, 31234, 31270, 32010, 32313, 32504; 34498,
        34541, 34895, 36388; 37083, 37349, 37838, 38035, 38246, 40280

Flooding and electron transport chain   B24024; 31234

Flooding and gas exchange 22830, 24183; 25780, 25992, 26749, 27248; 30551-2; 35049, 35636; 40083

Flooding and resistances to CO_2 and water vapour transfer 31270; 34651; 36974, 38310, 39005, 40169

Flooding and respiration 31234; 35636

Fluorescence, methods 22021, 22664, 23146, 23153, 23538, 23616, 25114; 29544, 30719, 31691, 31906; 32942, 33497, 35635, 36176, 36470

Fluorine see Pollution of air ...

Fraction I protein see Ribulose 1,5-bisphosphate carboxylase

Frost (hardiness) see Temperature, low ...

Fungus diseases see Phytopathological effects ...

Fusicoccin see Growth regulators ...

G

Gas exchange, general aspects see General aspects on CO_2 exchange ...

Gas exchange in algae
        21711, 21738, 21770, 21772, 21808, 21903, 21966, 21976, B22052, 22092,
        22104, 22109, 22137, 22211, 22265, 22287, 22303-4, 22363-4, 22387, 22407,
        22433, 22488, 22490, 22494, 22522, 22623, 22673, 22717, 22764, 22790-1,
        22828, 22842, 22846-8, 22861, 22944, 22961, 23018, 23043, 23099, 23133,
        23165, 23225, 23439, 23446, 23475, 23537, 23552-3, 23571-2, 23606, 23659,
        23685, 23718, 23729, 23884, 24013, 24021, 24029, 24043, 24053, 24072, 24134,
        24172, 24346-8, 24401-2, 24405, 24434, 24483, 24525, 24565, 24570, 24608,
        24612, 24637, 24667, 24679, 24708, 24724, 24735, 24800, 24862, 24865, 24867,
        24956, 24964-5, 24979, 24984, 24998, 25001, 25127
        25163, 25215, 25272, 25280, 25317, 25335, 25381-2, 25395, 25426, 25459,
        25501-3, 25536, 25543, 25548-9, 25559, 25598, 25601, 25635, 25651, 25706,
        25752, 25806, 25824-5, 25840, 25866-7, 25869, 25894, 25952, 25996, 26002,
        26023, 26035, 26089, 26108, 26148, 26167, 26170, 26211, 26214, 26313, 26367,
        26386, 26420, 26472, 26478, 26529, 26583-5, 26608, 26642-4, 26655, 26661,
        26698, 26781, 26794, 26807, 26822, 26829, 26896, 27156, 27177-8, 27185,
        27227, 27247, 27257-8, 27279, 27319-21, 27349, 27367-8, 27377, 27406, 27427,
        27429,·27438, 27463, 27490, 27511, 27513, 27534, 27548, 27567, 27613, 27668,
        27681, 27700, 27703, 27775, 27818, 27823, 27825, 27827, 27898, 27988, 27993-
        -4, 28008, 28076-7, 28088, 28124, 28128, 28140, 28152, 28170, 28174, 28223,
        28278, 28325, 28331, 28349, 28372, 28395, 28421, 28446
        28507, 28541, 28573, 28579, 28687, 28728, 28737, 28742, 28763-4, 28795, 28815,
        28846, 28879-80, 28889, 28900, 28921, 28939-40, 28979, 28981, 29118, 29140,
        29160, 29167, 29177, 29191, 29196, 29239, 29242, 29252-3, 29303, 29337, 29341,
        29378, 29416, 29429, 29451, 29500, 29503-4, 29527, 29582, 29601, 29629, 29649,
        29653, 29661, 29677, 29734, 29775, 29809, 29821, 29842, 29873, 29893, 29910,
        29928, 29945, 30023, 30026, 30124, 30137, 30147, 30151, 30172, 30185, 30193,
        30220, 30223, 30323, 30381, 30401, 30434-5, 30495, 30525, 30536, 30575, 30587,
        30622, 30624, 30630, 30757, 30768, 30845, 30848, 30944, 30962, 30968, 31051,
        31062, 31084, 31136, 31142, 31145, 31206, 31218, 31278, 31282, 31326-8, 31362-
                                                                (continued)

Gas exchange in algae (continued)
    -4, 31396, 31409, 31429-30, 31438, 31456, 31460, 31507, 31514, 31521, 31597,
    31651, 31676, 31679, 31681, 31700, 31704, 31723, 31757, 31772, 31835, 31904,
    31960, 31964, 31969, 32031-2, 32071, 32136, 32166, 32180, 32206-7, 32242,
    32344-5, 32356, 32441, 32448, 32470, 32517
    32622, 32633, 32687, 32714, 32716, 32787, 32789, 32798, 32813-4, 32818,
    32824, 32899, 32938, 32943, 33001, 33003, 33047, 33118, 33146, 33217-8,
    33243, 33258, 33292, 33360, 33384, 33390, 33466, 33482, 33496, 33509, 33544,
    33551, 33588, 33622, 33652, 33687, 33784, 33811, 33813, 33816, 33847, 33867,
    33901, 33912, 33915, 33922, 34070, 34143, 34157-8, 34188, 34222, 34328,
    34385, 34404-5, 34417, 34427, 34447, 34468, 34473, 34484, 34520, 34528,
    34555, 34558, 34562, 34568, 34570, 34589, 34643-4, 34741, 34801, 34819,
    34832-3, 34840, 34853, 34870, 34893, 34905, 34912, 34967, 34971, 34978,
    35015, 35024, 35069, 35072, 35126, 35138, 35153, 35175, 35179, 35182, 35191,
    35307, 35319, 35336, 35343, 35380-1, 35390, 35407, 35439, 35494, 35626,
    35688, 35715, 35755, 35761-3, 35780, 35793, 35802, 35804, 35878, 35935,
    35940, 35942, 35946, 35957, 36002, 36043, 36061, 36085, 36150, 36152-3,
    36155, 36179, 36214, 36320, 36411, 36446, 36455, 36506, 36510, 36540, 36550,
    36570, 36584, 36587
    36715, 36758, 36762, 36808, 36858, B36859, 36873, 36889, 36921, 36937, 36982,
    36992, 37074, 37076-7, 37097, 37142, 37162, 37188, 37196, 37210, 37246,
    37310, 37314-6, 37320, 37330, 37351, 37429, 37481, 37495, 37540, 37557,
    37566, 37606, 37639, 37659, 37715, 37737, 37757, 37785, 37787, 37818, 37861,
    37893, 37902, 37924, 37927, 37937, 37943, 38079, 38099, 38112, 38145, 38159-
    -60, 38202, 38251, 38274, 38296, 38326-7, 38337, 38374, 38394, 38417, 38426-
    -7, 38451, 38456-7, 38460, B38477, 38515, 38521, 38524, 38586-7, 38633,
    38659, 38686, 38693, 38759, 38794, 38956, 38974, 39025, 39102, 39104, 39108,
    39113, 39171, 39177, 39181, 39185, 39196, 39208, 39233, 39291, 39314, 39345,
    39347, 39361, 39370, 39425, 39455, 39533-4, 39538, 39648, 39697, 39728,
    B39759, 39787, 39790-2, 39823, 39826, 39859, 39931, 39960, 39970, 39992,
    40023, 40101, 40153, 40210, 40298, 40330, 40332

Gas exchange in isolated chloroplasts see Chloroplast, isolated, gas exchange by

Gas exchange in photosynthetic bacteria see Photosynthetic bacteria, gas exchange

Gas exchange, model see Model ...

Gas exchange of organs other than leaf
    24759; 25967, 26046, 26109, 27344, 28086; 29509, 29993, 30537, 31592, 32027,
    32057; 32965, 33132, 33566, 34153, 34523, 35030, 35074, 35121, 35161, 35349,
    35527, 36135; 36701, 37139, 37989, 38058, 38242, 39137, 39717

Gases, organic, and algae productivity 35929

Gases, organic, and carbon fixation pathways 26733

Gases, organic, and carotenoids 22323; 33010

Gases, organic, and chlorophyll
    21683, 22323; 25233; 29219, 29370, 29433, 31781, 31852; 33012, 35810, 35814;
    39587

Gases, organic, and chloroplast (chromatophore) 26733, 35813

Gases, organic, and ecosystem and plant productivity 24470, 33846; 36640, 37190

Gases, organic, and electron transport chain 22346; 31852

Gases, organic, and gas exchange 22434, 22830-1, 23151; 29219, 29239, 30502; 36064,
    36384; 37688, 37836

Gases, organic, and respiration 37836

                                                          (continued)

                                                              (continued)

Genetics of gas exchange (continued)
      32576, 32652, 32718, 32751, 32810, 32912, 33002, 33254, 33539, 33668, 33898,
      33921, 34137, 34310, 34702, 34782, 34792, 34881, 35034, 35634, 35775, 35807,
      35892, 36064, 36067, 36222, 36418-9, 36521, 36523
      36753, 36760, 37116, 37128, 37612, 37688, 37804, 38180, 38286-7, 38438,
      38603, 38768, 38783, 38798, -38868, 39147, 39421, 39572, 39980, 40103, 40313

Genetics of photorespiration
      23629, 25137; 26188, 26987-8, 28477-9; 31008, 32529; 32718, 32822;
      38287, 38438, 40345

Genetics of resistances to CO_2 and water vapour transfer
      25261; 28606, 30547, 31512; 32810, 33106, 33571, 34792, 36418, 36523; 37804,
      38075, 38847, 39147, 40313

Genetics of respiration
      22624, 25047; 25406, 26988, 27638; 29008, 31587; 34137, 35807; 38198, 38287,
      38438, 39421, 40045

Glycollate metabolism see Photorespiration ...

Glyoxysome see Peroxisome ...

Granum see Thylakoid ...

Gravimetric determination of photosynthesis see Dry-matter production ...

Gross photosynthetic rate
      21526, 21659, 21782, 22038, 22252, 22494, 22563, 22678, 22896-7, 23013,
      23091, 23145, 23278, 23298, B23407, 23464, 23553, 23603, 23632, 23894,
      23994, 24029, 24082, 24110, 24122, 24193, 24323-4, 24438, 24692, 24880,
      24915
      25195, 25307, 25592, 25654, 25666, 25801, 25810, 26037, 26046, 26420, 26487,
      26642, 26884, 27460, 28086-7, 28180, 28325, 28408, 28413
      28613, 28617, 28687, 28972, 29054, 29311, 29336, 29340, 29564, 29613, 29820,
      29842, 29848, 29887, 29967, 30011, 30081, 30381, 30476, 30500, 30622, 30643,
      30891, 31430, 31592, 31633, 31759, 31941, 32046, 32104, 32118, 32165, 32415
      32882, 32923, 32967, 33052, 33118, 33123, 33215, 33226, 33243, 33260, 33262,
      33379, 33428, 33496, 33561, 33566, 33741, 33915, 34037, 34101, 34485, 34535-
      -6, 34625, 34799, 34840, 35001, 35074, 35092, 35154, 35339, 35449, 35543,
      35611, 35664, 35673, 35761, 35791, 36067-8, 36133, 36198
      36602, 36672, 36748, 36775, 36804, 36817, 36940, 36983, 36985, 37122, 37348,
      37515-6, 37612, 37654, 37699, 37772, 37953, 37987, 37989, 38181, 38209,
      38215, 38676, 38710, 38828, 38897, 39057, 39211, 39325, 39395, 39567, 39628,
      39639, 39905, 39926, 39935, B39951, 40080, 40138, 40224, 40275

Growth analysis, methods
      21795, 21802, 21882, 22404, 22681, 22716, 23155, 23234, 23784, 24143, 24425,
      24509, 24564, 24602, 24672, 24726, 24971-2
      25469, 25475, 25531, 25569, 25575, 25707, 25910, 25921, 25972, 26534, 26789,
      27437, 27892, 28358, 28475
      28534, 29206, 29213, 29811, 30053, 30071, B30432, 30689, 30847, 30876, 31070,
      31224, 31500, 31947, 32191, 32263
      32891, 33232, 33432, B34075, 34090, 34784-5, 34913, 35174, 35902-3, 36391,
      36393
      36594, 36678, 37002, 37220, 37396, 37406, 37559, B37906, 37952, 38509, 39158,
      39561, 39899, 39985, 40060

Growth analysis, net assimilation rate, leaf area ratio, relative growth rate
      21523-5, 21621, 21803, 21853, 21919, 21951, 22074, 22076, 22112, 22123,
      22260, 22273, 22314, 22369, 22380, 22396, 22402, 22414, 22419, 22430-1,
      22449, 22479, 22516, B22572, 22593, 22603, 22676, 22787, 22806, 22829, 22855,
      22885, 22916, 22974, 22984, 22998, 23002, 23037-8, 23063-4, 23106, 23156-7,
                                                                          (continued)

Growth analysis, specific leaf area ... (continued)
        24267, B24271, 24291, 24295, 24321, 24324, 24430, 24436-8, 24470, 24474,
        24509, 24511, 24563, 24569, 24574, 24605, 24609, 24625, 24633, 24646-7,
        24661, 24668, 24670, 24689, 24705, 24752, 24758, 24773, 24780-1, 24794-5,
        24798, 24822, 24841, 24880, 24940, 25015, 25039, 25041, 25049, 25052, 25086,
        25103
        25167, 25169, 25189, 25203, 25264, 25303, 25336, 25412, 25454, 25482-3,
        25526, 25528, 25540, 25546, 25566, 25572, 25607, 25618, 25642, 25659-61,
        25673, 25681, 25686, 25772, 25774, B25775, 25792, 25909, 25943, 25948,
        25965, 25967, 25989, 26027, 26030, 26055, 26068, 26079, 26085-7, 26096,
        26133-4, 26164, 26197, 26202, 26204, 26213, 26263, 26308, 26337, 26372-3,
        26390, 26409, 26461, 26466, 26474, 26501, 26506, 26515, 26652, 26723,
        26747, 26782, 26787, 26798, 26803, 26828, 26878, 26903-4, 26906, 26912,
        26928, 26944-5, 26954, 26964, 27011, 27052, 27094, 27133, 27139, 27175,
        27188, 27195, 27210, 27216, 27228, 27234-5, 27242, 27269, 27294-5, 27325,
        27333, 27336, 27382, 27385-6, 27389, 27410, 27437, 27457, 27466, 27502,
        27507, 27563, 27587, 27625, 27630, 27664, 27686, 27701, 27710, 27720, 27734,
        27779, 27810, 27819, 27882, 27887, 27896, B27915, 27920, 27980, 28022,
        28068-70, 28087, 28089, 28092, 28097, 28123, 28195, 28197, 28231, 28247,
        28265, 28294, 28323, 28392, 28413, 28461
        28502, 28505, 28533, 28552, 28566, 28622, 28685, 28757, 28785, 28788, 28809,
        28864, 28899, 28941, 28989, 29003, 29007, 29009, 29022, 29060, 29064, 29068,
        29086, 29103, 29120, 29162, 29164, 29190, 29221, 29233, 29269, 29294, 29335,
        29362, 29372, 29427, 29443, 29475, 29492, 29496, 29550, 29559, 29565, 29574-6,
        29652, 29656, B29707, 29794, 29881, 29886-7, 29923, 29967, 29975, 29984-5,
        30011, 30031, 30057-8, 30070-1, 30131, 30163, 30167, 30195, 30203, 30221,
        30290, 30353, 30376, 30438, 30466, 30483, 30523, 30554, 30565, 30581, 30628,
        30633, 30663, 30674, 30677, 30699-700, 30711, 30733, 30802, 30945-6, 31019,
        31114, 31156, 31178, 31180, 31197, 31199, 31201, 31224, 31230, 31233, 31242,
        31250-2, 31265, 31348, 31356-8, 31392, 31463, 31494-5, 31500, 31509, 31517,
        31527, 31539, 31586, 31595, 31633, 31670, 31706, 31760-1, 31804, 31822, 31825,
        31840, 31875, 31895, 31925, 32006, 32012, 32046, 32096, 32111, 32115, 32120,
        32165, 32182, 32191, 32257, 32322, 32325, 32391, 32436, 32504, 32538, 32542
        32556, 32629, 32640, 32662, 32737-8, 32758, 32761, 32868-9, 32939, 32965,
        32973, 32986, 33019, 33022, 33096-7, 33123, 33184, 33195, 33222, 33242,
        33255, 33293, 33306, 33330-1, 33358-9, 33368, 33377, 33379, 33395, 33398,
        33433, 33435, 33449, 33495, 33508, B33510, 33542, 33547, 33565, 33572, 33669,
        33725, 33728, 33740, 33808, 33868, 33894, 33928, 33934, 33951, 33961-2,
        34000, 34037, 34048, 34076, 34091, 34135-6, 34151-2, 34181, 34206, 34228,
        34243, 34262, 34269, 34294, 34308-9, 34326, 34330, 34373, 34442, 34486,
        34500, 34622, 34676-7, 34690, 34720, 34732, 34782, 34794, 34808, 34829,
        34834, 34937, 34940, 34946, 34972, 35003, 35025, 35039, 35047, 35092, 35101,
        35103, 35154, 35158-9, 35169-71, 35202-3, 35262-3, 35273, 35303, 35340,
        35349, 35448, 35484, 35505, 35512, 35514, 35517, 35529, 35543, 35550, 35595,
        35607-8, 35632, 35660, 35662, 35691, 35775, 35790-1, 35808, 35870-1, 35882,
        35884, 35892-3, 35902, 35904-5, 35905, 35914, 35953-4, 35977, 36012, 36014,
        36027, 36068, 36075, 36076-7, 36120, 36133, 36136, 36225, 36242, 36252,
        36304, 36315, 36361, 36419, 36421, 36428, 36436, 36489, 36500-1, 36515,
        36521, 36528, 36548
        36593, 36599, 36602, 36617, 36691, 36700, 36703, 36726, 36736, 36776, 36845,
        36850, 36857, 36869, 36882, 36910, 36959, 36994, 37011, 37054, 37114, 37122-
        -3, 37125, 37171, 37173, 37176, 37198, 37268, 37298, 37302, 37334, 37412,
        37544, 37590, 37605, 37623, 37636, 37649, 37696, 37709, 37721, 37772, 37829,
        37847, 37890, 37895, B37906, 37953, 37983, 37987, 37992, 38043, 38071, 38123,
        38134, 38161, 38209, 38210, B38211, 38215, 38219, 38259, 38294, 38346, 38357,
        38436, 38438, 38455, 38500, 38520, 38535, 38563-4, 38584, 38616, 38629,
        38636, 38727, 38800, 38809, 38962, 38990, 38995, 39072-4, 39076,
        39080, 39089, 39112, 39134, 39138, 39158, 39280, 39442, 39510, 39541, 39565,
        39567, 39637, 39652, 39659-60, 39664, B39711, 39737, 39740, 39799, 39813,
        39821, 39843, 39852, 39875, 39904, 39946, 39976, 39998, 40043, 40070,
        40075, 40079, 40085, 40115, 40194, 40199, 40235, 40239, 40248, 40254, 40273-
        -5, 40304

Growth regulators and algae productivity 28537; 34474

Growth regulators and carbon fixation pathways
    21933, 22786, 23219, 23974, 24016, 25063; 25338, 25377, 25516, 26446, 27101,
    27408, 27651; 28662, 28981, 29092, 29567, 29800, 29850, 30817, 31096, 31977,
    32179; 32930, 33344, 33518, 33835, 33978, 34320, 34576, 34710, 34922; 37933,
    38126, B38238, 38517, 39079, 39546, 39689, 40238, 40351

Growth regulators and carotenoids
    21509, 21520, 21959, 22401, 23080, 23425, 23526, 23675, 23807, 23974, 24391,
    25142; 25432, 25877, 25905-6, 26291, 26634, 27101, 27165, 27372-3, 27443-4,
    27527; 28752, 28995, 29370, 30295-6, 30595, 31014, 31144, 31574; 32741,
    33463, 33518, 33629, 34869, 36331; 37062, 37428, 38517, 38819, 39429, 39585

Growth regulators and chlorophyll
    21509, 21520, 21536, 21701, 21781, 21948, 21959, 21961, 22067-8, 22290,
    22401, 22468, 22472, 22571, 22735, 22785, 23019, 23025, 23080, 23356, 23425,
    23526, 23535-6, 23638, 23674, 23883, 23974, 24037, 24084, 24127, 24168,
    24390-1, 24414, 24464, 24510, 24745-6, 24753, 25075, 25079, 25142
    25236, 25377, 25432-3, 25498, 25510, 25652, 25686, 25905-6, 26291, 26494,
    26581, 26586-7, 26602, 26634, 26701, 26814, 26918, 27101, 27205, 27311,
    27372-3, 27380, 27443-4, 27446, 27497, 27527, 27560, 27727, 27854, 28243
    28662, 28752, 28881, 28995, 29370, 29433, 29450, 29523, 29982, 30015, 30056,
    30173, 30240, 30296, 30431, 30992, 31014, 31096, 31144, 31601, 31855, 32001,
    32106, 32202, 32313, 32351, 32451
    32578, 32847, 32863, 32929-31, 33031, 33302, 33405, 33463, 33489, 33775,
    34209, 34225, 34267, 34320, 34322, 34869, 34922, 35021, 35023, 35065, 35133,
    35248, 35310, 35464, 35614, 35695, 35806, 36039, 36196, 36215, 36296, 36493,
    36520
    36610, 36614, 36759, 36896, 36899, 36920, 37061-2, 37349, 37428, 37523,
    B37532, 38121, 38163, 38803, 38853-4, 38936, 39062, 39206, 39254, 39429,
    39850, 39890-2, 40361

Growth regulators and chloroplast (chromatophore)
    21536, 22401, 22703, 22785, 24368, 24414, 24694, 24865, 24931, 25063,
    25075; 25187, 25333, 25377, 25551, 25995, 27160, 27373, 27443, 27497,
    27667, 27717, 27727, 28243, 28364; 28974, 28981, 29305, 29360, 29761,
    30295, 30300, 30467, 30595, 30646, 30992, 31424-5, 31541, 31768, 32240;
    32717, 32929-31, 33405, 33463, 33518, 33692, 34249, 34320, 34579, 34606,
    34922, 35065, 35384, 35602, 36296, 36370, 36488; 36690, 37062, 37227, 38273,
    38350, 38517, 38720, 38854, 39062, 39486, 39921, 39977, 40108, 40351

Growth regulators and ecosystem and plant productivity
    21560, 21630, 21781, 21847, 21928, 22096, 22663, 23431, 23863, 24127, 24246,
    24628, 24798, 25059, 25140; 25209, 25256, 25510, 25686, 26782, 27444, 27538,
    27821, 27990, 28207, 28246, 28339; 28618, 28662, 28873, 29046, 29848, 30427,
    30920, 31014-5, 31234, 31399, 31841, 31940-1, 32201, 32218, 32313, 32516;
    32567, 32847, 33079, 33270, 33318, 33449, 33831, 33935, 33970, 34266, 34275,
    34936, 35021, 35117, 35785, 35980; 36851, 37027, 37041, 37071, 37159, 37297,
    37629, 37807, 37848, 38385, 38498, 38609, 38834, 39031, 39070, 39141, 39249,
    39398, 39546, 39889, 39921, 39993, 40143-4, 40280, 40361

Growth regulators and electron transport chain
    21530, 21675, 21959, 23173, 23219, 23242, 23866, 24351, 25090-1; 25338,
    26291, 26766, 26902, 27408, 27442, 27532, 27553, 27562, 28246; 28730, 28887,
    28995, 29107, 29305, 29360, 29382, 30103, 30297, 30577, 30594, 31087, 31379,
    32179, 32309, 32483-4; 32862, 32930, 34093, 34693, 35009, 35134, 35843,
    36576; 37061-2, 38022, 38350, 38541, 38936, 40024, 40325

Growth regulators and gas exchange
    21536, 21786, 21869, 22189, 22782, 22785-6, 23219, 23242, 23244, 23430-1,
    23503, 23638, 23774, 23863, 23974, 24084-5, 24234, 24325, 24391, 24628,
    24865, 24868, 24919, 25063
    25256, 25338, 25516, 25642, 26237, 26536, 26824, 27101, 27311, 27380, 27408,
    27560, 27761

(continued)

Growth   regulators and gas exchange (continued)
          28981, 29046, 29092-3, 29296, 29350, 29363, 29486, 29800, 29849, 30103,
          30427, 30761, 30817, 30920, 30992, 31014, 31022, 31038, 31087, 31940-1,
          32049, 32201, 32483-4
          32929-31, 33079, 33096, 33288, 33388, 33489, 33835, 33969, 33978, 34125,
          34267, 34497, 34632, 34710, 34923, 34936, 35057, 35074, 35117, 35184, 35248,
          35504, 35759, 35895, 35939, 35980, 36196
          37041, 37184, 37217, B37532, 37807, 37933, 38163, B38238, 38252, 38465,
          38532, 38609, 38655, 39079, 39672, 39993, 40023, 40144, 40265

Growth   regulators and photorespiration
          24084-5; 27101; 31088, 31685, 31768, 31941; 35248; 37807, 38517

Growth regulators and resistances to CO_2 and water vapour exchange
          21713, 22964, 23218-9; 26237, 26775, 27356, 27574-5, 28321, B28340; 28509,
          28821, 29486, 30507-8, 30818, 31087, 32352; 33286, 33985, 34267; 36984,
          37184, 38465, 38532, 39212, 39662, 39890, 39892, 40265

Growth   regulators and respiration
          22785, 23080, 23863, 24084-5; 26536, 27101; 30296, 31768; 33978, 35117;
          38465

"Growth" respiration see Respiration, "growth"

H

H_2 evolution, photoreduction
          22305-6, 23202, 23667; 25370, 25915-6, 26171, 26544, 26640, 26659,
          26677, 26805, 27151-3, 27252, 27357, 27440, 27571, 28444
          28506, 28750, 28783, 28814-5, 28889, 29135, 29211, 29444, 29570, 29683-4,
          29734-6, 29796, 29940-1, 29974, 30265, 30308-9, 30399, 30416-7, 30422-3,
          30470, 30502, 30617, 30762, 30938, 31176-7, 31274, 31415, 31437, 31473, 31724,
          31730, 32082, 32479
          32789, 32860, 32897, 32937, 33027, 33325, 33442, 33459, 33505, 33563, 33710-
          -1, 33734, 33833, 34041, 34130, 34161-2, 34217, 34303, 34328, 34343, 34405,
          34417, 34466, 34476, 34801, 34858, 34863, 34902-4, 35075, 35191, 35207,
          35304-7, 35315, 35456, 35649, 35856, 36002, 36025, 36096, 36189, 36359, 36492,
          36513, 36536, 36541, 36578
          36605-7, B36769, 36843, 36848, 36858, 36878, 36928, 36930, 37177, 37402,
          37444, 37630, 37718, 37778, 37788, 37873, 37923, 38151, 38173, 38179,
          38318-9, 38321, 38352, 38393, 38424, 38506, 38526, 38565, 38649, 38709,
          38755, 38761, 38763, 38881, 38932, 39102-4, 39176, 39252, 39459, 39538,
          39542, 39719, 39885, 40298, 40368

H_2 isotopes see Deuterium ...

H^+ transport in chloroplast see Chemiosmotic hypothesis

Halobacterium photosynthesis
          21537, 21635, 21714-6, 22022, 22045, 22060, 22206, 22218, 23432, 23458,
          23495-6, 24195, 24211, 25110
          25191, 25296, 25298, 25337, 25351-2, 25369, 25393, 25461, 25547, 25561,
          25803, 25839, 25927, 26180, 26283, 26347, 26365, 26379, 26439, 26611,
          26636-7, 26728, 26819-21, 26850-1, 26963, 27305-7, 27441, 27545, 27599,
          27626, 27828-9, 27836, 27890, 27904-5, 28042-3, 28150, 28293, 28298-9,
          28378
          28722, 28826, 28831, 28848, 28943, 29028-9, 29034, 29084, 29095, 29099-100,
          29225, 29227-8, 29387-9, 29420, 29431, 29507, 29606-8, 29632, 29659-60, 29692,
          29742, 29797, 29847, 29854, 29900, 29926, 29929, 29935, 30055, 30063-5, 30192,
                                                                          (continued)

Halobacterium photosynthesis (continued)
        30239, 30258, 30284, 30383-4, 30390-1, 30408, 30471, 30519-21, 30614-5, 30632,
        30649, 30679, 30785, 31026, 31029, 31036, 31110-1, 31150, 31167, 31173-5,
        31286, 31309-10, 31458, 31491, 31497, 31773, 31838, 31860, 31997, 32167,
        32330, 32438, 32507
        32558-9, 32655, 32727, 32735, 32781, 32793, 32800, 32878-9, 32901, 32903,
        33046, 33063, 33075, 33099, 33228, 33267, 33275-7, 33313, 33326-8, 33376,
        33421, 33437, 33443-4, 33636-8, 33650, 33681-2, 33752, 33833, 33903, 33924,
        33947-9, 33966, 33972-3, 33993, 34015, 34028, 34077-8, 34083-5, 34112, 34221,
        34263, 34268, 34270, 34345, 34391-2, 34398, 34449, 34516, 34590-1, 34662,
        34667-8, 34748, 34900, 34977, 35120, 35135, 35185, 35189, 35199, 35208,
        35228-9, 35253, 35312, 35486, 35551, 35565, 35568, 35718, 35722-3, 35782,
        35805, 35817, 35829-31, 35913, 35926, 35951, 36006-8, 36042, 36131, 36160-1,
        36317-8, 36321-2, 36349, 36374-5, 36380
        36743, 36765, B36769, 36915, 36954, 36999, 37016, 37214, 37231, 37283-4,
        37345, 37353, 37403-5, 37424, B37493, 37508, 37538, 37583, 37608, 37625,
        37720, 37816, 37853-4, 37865, 37910-1, 38032, 38184, 38197, 38278, 38282-3,
        38330, 38368, 38403-5, 38432, 38491, 38494, 38515, B38528, 38595, 38712,
        38815, 38817, 38821, 38937-8, 38986-7, 39011, 39021-2, 39028, B39220, 39221,
        39228, 39302, 39320, 39340, 39362, 39396, 39470, 39495-6, 39516, 39575,
        39671, 39685, 39729, 39735, 39780, 39782-3, 39878-9, 39914-6, 39978, 40094,
        40110, 40148, 40150, 40171, 40370

Hatch-Slack cycle see C$_4$ pathway ...

Herbicides see Pesticides, herbicides ...

Heterogeneity of leaf blade (organ) and carbon fixation pathways
        21785, 21814, 22071; 22399, 22406, 22635, 22698, 22839, 22983, 23030-3,
        23231, 24164-5; 25768, 25987, 26057, 26165, 26441, 26445, 26598, 27008,
        27091, 27202, 27361, 27596, 28351; 28957, 29669, 29865, 30233, 31266, 31578,
        32462; 34473, 34618, 34749, 35427-8, 35433, 35467; 37215, 37242, 38588,
        39235, 39761

Heterogeneity of leaf blade (organ) and carotenoids
        22395, 23026, 23213, 23400, 23810, 24163; 30313; 35400, 35768; 38561

Heterogeneity of leaf blade (organ) and chlorophyll
        21790, 22027, 22328-30, 22394-5, 22698, 22799, 22815-6, 22930, 23213, 23586,
        23698, 23810, 24163, 24715, 24817; 25204, 25310, 25877, 25913, 26099, 26444,
        26666-7, 26880, 26994, 27526, 27581-2, 27622, 28205; 28529, 28655-6, 28939,
        29534, 29594, 29669, 29833, 29876, 29903, 30562, 31243, 31578, 31735; 33108,
        33523, 33630, 34043, 34045, 35400, 35427, 35658, B36191; 36616, 37101, 37119,
        37215, 37280, 37348, 37539, 37723, 38561-2, 39237, 39281

Heterogeneity of leaf blade (organ) and chloroplast
        21613, 22033, 22156, 22281, 22329, 22395, B22572, 22712, 22815-6, 23001,
        23092, 23094, 23689, 23743, 24157, 24159, 24546, 24615, 24772, 24817;
        25428, 25431, 25512, 25530, 26665, 27106, 27157, 27164, 27351, 27621, 28357;
        28655, 28909, 29876, 30016, 30313, 30557, 30562, 30690, 30747, 30826, 30871,
        31243, 31325, 31568, 32061; 32859, 33170, 33498, 33502, 33578, 34044, 34155,
        34705, 34956, 35604, 35658, B35862, B36191; 36696, 36927, 36998, 37101,
        37119, 37336, 37393, 38214, 38260-1, 38425, 38747, 40354

Heterogeneity of leaf blade (organ) and electron transport chain
        21790, 21804, 22816, 22834, 22973, 23351, 23851, 24159-60, 24163, B24176,
        24850; 25768, 25913, 26442-3, 26872, 27552, 27580-1, 27921, 28420; 28529,
        28655, 29534, 30558, 30866, 31418, 31551; 33170, 33869, 34043, 35059, 35427,
        35433; 37242, 38588, 39235, 39237, 40124

Heterogeneity of leaf blade (organ) and gas exchange
        21677, B21833, 22836, 22962, 23213, 23418, 23479, 24324, 24336, 24715, 24821,
        24850; 25821, 25987, 26078, 26112, 26316, 27007, 27389, 28205; 28656, 28939-40,
                                                                                (continued)

Heterogeneity of leaf blade and gas exchange (continued)
        29421, 29594, 29775, 30232, 30562, 30748, 31266, 31449, 31451-2, 31734, 32226;
        32826, 33630, 33869, 34473, 34835, 35059, 35504; 37348, 37857, 38588, 40210

Heterogeneity of leaf blade (organ) and photorespiration
        22071, 23231, 23877; 25711; 31266, 31451; 34256

Heterogeneity of leaf blade (organ) and photosynthates
        24275; 26489

Heterogeneity of leaf blade (organ) and resistances to CO_2 and water vapour transfer
        26993; 30991, 32226; 34334

Heterogeneity of leaf blade (organ) and respiration
        27389, 29421

Heterotrophy see Carbon metabolism types ...

Hill reaction see Photosystem 2 activity ...

Hill reaction, methods see Photosystem 2 activity, methods

Humidity of air and carbon fixation pathways
        33130; 36997, 39605, 40213, 40238

Humidity of air and chlorophyll
        21536, 22911; 25898, 27040; 29391, 31991, 32232; 33014, 36039, 36235; 36920,
        38865

Humidity of air and chloroplast 21536, 21873; 36487

Humidity of air and ecosystem and plant productivity
        22351, 22485, 22584, 22999, 23250, 23678, 23733, 24645; B25775, 25785, 25929,
        26920, 27133, 27235, 27656, 28134, 28427; 30820, 30917, 31496, 32033; 33020,
        33060, 33073, 33330, 35007, 36388; 37368, 37696, 38182, 39130, 39549, 39911,
        40240

Humidity of air and electron transport chain   36235

Humidity of air and gas exchange
        21534, 21773, 22302, 22314, 22344, 22651-2, 22950, 22999, 23015, 23069,
        23105, 23257, 23298-9, 23330, 23608, 23814, 24183, 24376, 24593, 24826,
        24835, 24988; 25167, 25415, 26227, 26397, 26580, 26646-7, 26905, 27768,
        27888, 28133-4, 28143; 29020, 29102, 29494, 29409, 29572, 29793, 30898,
        31585; 32810, 32824, 33060, 33441, 33984, 34260, 34646, 34966, 35395, 35624,
        35886, 36069, 36358; B36781, 37287, 37799, 37803, 38400-1, 38436, 38529,
        38869, 38996, 39501, 39552, 39911, B39951

Humidity of air and resistances to CO_2 and water vapour exchange
        21549, 22116, 22403, 22584, 22650-2, 22920, 22938, 23299, 24377, 24826
        25251, 25604, 25742, 25999, 26015, 26265, 26397, 26532, 26612, 26646, 26865,
        26947, 27133, 27432, 28089, 28125, 28143
        28732, 29020, 29130, 29319, 29394-5, 29467, 29496, 29793, 29881, 30069, 30371,
        30387, 30392, B30477, B30511, 30643, 30738, 31268, 31762, 31769-71, 31867,
        31888, 31956, 32051, 32110, 32369
        33000, 33020, 33050, 33451, 33547, 33984-5, 34636, 34646, 34829, 35537,
        35594, 36069, 36075, 36358
        36732, 36907, 37368, 37413-4, 38075, 38400-1, 38436, 38529, 38534, 38915,
        38996, 39036, 39424, 39501, 39552, 39737, 39911, 39991, 40213

Humidity of air and respiration
        23764, 24183; B30477

Humidity of air, methods (*cf*. also Infra-red analyser for water vapour)
  21801, 22189, 22505, 22686, 23299, 23330, 23457, 24677; 26264, 26577, 26592,
  26635, 26708, 27826, 28245; B30325, B30432, 30860, 31968, 32015; 32877;
  36900, 37567, 38713

Hydration level of leaf and biliproteins 30481

Hydration level of leaf and carbon fixation pathways
  23296, 23481, 24676, 24808; 25520, 25772, 26446, 26689, B26846, 26864,
  27360, 27363, 27793; 28717-8, 29204, 29817, 30174, 30272, 30332, 30543-4,
  30665, 30823, 31165, 31798, 32378; 33654, 34045, 34368, 34535, 34949,
  35326; 36916, 37208, 37400, 37949, 38399, 39275, 39963

Hydration level of leaf and carotenoids
  23050, 23263, 23375, 23753; 25716, 25905-6, 27855; 31851; 34480, 34483,
  35962; 39736

Hydration level of leaf and chlorophyll
  21537, 21908, 22509, 23050, 23263, 23450, 23753, 24389, 24775, 24794, 25050;
  25173, 25507, 25716, 25905-6, 25991, 26338, 26437, 27163, 27496, 27855,
  28266; 28529, 28718, 28789, 29554, 29727, 30272, 30481, 30825, 31052, 31581,
  31851; 31903; 32805, 33187, 33292, 33405, 33654, 34045, 34464-5, 34480,
  35962, B36081, 36439; 36794, 37378, 37400, B37532, 38399, 38685, 38695,
  B38740, 39736

Hydration level of leaf and chloroplast
  21861, 22148, 22415, 23260, B23262, 23263, 24018, 24269, 24636, 25050;
  25906, 26090, 26128, 26141, 27454, 28266; 29360, 30111, 30879, 31072; 32838,
  33138, 33405, 33524, 33659, 33812, 33839-40, 34129, 34440, 34480-3, 34663,
  35094, 35603, 35605, 36526; 36794, 37270, 38026, 39841

Hydration level of leaf and ecosystem and plant productivity
  21623, 21823, 21847, 21854, 21878, 21908, 22380, 22974, 23027, B23302,
  23305, 23308, 23504, 24000; 25173, 25238, 25520-1, 25860, 25885, 25992,
  26409, 26437-8, 26484, 26537, 26759, 26970, 27749, 28456; 28566, 28809,
  28986, 30272, 30310, 30524, 30739, 30808, 30820-1, 30903; 33020, 33222,
  33928, 34181, 34831, 35194, 35904-5, 36205, 36325; 36637, 37108, 37334,
  37492, 37499, 37544, 37591, 37768, 38100, 38134, 38206, 38299-300, 38823,
  38979, 39097, 39130, 39549, 39560-1, 39815, 39989

Hydration level of leaf and electron transport chain
  22415, 23079, 23296, 23673, 24147-8, 24312, 24537, 25050; 25520, 26007,
  26335, 27163, 27274, 27496; 28529, 28717-8, 28789, 29360, 30111, 30386,
  30892, 30919, 31581; 33187, 33659, 34663, 36439; 38026, 38136, 38335, 40321-2

Hydration level of leaf and gas exchange
  21517, 21648, 21670, 21765, 21878, 21908, 21910, 21947, 22031, 22344, B22358,
  22666, 22830-1, 22948, 22974, 22993, 23027, 23069, 23090, 23151, 23218-9,
  23260, 23296, 23298-9, B23302, 23306-8, 23329-30, 23465, 23587, 23608-9,
  23628, 23665, 23673, 23696, 23887, 24017, 24033, 24063, 24175, 24215, B24271,
  24302, 24389, 24437, 24444, 24507, 24537, 24649, 24804, 24820, 24821, 25021
  25173, 25258, 25283, 25407, 25519, 25520-1, 25549, 25592, B25637, 25642,
  25653, 25755, 25772, 25774, B25775, 25828, 25860, 25902, 25928, 25991,·
  25993, 26121, 26296, 26300, 26348, 26409, 26437-8, 26484, 26494, 26497,
  26537, 26580, 26685, 26689, 26759, B26846, 26863, 26970-1, 27135, 27139,
  ·27163, 27302, 27348, 27363, 27409, 27411, 27768, 27782, 27793, 27998, 28064,
  28066-7, 28342
  28500, 28504, 28688, 28717-8, 28731, 28809,
  28919, 29020, 29071, 29122, 29156, 29243, 29294-5, 29367, B29468, 29721,
  29766, 29769, 29843, 29931, 30168, 30249, 30270, 30272, 30285-6, 30310,
  B30511, 30515, 30523, 30542-3, 30608, 30633, 30723, 30808, 30821, 30825,
  30861, 30892, 30903, 30919, 30944, 31018, 31065, 31090, 31165, 31258, 31293,
  31321, 31536, 31899, 31983, 32197, 32252, 32378, 32541

(continued)

Hydration level of leaf and gas exchange (continued)
        32777, 32837, 32997, 33002, 33029, 33050,
        33249, 33292, 33441, 33659, 33812, 33839, 34129, 34181, 34288, 34295, 34302,
        34497, 34518, 34535, 34547, 34649, 34792, 34886, 35370, 35476, 35623, 35895,
        36050, 36060, 36069, B36081, 36090, 36135, 36205, 36207
        37002, 37021, 37049, 37184, 37208, 37243, 37287, 37383, 37400, 37476, 37492,
        37516, B37532, 37701, 37709, 37768, 37915, 38026, 38077, 38132, 38312,
        38358, 38389, 38399, 38411, 38429, 38444, 38493, B38739-40, 38814, 38902,
        38997, 39058, 39270, 39381, 39501, 39815, 39867, 39935, 39945

Hydration level of leaf and photorespiration
        23329; 25520, 25592, 25755, 26689, 26863; 30270, 30542-4, 31018, 31293;
        34535, 35324, 35370; 37516, 38429

Hydration level of leaf and resistances to CO_2 and water vapour transfer
        21713, 21787, 22283, 22316, 22380, 22662, 22676, 22920, 22964, 23167, 23218,
        23329-30, 23608-9, 23648, 23679, 23696, 23813, 23887, 24063, 24302, 24377-8,
        24437, 24826
        25173, 25283, 25327, 25454, 25572, 25592, 25604, 25742, 25843, 25999,
        26114, 26265, 26438, 26532, 26843, B26846, 26865, 26970-1, 27109, 27289,
        27363, 27433, 27749, 27882, 27897, 27998, 28064, 28067, 28123, 28321, 28341,
        28367
        28500-1, 28504, 28688, 28717-8, 28731, 28987-8, 29020, 29162, 29203, 29394-5,
        29467, 29725, 29766, 29881, 30069, 30163, 30249, 30270, B30511, 30738-9,
        30903, 30919, 30931, 31165, 31770, 31899, 32051, 32110-1, 32252, 32352, 32369,
        32485, 32539
        32777, 33002, 33022, 33286, 33451, 33527, 33659, 33928, 33984-5, 34109,
        34181, 34535, 34541, 34651, 34792, 34851, 34886, 35354, 35369-70, 35476,
        35492, 35529, 35594, 35963, 36050, 36069, 36206, 36431
        36637, 36821, 37177, 37184, 37492, 37496, 37516, 37799, 37915, 38075, 38077,
        38134, 38389, 38436, 38529, 38534, B38740, 38814, 39212, 39224, 39270,
        39328, 39424, 39492, 39501, 39560-1, 39600, 39709, 39991

Hydration level of leaf and respiration
        22344, 22830-1, 23069, 23079, 23090, 23298, 23306, 23329, 24175, 24437;
        25173, 25520-1, 25549, 25592, 25991, 26437, 26863, 26971, 27163; 29843,
        30270, 30515, 30808, 30825, 31018, 31065, 31293, 31882; 32837, 33925, 34440,
        35370, 36205; 37002, 37049, 37768, 38132, 38335, 38358, 38429, B38740,
        38814, 39935

Hydrogen see H_2 ...

Hydrogenase see O_2 evolution mechanism and kinetics; H_2 evolution ...

Hygrometer see Humidity of air, methods; Infra-red analyser for water vapour

I

Ideotype see Model ...

Immobilization of chloroplasts and photosynthetic systems see Photosystems stabi-
        lization ...

Induction phenomena see Transient phenomena ...

Infra-red analyser for CO_2
        21801-2, 21838, 22005, 22143, 22152, 22486, 23059, 23299, 23979, 24033,
        24156, 24275, 24294, 24677, 24977, 25032
                                                                        (continued)

                                                                    (continued)

Inhibitors of electron transport chain (continued)
        28519, 28524-5, 28541, 28560, 28562, 28571, 28588, 28592, 28594, 28604,
        28633, 28678, 28681-2, 28694-6, 28705, 28738-9, 28742, 28745, 28766, 28769,
        28772, 28793, 28812, 28814-5, 28827, 28847, 28867, 28875, 28879, 28884, 28886,
        28893, 28903, 28935-6, 29017, 29032, 29059, 29097, 29101, 29114, 29143, 29183,
        29189, 29191, 29258-9, 29261, 29270, 29304-5, 29309, 29328, 29343, 29353,
        29356, 29418, 29435-6, 29471-2, 29504, 29517, 29525, 29528, 29531, 29533,
        29537, 29601, 29634-5, 29637, 29640, 29647, 29649, 29662, 29665, 29684, 29697,
        29715, 29731, 29735, 29737, 29750, 29773, 29784-6, 29831, 29856, 29889, 29904,
        29906, 29939, 30001-2, 30012, 30035, 30108, 30112, 30115, 30118-9, 30127,
        30136, 30154-5, 30158, 30160, 30183, 30188, 30215, 30219-20, 30223,
        30243, 30245, 30291, 30294, 30324, 30355, 30365, 30403, 30430, 30437, 30468,
        30470, 30502, 30536, 30539, 30589, 30594, 30596, 30599-600, 30635, 30681,
        30701, 30710, 30736, 30749, 30771, 30773, 30781-2, 30803-6, 30834, 30837,
        30841, 30853, 30878, 30970, 30983-5, 30993, 31007, 31042, 31050, 31098, 31109,
        31146, 31169, 31195, 31207, 31211, 31246, 31274, 31291, 31342, 31344, 31409,
        31410, 31416, 31420, 31447, 31449-51, 31456, 31461, 31465-6, 31483, 31503,
        31518, 31521-2, 31524, 31526, 31541, 31551, 31567, 31582, 31597, 31612, 31626,
        31660-1, 31663-4, 31667-8, 31686, 31692, 31717, 31723, 31731, 31733, 31743,
        31746-7, 31772, 31802, 31809, 31831, 31872-3, 31877-8, 31882, 31897-8, 31918,
        31942, 31950, 31957, 31959, 31969, 31998, 32036, 32039-40, 32042, 32082,
        32092, 32135, 32141, 32161, 32178, 32207, 32233-5, 32247, 32250, 32256, 32262,
        32316, 32357-8, 32384, 32414, 32418, 32432, 32451, 32497-9, 32512, 32536
        32590, 32601, 32609, 32631, 32636, 32684, 32715, 32719, 32734, 32747, 32755,
        32767-8, 32789, 32791, 32798, 32820, 32822, 32834, 32848, 32853, 32860, 32876,
        32900, 32942, 32954, 32971, 33137, 33144, 33154, 33157, 33273, 33309, 33323,
        33329, 33350, 33460, 33462, 33482, 33496, 33499, 33509, 33606, 33655, 33716-
        -7, 33743-4, 33749, 33802, 33850, 33883, 33886, 33895, 33906, 33935, 33980,
        33990-1, 34025, 34027, 34053, 34103, 34106, 34117, 34127, 34173, 34259,
        34266, 34275-6, 34280, 34286, 34299, 34305, 34344, 34355, 34372, 34377,
        34400, 34405, 34417, 34509, 34520-1, 34531, 34539, 34560-1, 34585, 34643,
        34657, 34665, 34671, 34675, 34703, 34728-9, 34766, 34773, 34817, 34843-5,
        34848, 34866-7,   34891, 34964, 34967, 34984, 34998, 35009, 35024, 35058,
        35068, 35072, 35104, 35136, 35144, 35187, 35204, 35212-3, 35245, 35305, 35307,
        35327-9, 35343, 35362, 35367-8, 35384, 35391, 35401-2, 35404, 35413, 35418,
        35444, 35447, 35468, 35471-2, 35485, 35495, 35511, 35516, 35519, 35524,
        35555, 35557, 35566, 35591, 35606, 35635, 35669, 35681-2, 35684-5, 35692,
        35697, 35699, 35719, 35736, 35748, 35756, 35858, 35894, 35915-6, 35918,
        35935, 35955, 35982, 35985, 36023, 36035, 36043, 36099, 36122, 36173-5,
        36185, 36200, 36214, 36236, 36250, 36276, 36316, 36366, 36389, 36395-7,
        36422, 36439, 36453, 36460-2, 36488,   36495-6,   36502-3, 36508, 36511,
        36541-2, 36557, 36561
        36653, 36658, 36699, 36710, 36731, 36745, 36749, 36754, 36782-3, 36815,
        36833, 36841-2, 36864, 36877, 36884, 36892, 36899, 36935, 36939, 36968,
        36982, 36993, 37075-7, 37107, 37115, 37129, 37140, 37143-5, 37186, 37194,
        37222, 37226, 37261, 37267, 37273, 37280-1, 37344, 37391, 37400, 37548,
        37563, 37576, 37629, 37640, 37642, 37668, 37705, 37708, 37723, 37754-5,
        37764, 37786, 37822, 37842, 37860, 37866, 37875, 37901-2, 37907, 37913,
        37919, 37929, 37968, 37979, 38002, 38026, 38029, 38049, 38069, 38113,
        38122, 38126, 38129, 38231, 38250, 38275, 38338, 38377, 38394, 38430, 38458-
        -60, 38466, B38477, 38510, 38588, 38592, 38619, 38672, 38698-9, 38707,
        38757-8, 38767, 38773, 38797, 38802, 38810, 38838, 38844, 38888-9, 38928-9,
        38939-40, 38946, 38950, 38953, 38970, 38982-3, 39025, 39038-9, 39048, 39102,
        39123, 39125, 39177, 39194, 39225-6, 39235, 39247, 39288, 39295, 39299,
        39301, 39316, 39329-30, 39338, 39345, 39357, 39373, 39379, 39390, 39401,
        39427, 39448, 39458, 39473-4, 39494, 39498, 39538, 39607, 39627, 39630,
        39645, 39648, 39674-5, 39702, 39708, 39811, 39826, 39839, 39880, 39887,
        39915, 39929-30, 39936, 39955-7, 39973-4, 40005, 40014, 40037, 40046, 40058,
        40068, 40076, 40082, 40112, 40124, 40132, 40164, 40192, 40247, 40251, 40259,
        40265, 40301

Insertion see Ontogeny ...

Intercellular spaces, CO_2 concentration inside
 21757, 21767, 21773, 22041, 22652, 23928, 24239-40; 25261, 25437, 25611,
 25780, 26265, 26600, 27180, 27370, 27533; 28614, 29054, 29130, 29486,
 30139, 30455, 30499, 30643, 30699, 31067, 31214, 31232, 31256, 31470; 32715,
 32897, 33290, 33305, 33358-9, 33388, 33435, 33505, 33742, 34097, 34219,
 34375, 34451, 34636, 34646, 34945, 35097, 35374, 35441, 35450, 35460, 35766,
 36472; 37021, 37241, 37515, 37649, 37752, 37774, 38508, 39057, 39094, 39725,
 40265

Intermediates of carbon fixation pathways, methods
 22710, 25067; 29113, 30564, 31313, 31452, 32075; 33272, 34455, 34742; 37634,
 39150, 39618

Intracellular resistance see Resistance, intracellular (mesophyll)

Ionizing radiation (gamma, X, cosmic, etc.) and carbon fixation pathways   25116

Ionizing radiation (gamma, X, cosmic, etc.) and carotenoids 24751; 25297, 27885;
 34869; 36591

Ionizing radiation (gamma, X, cosmic, etc.) and chlorophyll
 21603, 21831, 22911, 24141, 24388; 25297, 25715, 25864, 27201, 27429, 27477,
 27885, 27902-3; 29499, 30404, 30548; 34050, 34261, 34869, 35310, 36004,
 36157, 36268; 37519, 39159, 39701, 40107

Ionizing radiation (gamma, X, cosmic, etc.) and chloroplast (chromatophore)
 27054, 28149; 31333; 33475, 35989; 38545, 39159

Ionizing radiation (gamma, X, cosmic, etc.) and ecosystem and plant productivity
 23763, 25126; 25239, 27821; 29175; 37221, 37440, 38658, 39159, 39570

Ionizing radiation (gamma, X, cosmic, etc.) and electron transport chain
 30548; 36568; 38658, 39701

Ionizing radiation (gamma, X, cosmic, etc.) and gas exchange
 23850, 24612; 27429, 27477, 28487; 28702, 29175, 32219; 33360, 34454, 34475,
 35989, 36568; 37320, 38635, 38658, 39159, 39701

Ionizing radiation (gamma, X, cosmic, etc.) and photorespiration 36158

Ionizing radiation (gamma, X, cosmic, etc.) and resistances to CO_2 and water vapour
 transfer 37440

Ionizing radiation (gamma, X, cosmic, etc.) and respiration 23850, 24001

Irradiance, compensation see Compensation irradiance

Irradiance, flash, and algae productivity 28172

Irradiance, flash, and biliproteins 35133

Irradiance, flash, and carotenoids 23115, 23399; 27071, 28029; 28651,29266, 29760,
 31219; 39119

Irradiance, flash, and chlorophyll
 21507, 21532, 22058, 22359, 22439, 22600-1, 22780, 22795, 22856-7, 22881-2,
 22955, 23403, 23520, 23538, 23639, 23823, 23874, 23965, 24189-90, 24524,
 24653-4, 24901-3, 24916
 25544, 25630, 25645, 25840, 25861, 25988, 26459, 26481-2, 26545, 27071,
 27452, 27867, 28028-9, 28172-3, 28258

                                                                    (continued)

Irradiance (PhAR) and biliproteins
        23083; 25867, 25869, 26168, 26608, 27568, 27901; 29378, 30531, 31499, 31584,
        31757; 33657, 34596, 35147; 38160, 38290, 38367, 38899, 38948, 39883, 40034-5

Irradiance (PhAR) and carbon fixation pathways
        21576, 21616, 21711, 21724, 21810, 21975, 22157, 22274, 22398, 22507, 22530,
        22552, 22868, 22905, 22980-1, 23013, 23060, 23195-7, 23286, 23419, 23451,
        23454, 23668, 23772, 23938, 24016, 24160, 24201, 24265, 24348, 24380, 24397,
        24553, 24708, 24843
        25213, 25221, 25261, 25431, 25529, 25672, 25875, 25913, 26251, 26526, 26598,
        26624, 26760, 26858, 26888, 26940, 27182, 27247, 27272, 27524, 27709, 27823,
        27901, 27911, 27928, 28116, 28405
        28775, 28835, 29015, 29267, 29313, 29316, 29462, 29533, 29772, 29808, 29889,
        29922, 29951, 30333, 30887, 31089, 31109, 31620, 31910
        32585, 32720, 32768, 32843, 32929, 32968, 33130, 33148, 33198, 33513, 33518,
        33658, 33673, 33809, 34064-5, 34173, 34219, 34253, 34276, 34657, 34861,
        34906, 35139, 35183-4, 35192, 35241, 35762, 36159, 36192, 36234, 36451-2,
        36533
        36624, 36663-4, 36676, 36712, 36725, 36800, 37331, 37400, B37531, 37561,
        38033, 38115-6, 38126, B38139, 38146, 38552, 38560, 38586, 38643, 38872,
        39275, 39351, 39466-7, 39505, 39603, 39620, 39876, 39963, 39972, 40201,
        40203, 40213-4, 40236

Irradiance (PhAR) and carotenoids
        21509, 22401, 22450, 22507, 22518, 22619, 22644, 22841, 23116, 23393, 23471,
        23804, 24129, 24146, 24228, 24317-8, 24451, 24489, 24621, 25077-8
        25553, 25696, 25969, 26168, 26707, 26740, 26975, 27066, 27388, 27511, 27586,
        27677, 27711, 27805, 27823, 27850, 27864, 27946, 28217, 28239-40, 28357
        28713, 28716, 28947, 29121, 29456, 29573, 30230, 30313, 30395, 30581, 31188,
        31236, 31276, 31330, 31349, 31430, 31673, 31757, 31810, 31831, 31851, 32142-3,
        32424
        32595, 32763, 32914-5, 33010, 33177, 33518, 33901, 34175, 34399, 34435,
        34441, 34596, 34744, 34791, 34948, 35237, 35381, 35596, 35617, 35854, B36191,
        36411, 36587
        36620, 36747, 37095, 37309, 37346, 37568, 38290, 38292, 38332, 38478, 38874,
        38899, 39044, 39400

Irradiance (PhAR) and chlorophyll
        21509, 21513, 21537, 21558, 21701, 21734, 21842, 21870, 22001, 22011, 22056,
        22068, 22093, 22102, 22230, 22312, 22317, 22337, 22364, 22401, 22440, 22449-
        -50, 22453, 22475, 22506-7, 22518, 22786, 22789, 22799, 22816, 22827, 22841,
        22858, 22905, 22911, 22980-1, 22988, 22994, 23029, 23042, 23083, 23091,
        23108, 23118, 23132, 23291, 23318, 23361, 23393, 23395, 23402, 23429, 23462,
        23488-9, 23560, 23581, 23614, 23639, 23751-2, 23759, 23804, 23812, 23830,
        23849, 23875, 23882, 23896, 23959, 23969, 23993, 24016, 24037, 24086, 24103,
        24177, 24208, 24228, 24321, 24330, 24371, 24402, 24464-5, 24485, 24545,
        24617, 24621, 24713, 24753, 24792, 24817-8, 24895, 24912, 24930, 24986,
        24993-4, 24996, 25011, 25015, 25029, 25077-8
        25186, 25237, 25248, 25268-9, 25287, 25427, 25440, 25577, 25596, 25696,
        25748, 25764, 25852, 25867, 25869, 25887, 25898, 25900, 26070-1, 26094,
        26099, 26168, 26192, 26215, 26310, 26338, 26399, 26481, 26486, 26608,
        26617, 26706-7, 26727, 26737, 26842, 26890, 26910, 26930, 26937, 26940,
        26976, 27077, 27216, 27272, 27287, 27291, 27368, 27388, 27407, 27497,
        27500, 27511, 27520, 27536, 27567-8, 27622, 27659, 27677-8, 27703, 27718,
        27795, 27805, 27823, 27850, 27854, 27857, 27864, 27885, 27901, 27997,
        28017, 28156, 28166, 28173, 28217, 28221-2, 28239-40, 28357
        28637-8, 28713, 28719, 28752, 28775, 28817, 28828, 28835, 28866, 28896, 28902,
        28925, 28947, 28989-90, 29067, 29077, 29107, 29121, 29139, 29155, 29163,
        29249, 29329, 29341, 29354, 29378, 29392, 29461, 29505, 29523, 29573, 29682,
        29704, 29835, 30006, 30199-200, 30230, 30301, 30331, 30580-1, 30594, 30602,
        30605-6, 30611, 30633, 30696-7, 30722, 30741, 30850, 30881, 31094, 31110-1,
        31155, 31178, 31188, 31231, 31236, 31238, 31254, 31276, 31330, 31349, 31364,
                                                                        (continued)

                                                                      (continued)

Irradiance and ecosystem and plant productivity (continued)
        34959, 35052, 35062, 35102, 35151, 35170, 35184, 35261, 35263, 35301, 35331,
        35351-2, 35416, 35455, 35619, 35849, 35882, 36070, 36097, 36112, 36143,
        36163, B36191, 36288, 36290, 36304, 36348, 36388, 36394, 36494
        36602, 36721, 36742, 36746, 36776, 36875, 36882, 37010-1, 37017-8, 37114,
        37121-2, 37169, 37173, 37191, 37199, 37298, 37302, 37491, 37591, 37623,
        37636, 37687, 37696, 37796, 37845, 37886, 37935, 37974, 37983, 38455,
        38501, 38535, 38627, 38658, 38760, 38873, 38886, 39043, 39072-3, 39130,
        39141-2, 39187, 39342-3, B39523, 39549, 39560-1, 39637, 39752, 39755, B39903,
        39989, 40035, 40062, 40092, 40142, 40144, 40207, 40239

Irradiance (PhAR) and electron transport chain
        21528, 21561, 21644, 21744, 21810, 21825, 22127, 22179, 22310, 22349-50,
        22376, 22400, 22588, 22594, 22613, 22619, 22721, 22811, 22853, 22883-4,
        23076, 23241, 23351, 23365, 23403, 23591, 23655, 23704, 23889, 23921, 24107,
        24206, 24209, 24321, 24373, 24418, 24433, 24490, 24545-6, 24568, 24760-2,
        24808, 24891, 24904, 25029, 25102
        25183, 25212, 25335, 25399, 25417, 25494, 25528, 25589, 25632, 25668, 25696,
        25700, 25918, 26146, 26310, 26335, 26399, 26424, 26486, 26703, 26766, 26872,
        26910, 26925, 27093, 27104, 27126, 27184, 27279, 27332, 27354-5, 27403,
        27496, 27504, 27590, 27659, 27699, 27856-7, 27921, 27977, 28002, 28221,
        28297, 28332
        28508, 28587, 28595, 28655, 28817, 28835, 28879, 28990, 29070, 29081, 29238,
        29244, 29275, 29285, 29341, 29353, 29596, 29730, 29847, 29949-50, 30013,
        30085, 30214, 30251, 30329, 30355, 30551, 30558, 30579-80, 30594, 30720,
        30777, 30832, 30837, 30881, 30909, 30964, 31053-4, 31128, 31146, 31152-3,
        31194, 31238, 31588, 31683, 31690, 31726, 31737, 31871, 31874, 31877-8,
        32198, 32248, 32408, 32418
        32591, 32595, 32604, 32673, 32733, 32834, 32867, 32889, 33004-5, 33323,
        33389, 33482, 33502, 33513, 33551, 33582, 33714, 33749, 33761, 33869, 33901,
        34219, 34234, 34300, 34432, 34441, 34521, 34588-9, 34607, 34746, 34769,
        34774, 34805, 34841, 34868, 35143, 35157, 35187, 35240-1, 35269, 35273,
        35410, 35413, 35530, 35617-8, 35773, 35780, 35819, 35879, 36122, 36139,
        36234, 36244, 36250, 36310, 36397, 36569
        36673, 37118, 37147, 37371, B37531, 37615, 37706, 37891, 38055, 38080,
        38160, 38276, 38510, 38552, 38901, 38928-9, 38960, 38994, 39152, 39193,
        39379, 39796, 39845, 39845, 40192, 40201, 40302

Irradiance (PhAR) and gas exchange (*cf.* also Irradiance {PhAR} and gas exchange,
        analysis of light curves)
        21517, 21525-6, 21538, 21572, 21629, 21633-4, 21662, 21709, 21711, 21717,
        21748, 21770-1, 21773, 21779, 21803, 21810-3, 21815, 21825, 21841, 21903,
        21946, 21967, 22013, 22031, 22038-41, B22052, 22053, 22066, 22075, 22082,
        22104, 22109, 22131, 22139, 22152-3, 22157, 22165, 22197, 22209, 22211,
        22234, 22284, 22295, 22309-10, 22321, 22351, 22363-4, 22369, 22413, 22434,
        22449, 22453, 22475, 22481, 22485, 22487-8, 22494, 22521, 22544, B22572,
        22588, 22608, 22639, 22649, 22678, 22686, 22717, 22725, 22764, 22786, 22791,
        22794, 22796, 22819, 22828, 22858-9, 22873, 22875, 22897, 22937, 22944,
        22984-5, 22992-4, 23015, 23040, 23069, 23105, 23121, 23128, 23151, 23165,
        23172, 23195, 23219, 23257, 23286, 23295, 23298, 23300, 23304, 23307-8,
        23334, 23369, 23371, 23402, B23407, 23439, 23443-4, 23462-4, 23467, 23493,
        23503, 23522, B23540, 23556, 23571-2, 23578, 23587, 23596, 23645, 23648-9,
        23659, 23665-6, 23668, 23671, 23673, 23689, 23692, 23712, 23745, 23751,
        23756-7, 23761, 23771, 23773, 23775, 23780, 23783-4, 23814, 23827, 23845,
        23848-50, 23873, 23895-6, 23901, 23922, 23938, 23942, 23972, 23976, 23978,
        23980, 23988, B24024, 24025, 24029, 24046, 24063, 24072, 24082-3, 24086,
        24100, 24110, 24142, 24172, B24176, 24183, 24215, 24229, B24271, 24272,
        24280, 24291, 24302, 24321, 24323-4, 24378, 24397, 24401-2, 24437, 24440,
        24483, 24525, 24535, 24556, 24568, 24578, 24593, 24596, 24598, 24601, 24608,
        24629, 24662, 24688-9, 24699, 24708, 24752, 24758-9, 24787, 24797, 24804,
        24821, 24835, 24863, 24867, 24881, 24915, 24918, 24931, 24934, 24956, 24979,
        24984, 24988, 24998, 25000, 25021, 25025, 25029, 25032, 25048-9, B25064,
        25069, 25077-8, 25123, 25136, 25160

                                                              (continued)

Irradiance (PhAR) and gas exchange (continued)
                                                                        (continued)

Irradiance (PhAR) and gas exchange (continued)
        37772, 37774, 37784, 37818, 37874, 37895, 37929, 37943, 37987-8, 38059,
        38068, 38079, 38082, 38085, 38098, 38099, 38141, 38143, 38174, 38180, 38209,
        38222, 38251-2, 38264, 38326, 38337, 38360, 38374, 38401, 38411, 38416-7,
        38435, 38444, 38448, 38465, 38472, B38477, 38492-3, 38508, 38515-6, 38543,
        38589, 38603, 38631-2, 38658-9, 38686, 38703, 38710, B38739, 38794, 38868,
        38873, 38886, B38892, 38896, 38934, 38942, 38960, 38967-8, 38996, 39037,
        39057, 39060, 39063, 39072-3, 39087, 39102, 39105, 39149, 39178-9, 39181,
        39185-6, 39247, 39252, 39259, 39270, 39291, 39309, 39342, 39370, 39381, 39391,
        B39394, 39402, 39423, 39455, 39461, 39500, 39503. 39533-4, 39567, 39584,
        39620, 39637, 39639-40, 39648, 39697, B39711, 39728, 39759, 39784, 39794,
        39823, 39828, 39852, 39869, 39873-5, 39882, 39926, 39932, B39951, 39960,
        39971-2, 39976, 39981, 39996, 40033, 40075, 40077, 40080, 40089, B40105,
        40126, 40144, 40153-4, 40166, 40201, 40210-1, 40217, 40239, 40254, 40293,
        40298, 40330, B40331, 40332

Irradiance (PhAR) and gas exchange, analysis of light curves
        21517, 21525, 21633-4, 21662, 21686, 21709, 21717, 21782, 21802, 21812,
        21946, 22039, 22109, 22134, 22234, 22252, 22321, 22367, 22649, 22652, 22728,
        22875, 22890, 22951, 22985, 23007, 23059, 23286, 23299, B23407, 23443, 23464,
        23522, 23545, 23578, 23757, 23780, 23847-8, 23896, 23928, 23980, 24000,
        24046, 24086, 24161, 24229, 24272, 24291, 24324, 24438, 24526, 24647, 24662,
        24680, 24692, 24759, 24787, 24798, 24804, 24813, 24821, 24863, 24880, 24934,
        25032, 25086
        25186, 25195, 25224, 25294, 25367, 25476, 25618, 25642, 25674, 26001, 26078,
        26085, 26116, 26164, 26300, 26370, 26426, 26476, 26661, 26718, 26720, 26970,
        27035, 27069, 27139, 27180, 27210, 27227, 27270, 27317, 27348, 27399, 27422,
        27469, 27490, 27630, 27728, 27768, 27920, 27980, 27995, 28077, 28087, 28102,
        28106-7, 28110, 28192-3, 28229-30, 28247, 28287, 28300, 28337, 28413
        28552, 28616, 28685, 28783, 28795, 28810, 28818, 28866, 28896, 28898, 28901,
        28990, 29068, 29087, 29090, 29188, 29191, 29242, 29289, 29333, 29337, 29373,
        29383, 29524, 29563, 29575-6, 29675, 29790, 29848, 29975, 29985, 30047,
        30172, 30185, 30457-8, 30499, 30556, 30685, 30699, 30721-2, 30761, 30916,
        30945-6, 31012, 31040-1, 31077, 31119, 31232, 31359, 31372, 31411, 31470,
        31586, 31704, 31712, 31759, 31836, 31866, 31916, 31925, 32031, 32046, 32093,
        B32150, 32175, 32231, 32287, 32294, 32296, B32515
        32687, 33124, 33175, 33244, 33513, 33694, 33741, 33759, 33798, 33879, 34070,
        34101, 34123, 34219, 34330, 34373, 34467, 34792, 34899, 35092, 35105, 35129,
        35140, 35158, 35184, 35200, 35281, 35314, 35664, 35674, 35773, 35807, 36067-
        -8, 36077, 36132, B36191, 36195, 36203, 36417, 36569-70
        36602, 36683, 37289, B37299, 37348, 37389, 37614, 37700, 37924, 38098,
        38174, 38472, 38515, 38745, 38891, B38892, 38996, 39461, 39567, 39748,
        39867

Irradiance (PhAR) and photorespiration
        21812, B21833, 22071, 22131, 22982, 23714, 25160; 25653, 26037, 26785, 27493,
        27754, 27901, 27929, 28353; 28902, 29693; 33324, 33785, 34375, 34718, 35044,
        35334, 36280, 36562; 36687, 36757, 37067, 37084, 37105, 37754, 37881, 38020,
        38552, 39073, 39391, 39710, 39716, 39869, 40114

Irradiance (PhAR) and resistances to CO_2 and water vapour transfer
        21549-50, 21943, 22157, 22263, 22314-6, 22321, 22403, 22650, 22676, 22761-2,
        22920, 22937, 22964, 23295, 23609, 23648, 23653, 23679, 23757, 23780, 23801,
        23978, 24183, 24302, 24688, 24725, 24880
        25327, 25430, 25604, 25620, 25653, 25843, 25947, 25999, 26065, 26237, 26240,
        26389, 26514, 26843, 26865, 26947, 26970-1, 27180, 27187, 27329, 27360,
        27432, 27595, 27939, 28067, 28125, 28341, 28350, 28367, 28462
        28500, 28732, 28803, 28810, 28821, 28898, 28925, 29020, 29130, 29247, 29319,
        29373, 29394-5, 29425, 29793, 29795, 30078, 30212, 30249, 30457-8, B30511,
        30643, 30738, 30808, 30903, 30991, 31193, 31268-9, 31627, 31885, 31888, 31956,
        32051, 32194, 32369

(continued)

Irradiance (PhAR) and resistances ... (continued)
    32573, 32690, 32777, 33000, 33244, B33281, 33395, 33451, 33508, 33558, 33879,
    33916, 33976, 33985, 34038, 34193, 34633, 34647, 34676, 34743, 34747, 34814,
    34954, 35261, 35340, 35369, 35460, 35634, 35773, 35963, 36069, 36120, 36135,
    36190, 36207, 36358, 36381, 36457, 36472
    36683, 36732, 36742, 36821, 37171, 37173, 37177, 37325, 37413-5, 37514,
    37752, 38401, 38435, 38472, 38529, 38534, 38737, 38847-8, 38915, 38934,
    38996, 39005, 39036, 39072-3, 39270, 39533, 39556, 39567, 39661, 39709,
    39737, 39828, 39875, 39891-2, 39991, 40239, 40265

Irradiance (PhAR) and respiration
    22131, 22157, 22819, 23464, 23467, 23714, 24001, 25032, 25047; 25653,
    26046, 26299, 26578, 26744, 26746, 27687, 27873; 28685, 31077,
    31470, 32115, 32175; 32975, 33565, 33891, 34219, 34281, 34810, 35261; 36742,
    36776, 36799, 37124, 37325, 37576, 37844, 37886, 39342, 40270

Irradiance (PhAR, total) measurement
    21539, 21693, 21801, 21881, 21956, 22302, 22680, 22769, 22788, 23426, B23688,
    24641, 24677, 24913; 25552, 25676-7, 26000, 26177, 26564, 27080, 27096,
    27770, 28051-2, 28188, 28208, 28210, 28233; 28735, 28806, 30071, 30129,
    30315, B30325, 30407, B30432, 30553, 30634, 30882, 30956, 31501, 31547,
    31886, 32096, 32371; 32721, 32812, 33362, 33375, 33666, 33724, 33878, 34090,
    34402, 34806, 34943, 35116, 35168, 35588-9, 35850, 36125, 36138, 36248,
    36286; 36596, 36893, 36941-2, 36949, B37567, 37605, 37648, 37839, 37912,
    37990, 38086, 38147, 38171, 38216, 38279-80, 38343, 38792, 39071, 39341,
    39454, 40215, 40276

Irradiance, spectral composition and algae productivity
    22261, 23133, 24061; 25390, 25534, 25849, 25894, 26217, 27144, 28147; 30226,
    32446; 34715; 37673

Irradiance, spectral composition and biliproteins
    21836, 24254; 25867, 26168, 27177-8, 27568; 29866, 30552, 32053, 32288;
    33758, 35015, 35706, 36294; 37030, 37040, 37622, 37785, 38947, 39484

Irradiance, spectral composition and carbon fixation pathways
    22308, 22997, 23060, 23674, 24397, 24931, 24933; 25515, 26077, 28009, 28285;
    28883, 29567-8, 29603, 30893-5, 31417, 31620, 32236, 32317, B32515; 32792,
    32959, 33271, 33439, 34682, 34749, 34905-6, 34952, 35139, 35359, 35975;
    B37531, 37697, 37725, 38758, 40104, B40105, 40111

Irradiance, spectral composition and carotenoids
    21594, 21719, 21891, 22618, 23398-401, 23440, 23691, 24166, B25064; 25778,
    26168, 26491, 26779, 26855, 27239, 27513, 27848; 28713, 29208, 29516, 30298,
    30446, 30595, 31937, 32288; 33174, 34163, 34606, 35237; 36747, 37522, 37785,
    39436, 39557, 39958, 40111, 40244

Irradiance, spectral composition and chlorophyll
    21891, 21993, 22011, 22067, 22107, 22307, 22410, 22439, 22472, 22559, 22618,
    22672, 22691, 22818, 22857, 22902, 22988, 22930, 23024, 23029, 23035, 23398,
    23400-1, 23440, 23528, 23564, 23674, 23691, 23794, 23880, 23882, 24158,
    24168, 24353, 24510, 24533, 24818, 24910, B25064, 25079
    25208, 25220, 25268-9, 25688, 25829, 25867, 25892, 26168, 26285-6, 26415-6,
    26459, 26520, 26588, 26603-4, 26779, 26813, 26855, 26935-6, 27041, 27064,
    27078, 27177-8, 27238-9, 27260, 27499, 27513, 27568, 27848, 27854, 27899,
    27947, 27997, 28218, 28476
    28589, 28713, 28883, 28933, 28942, 28958, 29329, 29386, 29392, 29603, 29623,
    29743, 29894, 30080, 30101, 30137, 30199, 30201, 30298, 30318-9, 30330, 30389,
    30552, 30862, 30893-4, 31044, 31064, 31254, 31324, 31429, 31719, 31744, 31829,
    31856, 31937, 32001, 32053, 32084, 32086, 32155, 32237, 32288, 32298-9, 32385,
    32433
    32594, 32704, 32765, 32792, 32883, 32974-5, 33174, 33298, 33302, 33392, 33761,
    34094, 34198, 34606, 34728, 34920-1, 35015, 35132-3, 35144-5, 35237, 35483,
    35602, 35706, 35754, 35806, 36038, 36101, 36137, 36164, 36306, 36316, 36454,
    36586

(continued)

Leaf anatomy (*cf.* also Leaf thickness)
        21535, 21633, 21775, 21827, B21833, 21863, 21926-7, 22009, 22016, 22039,
        22042, 22071, 22075, 22082, 22085, 22097, 22114, 22156-7, 22165, 22208, 22251,
        22489, 22509, 22516, 22624, 22700, 22719, 22837, 22868, 22917, 22958,
        23001, 23092, 23443, 23502, 23521-2, 23545, 23634, 23713, 23768, 23791,
        23848, 23929, 23937-8, 23986, 24216, 24261, 24309, 24319, 24346, 24359,
        24434, 24469, 24554, 24804, 24817, 24831, 25018, 25028, 25042
        25264, 25431, 25445, 25512, B25603, 25612, 25754, 25767, 25813, 25878, 25924,
        25986, 26078, 26086, 26116, 26243, 26317, 26319, 26428, 26446, 26550, 26628,
        26776, 26816-7, 26872, 26951, 27012, 27060-1, 27141, 27158, 27270, 27361,
        27364, 27431, 37433, 27506, 27585, 27928, 27930, 28114, 28179-81,
        28195, 28205, 28288-9, 28346, 28400, B28482
        28622, 28625, 28703, 28733, 28748, 28989, 29047, 29052, 29068, 29110-1,
        29181, 29395, 29407, 29470, 29525, 29565, 29572, 29704, 29865, 30078, 30152,
        30273-4, 30283, 30721, 30733, 30764, 30895, 30946, 31066-7, 31168, 31230,
        31232, 31239, 31243, 31252, 31394, 31568, 31610, 31627, 31769, 31801, 31836,
        31887, 32065, 32126, 32188, 32428, 32439
        32568, 32572, 32653, 32717, 32929, 32931, 33080, 33093, 33158, 33168, 33280,
        33429, 33500, 33513, 33524, 33534, 33549, 33602, 33695, 33759, 33937, 33974,
        34052, 34260, 34310, 34416, 34471, 34510, 34548, 34597, 34648, 34730, 34827,
        34924-6, 35031, 35045, 35051, 35091, 35096, 35122, 35200, 35261, 35429-30,
        35450-1, 35460, 35604, 35659, 35708, 35842, 35924, 35931, 35936, 36057, 36190,
        B36191, 36217, 36432, 36521, 36552
        36691-4, 36696, 36726, 36736, 36757, 36822, 36827, 36917, 37052, 37093-4,
        37114, 37123, 37265, 37398, 37569, 37614, 37627, 37695, 37731, 37747, 37984,
        38065, 38098, 38237, 38313, 38372, 38472, B38477, 38478, 38493, 38519, 38632,
        38679, 38793, 38854, 39008-9, 39126, 39145, 39165, 39235, 39265, 39275, 39284,
        39380, B39394, 39442, 39460, 39690, 39695, 39765, 39775, 39828, 39881, 39900,
        39972, 39977, 39981, 40062, 40085, 40119, 40127, 40146, 40165, 40177, 40201,
        40238, 40345

Leaf and plant development and ageing, morphology (*cf.* also Ontogeny ...)
        21593, 21696, 21872, 21882, 21925, 22015, 22113, 23048, 23468, 23490, 23536,
        23604, 23756, 24016, 24115, 24127, 24231, 24267; 25737, 25832, 25857, 25955,
        25986, 26213, 27108, 27213, 27981; 28791, 29254, 29461, 29795, 30916, 30929,
        30979, 31041, 31563, 31840-1; 32696, 33835, 34176, 34243, 34279, 34318,
        34552, 34937, 35021, 35106, 36109, 36196, 36345, 36478-9, 36509; 36746,
        37141, 37227, 37267, 37289, 37589, 37871, 37874, 38590, 38975, 39080, B39523,
        39775, 39979, 40062, 40225, 40241

Leaf area duration see Growth analysis, specific leaf area ...

Leaf area index see Growth analysis, specific leaf area ...

Leaf area measurement
        21658, 21998, 22128-9, 22229, 22255, 22284, 22437, 23158, 23233, 23778,
        23985, 24275, 24604, 24977, 25026; 25250, 25643, 25702, 26133, 26245, 26318,
        26625, B26846, 27102, 27296, 27799, 28044; 29320, 29408, 29455, 30379,
        B30432, 30692, 31267, 31729, 31905; 32846, 33051, 33573, 34088, 34737, 35707,
        36218, 36221, 36237, 36313; 36594, 36881, 37087, 37126, 37342, 37940, 37991,
        38047, 38265, 38533, 38931, 39139, 39195, 39461, 39765

Leaf area ratio see Growth analysis, net assimilation rate ...

Leaf chamber see Assimilation chamber

Leaf dimensions see Extension growth, leaf dimensions

Leaf epidermis, anatomy
        21633, 21695, 21713, 21748, 21822, 21887, 22084-5, 22087, 22263, 22283,
        22315, 22484, 22508, 22624, B23006, 23547, 23628, 23696, 23703, 23757, 23784,
        23873, 23937, 24261, 24436, 24449, 24636, 24704, 24795, 24853, 25031, 25045
                                                                        (continued)

Leaf movements
        21518, 22963; 27820, 28051; 32329; 33627, 34551-2; 39825

Leaf optical properties (*cf.* also Carotenoids absorption spectra *in vivo*; Chlorophyll
        absorption spectra *in vivo*)
        21629, 21695, 22247, 22296, 22302, 22333-5, 22508, 23368, 23523, 23692,
        23992, 24436, 24775, 24818; 25176, 25329, 25560, 25922, 25924, 26117,
        26285-6, 26476, 26483, 26515, 26898, 27036, 27370, 27388, 27460, 27630,
        27728, 27981, 28141, 28188, 28197; 28616, 28866, 28976, 29373, 29398-402,
        29574, 29616-7, 30722, 30881, B31140, 31197, 31547, 31801, 32103, 32175,
        32186-9, 32552; 32665, 32956, 33174, 33432, 33434, 33436, 33471, 33540,
        33998, 34550, 34784, 35449, 35491, 35842, 36011, 36058, 36138, B36191, 36203;
        37396, B37906, 37912, 38215, 38790, 39126, 39380, B39611, 39652, 39928, 39985,
        40293

Leaf resistance   see   Resistances to water vapour ...; Resistance, stomatal ...

Leaf, sun- and shade leaf   see   Leaf anatomy

Leaf temperature (methods and results)
        21569, 21695, 21773, 22505, 22507, 22652, 22875, 23297, 23368, 23447, 23791,
        23992, 24045, 24627, 24718, 24739, 24889, 24934; 25326, 25367, 25450, 25885,
        26114, 26265, 26375, 26397, 26474, 26708, 26854, 26898, 26941, 26970, 27107,
        27154, 27180, 27322-3, 27826, 28085, 28087; 28976, B29023, 29722, 30070,
        30343, B30432, 30455, 30457, 30563, 30583, 30643, 30775, 30943, 31994,
        32094, 32120, 32296; 33436, 33438, 33695, 33984, 34219, 34333-4, 34451,
        34485, 34511, 34747, 34755, 34814, 35096, 35344, 35529, 35595, 35930, 36050,
        36069, 36348, 36403, 36545; 37544, 37808, 37828, 38217, 38508, 38534, 38829-
        -30, 38915, 39090, 39549, 39552, 39813, 39869, 39911, 40079, 40318

Leaf temperature measurement   see   Leaf temperature (methods and results)

Leaf thickness
        21633, 21801, 22039, 22082, 22084, 22157, 22509, 22516, 22787, 22885, 22937,
        23521, 23787, 23848, 23896, 24076, 24299, 24411, 24601
        25404, 25406, 25654, 25902, 27012, 27270, 27386, 27410, 27488-9
        28546, 28622, 28785, 28866, 28989, 29068, 29190, 30187, 30283, 30643, 30733,
        30764, 30861, 30881, 30932, 31067, 31178, 31230, 31251, 31627, 31839, 31904,
        32126, 32188, 32196, 32364
        32652, 33107, 33158, 33368, 33513, 33694-5, 34265, 34310, 34597, 34730,
        34783, 34827, 35045, 35065, 35091, 35096, 35122, 35261, 35350, 35514, 35708,
        35842, 35931, B36191
        36726, 36732, 36736, 37031, 37114, 37122-3, 37569, 37589, 37618, 38087,
        38098, 38372, 38519, 38847, B39394, 39442, 39852, 39881, 39972, 40062,
        40274, 40291

Leaf volume, thickness and internal area measurement (*cf.* also Volume changes in
        leaf ...)   29493; 39690

Light   see   Irradiance ...; Canopy, radiation ...

Lighting system   see   Irradiation, illumination equipment and systems

Linear dichroism   see   Dichroisms ...

Lipids, fatty acids, and carbon fixation pathways
        22411, 23036; 25471, 26693; 29888-9, 30817, 30978, 31676; 33796, 33939,
        34182, 34820, 35005, 35851; 36884, 37249, 37851

Lipids, fatty acids, and carotenoids   34941; 39540, 39862

Lipids, fatty acids, and chlorophyll
        24284; 26681, 27561; 28953; 32876, 32890, 33308, 35284, 35754; 38186, 38561,
        38591, 39146, 39363-4, 39818, 39862, 40135, 40173

Lipids, fatty acids, and chloroplast (chromatophore)
        21575, 21653-4, 22002-3, 22620, 22673, 22694, 22712, 22779, 22820, 23116,
        23228, 23323, 23402, 23459, 23482, 23730, B23933, 24468, 24827, 24878-9,
        B25064, 25092-3
        25263, 25287-9, 26153, 26167, 26378, 26399, 26553, 26769, 26893, 27168,
        27205, 27288, 27468, 27665, 27674, 27784, 27816, 28203
        28650, 28723, 28828, 28840, 29218, 29300, 29318, 29527, 29529, ·29627, 29698,
        29765, 29948, 30150, 30177, 30182, 30449, 30562, 30801, 30814, 30826, 30859,
        30900, 30971-2, 30977, 30986, 31028-9, 31109, 31120-1, 31366, 31413-4, 31557,
        31719, 31779, 31814, 31893, 31937, 32029, 32084-6, 32204, 32267
        32632, 32668, 32858-9, 32876, 32893, 32917, 33113, .33373, 33387, 33481,
        33939-40, 33964, 33994-5, 34180, 34297, 34307, 34426, 34446, 34820, 34856,
        34928-9, 34931, 35054, 35066, 35296, 35400, 35438, 35749, 36101, 36177,
        36238, 36262
        36751, 36816, 36884, 36923, 37062, 37202, 37235-6, 37301, 37335-6, 37371,
        37388, 37524, 37526-7, 37741, 37747, 37758, 37849-51, 37925, 38093, 38186,
        38226, 38383, 38460, B38477, 38479, 38561, 38671, 38766, 38840, 38857,
        38898, 38951, 39001, 39146, 39197, 39227, 39241-2, 39305-6, 39363-4, 39450,
        39508, 39623, 39958, 40021, 40078, 40157, 40295, 40319-20

Lipids, fatty acids, and electron transport chain
        22711, 23227, 23935, 24494; 28693, 30986, 31120, 31813, 31816; 34232-3,
        36240; 37922, 38844, 39306, 39627

Lipids, fatty acids, and gas exchange 26229; 33521, 36557; 38843, 38845, 39363,
        39627

Lipids, fatty acids, and photorespiration 30801

Lutein see Carotenoids ...; Xanthophylls ...

M

"Maintenance" respiration see Respiration, "growth" and "maintenance"

Malate dehydrogenase, methods see Malic enzyme, malate dehydrogenase, methods

Malic enzyme
        21580, 21814, 22054, 22157, 22239, 22381, 22397, 22399, 22635, 22697-8,
        22983, 22997, 23025, 23030-1, 23060, 23231, 23289, 23586, 23797, 23877,
        23938, 24008, 24122, 24126, 24164-5, 24311, 24615, 24734, 24795, 24804,
        25065-6
        25211, 25338, 25409, 25431, 25436, 25768, 25862, 25867, 26004, 26181, 26249,
        26280, 26316-7, B26551, 26623, 26691-2, 26806, 26872, 26921, 27007-8, 27183,
        27198, 27358-9, 27435, 27542, 27621, 27651, 27709, 28040, 28142, 28219,
        28355, 28357, 28399
        28613, 28786-7, 28989, 29085, 29098, 29104, 29312-3, 29462, 29772, 29833,
        29846, 29850, 29860, 29862, 30232, 30234, 30332, 30431, 30537-8, 30783-4,
        30823, 30936, 30951, 31034, 31440-1, 31449, 31451-2, 31611, 32364, 32435,
        32477
        32867, 32912, 33099, 33148, 33554-5, 33612, 33849, 33912-3, 34147, 34253,
        34256, 34259, 34353, 34525, 34549, 34684, 34950, 34952, 35192, 35430, 35434,
        35442, 35467, 35471, 35675, 36019, 36121, 36241, 36366, 36426, 36443, 36533
        36664, 36719-20, 36725, 37034, 37120, 37197, 37242, 37317, 37387, 37841,
        37934, 38125, 38175, 38235-6, 38298, 38428, 38560, 38788, 39153, 39156,
        39201, 39217, 39235, 39237, 39264-6, 39284, 39366, 39416, 39466, 39500,
        39604, 39724-5, 39809, 39876, 39953, 40256, 40360

Malic enzyme, malate dehydrogenase, methods 22381, 25065-6; 29846, 32338; 34631;
        36720, 38566

Mass culture of algae   see   Algae mass cultures ...

Maximum photosynthetic rate   see   Potential photosynthetic rate ...

Mehler reaction   see   Photosystem 1 activity ...

Membrane transport of CO_2 see CO_2 transfer across membranes

Mesophyll resistance   see   Resistance, intracellular (mesophyll)

Microbody   see   Peroxisome

Microelements   see   Mineral elements (other than N,P,K) ...

Mineral elements (N,P,K) and algae productivity
        21735, 22132, 22417, 22551, 22976, 23301, 23606, 23932, 24248, 24349, 24687,
        24764, 25130; 25389, 25580, 26212, 26273, 26358, 26407, 26510, 26543, 26882,
        27227, 27652, 27913, 27960, 28017, 28025, 28317; 28536-7, 28830, 28996,
        29226, 29375, 29466, 29653, 29825, 29917, 29952, 30281, 30572, 31382, 31513,
        32089, 32268, 32308, 32404, 32493, 32518; 32696, 33126, 33238, 33340, 33634,
        33656, 33856, 33896, 33971, 34592, 35063, 35070, 35584, 35690, 35911, 36028,
        36216, 36369, 36553; 36868, 37058, 37204, 37436-7, 37450, 37453, 37483,
        37673, 38031, 38409, 38973, 39108, 39663, 39696, 39787, 39877, 39908, 40089,
        40224

Mineral elements (N,P,K) and biliproteins
        22496; 30531, 31423, 32102; 37022, 38212, 38899, 39046, 39829, 40034

Mineral elements (N,P,K) and carbon fixation pathways
        21672, 21785, 22150, 22552, 22685, 22837, 23333, 23386, 23480, 23660, 24064,
        24442, 24734, 24752, 24854, 24982, 25009, 25035-6; 25221, 25372, 25449,
        25772, 26525, 26539, 26866, 27048, 27358, 27464, 27593, 28146, 28183, 28220,
        28303; 29153, 29581, 29891, 29945, 29983, 30037, 30584, 30844, 30952, 30988,
        31315, 32426; 32742-3, 32835, 32913, 32996, 33041, 33654, 33944, 34972,
        35183-4, 35197, 35326, 35592-3, 35996, 36116, 36146, 36148, 36331, 36365,
        36412; 36728, 36874, 36917, 37216, 37331, 37470, 37866, 37928, 37931, 38103,
        38126, 38192, 38643, 38738, 38771, 39099, 39213, 39239, 39275, 39351, 39603,
        39800, 40123

Mineral elements (N,P,K) and carotenoids
        21697, 22518, 22565, 23375, 23550, 23853, 24145-6, 24540, 24597, 24778,
        24978; 25479, 25553, 25871, 27022, 27396, 27715, 28185; 29456, 29762, 30400,
        30452, 30802, 31276, 31384, 31406, 31927, 32016, 32103, 32212; 32601, 33322,
        33572, 35364, 35890, 36146, 36456, 35680; 37620-1, 38272, 38439, 38581,
        38899, 39192, 39399, 39829, 39995, 40328

Mineral elements (N, P, K) and chlorophyll
        21544, 21697, 21767, 21877, 21949, 22012, 22064, 22197, 22247, 22496, 22512,
        22516, 22518, 22565, 22779, 22808, 22863, 22893, 23258, 23549-50, 23593,
        23663, 23814, 23853, 24233, 24407, 24540, 24571, 24597, 24686, 24732, 24752,
        24778, 24857, 24896, 24956, 24965, 24978, 25108, 25117, 25130
        25316, 25333-4, 25479, 25518, 25680, 25764, 25871, 25990, 26050, 26543,
        26605, 26608, 26797, 27129, 27227, 27258, 27396, 27652, 27715, 27973,
        28017, 28025, 28124, 28185, 28280, 28312, 28317
        28661, 28762, 29127, 29155, 29163, 29237, 29338, 29457, 29762, 29945, 30350,
        30400, 30452, 30620, 30693, 30802, 30933, 31083, 31139, 31276, 31406, 31495,
        31533, 31602, 31927, 32103, 32212, 32493
        32601, 32729, 32742, 33041, 33094, 33121, 33322, 33572, 33608, 33641, 33654,
        33856, 34047, 34100, 34242, 34465, 34630, 34693, B34706, 34782, 34838,
        34846, 34956, 34972, 35017, 35035, 35042, 35184, 35337, 35364, 35377, 35389,
        35577, 35637-8, 35890, 35958, 35973, 36116, 36146, 36215, 36332, 36390,
        36456, 36493

                                                              (continued)

Mineral elements (N,P,K) and gas exchange (continued)
      25163, 25221, 25223, 25317, 25372, 25404, 25642, 25706, 25772, 25809, 25824,
      25950, 25996, 26020, 26106, 26166, 26471, 26493, 26499, 26566, 26651, 26685,
      26742, 26812, 26896, 26971, 27009, 27129, 27227, 27267, 27327, 27347, 27366,
      27396, 27416, 27464, 27466, 27595, 27672, 27920, 27980, 27994, 28090, 28124,
      28146, 28167, 28169, 28183, 28206, 28301, 28375, 28421, 28461
      28527, 28556, 28783, 28899, 29102, 29155, 29331, B29468, 29726, 29945, 29975,
      30047, 30131, 30243, 30377, 30523, 30642, 30706, 30844, 31156, 31218, 31249,
      31283, 31292, 31320, 31492-4, 31525, 31527, 31570, 31939, 31978, 32104, 32115
      32177, 32206, 32343, 32545
      32627, 32742-3, 32787, 32818, 32832, 32835-6, 32867, 32913, 32923, 32996,
      33002, 33041, 33107, 33121, 33193, 33532, 33729, 33808, 33944, 34100, 34287,
      34475, 34520, 34630, 34693, 34782-3, 34833, B34849, 34888, 34946, 34956,
      34967, 34972, 35155, 35337, 35373, 35377, 35390, 35402, 35459, 35470, 35543,
      35554, 35624, 35636-7, 35659, 35761, 35886, 35954, 35973, 35996, 36002,
      36014, 36060, 36253, 36332, 36359
      36597, 36698, 36874, 36877, 37188, 37259, 37470, 37472, B37532, 37701-2,
      37753, 37768, 37786, 37801, 37863, 37928-9, 37931, 38018-9, 38103, 38296,
      38417, 38421, 38434, 38441, 38493, 38496, 38518, 38649, 38657, 38659,
      B38739, 38753, 38761, 38770, 38783, 38804, 38844, 38924, 38984, 39019,
      39025, 39052, 39079, 39087-8, 39099, 39276, 39351, 39395, 39566, 39627,
      39665, 39672, 39751, 39764, 39787, 39794, 39808, 39852, 39946, 39980-1,
      39995, 40008, 40082, 40103, 40166, 40217, 40265, B40331, 40368

Mineral elements (N,P,K) and photorespiration
      21572, 21587, 24627; 27416, 28183, 28220; 28899, 31678, 32088, 32104;
      34410, 36116; 37216, 37753, 38103, 38496, 39099, 39401

Mineral elements (N,P,K) and resistances to CO_2 and water vapour transfer
      22964, 23134; 25604, 26971, 27595, 28220, 28461; 28899, 28914, 29053, 29331,
      31156, 31212; 32913, 33041, 35335; 37108, 37470, 37753, 38018-9, 38848,
      38920, 39099, 39134, 39224, 39751, 40265

Mineral elements (N,P,K) and respiration
      24001, 25048-9; 35155, 35636, 35761, 35996, 36116, 36303; 37702, 37753,
      37866, 39099, 39395, 39401, 39808, 39946, 40103

Mineral elements (other than N,P,K) and algae productivity
      22132, 22417, 22457, 22976, 23099, 24349, 25130, 25154; 25272, 26089, 26212,
      26273, 27224; 28536-7, 28829, 29937, 30501, 30827, 30843, 32108, 32268;
      33446, 33619, 34272, 35336; 37246, 38782, 39291, 39325, 39908, 39959

Mineral elements (other than N,P,K) and biliproteins   24345;29498,31423;37022,39497

Mineral elements (other than N, P, K) and carbon fixation pathways
      21592, 21604, 21655, 21671, 21722, 22069, 22077, 22080, 22119, 22318, 22671,
      22808, 22975, 23343, 23492, 24255, 24730, 24843, 25148
      25284-5, 26018, 26275, 26549, B26551, 26560, 26958, 27063, 27074, 27435,
      27492, 27583, 27777, 27797, 27860, 28303, 28389-90
      28652, 29863, 29892, 30707, 30856, 30952, 30960, 31264, 31343, 31806, 32244,
      32334, 32426, 32536
      32720, 32767, 32968, 33154, 33648, 33772, 33943-4, 33981, 34062, 34573,
      34657, 34817, 34950, 34972, 35247, 35326, 35428, 35461, 35940, 36412, 36415-
      -6, 36572
      36663, 36719, 36754, 36917, 36922,37022, 37024, 37149, 37170, 37282, 37471,
      37930, 37934, 37947, 38036, 38247, 38289, 38356, 38487, 38560, 38641, 38643,
      39006, 39009, 39204, 39338, 39410, 39475, 40020, 40090, 40203, 40205

Mineral elements (other than N,P,K) and carotenoids
      21948, 22565, 23174-6, 23259, 23550, 23568, 23781, 23853, 23918; 26016,
      26019, 26928, 27487, 27715, 27850; 29498, 29947, 30494, 30549, 31171, 31276,
      31403, 32016; 32769, 32829, 33224, 33269, 35321, 36456; 37620-1, 38363,
      38443, 39114, 39408-10, 39827, 39995, 40020, 40289

Mineral elements (other than N, P, K) and chlorophyll
    21533, 21607-8, 21679, 21694, 21701, 21948, 22094, 22107, 22181, 22194,
    22512, 22565, 22623, 22786, 22808, 22843, 22936, 23052, 23054, 23174-6,
    23208, 23246, 23259, 23428-9, 23549-50, 23568, 23614, 23656, 23781, 23793,
    23853, 23918, 23954, 24012, 24047, 24051, 24075, 24077-9, 24103, 24233,
    24236, 24266, 24345, 24539, 24571, 24615, 24713, 24728, 24747, 24845, 24911,
    24956, 24964-5, 24985-6, 25083, 25108, 25130
    25246, 25297, 25316, 25329-30, 25368, 25435, 25456, 25501, 25700, 25819,
    26016, 26019, 26338, 26351, 26479, B26551, 26682, 26684, 26813, 26822, 26852,
    26926-30, 27074, 27085, 27140, 27317-8, 27423, 27429, 27479, 27487-9, 27671,
    27685, 27715, 27758, 27850, 28104, 28138, 28176, 28270, 28276-7, 28309-13,
    28318, 28420, 28467
    28585-6, 28595, 28676, 28678, 28771, 28886, 29127, 29137, 29163, 29233, 29268,
    29498, 29542, 29562, 29635, 29755, 29906, 29947, 30013, 30354, 30494, 30518,
    30541, 30549, 30569, 30576, 30610-2, 30620, 30644, 30827, 30875, 30950, 30971,
    31043, 31116, 31171, 31187, 31209, 31213, 31264, 31276, 31403, 31544, 31657,
    31692, 32072, 32108, 32302, 32380
    32574, 32603, 32659-61, 32676, 32725, 32746, 32769, 32829, 33025, 33143,
    33224, 33268-9, 33446, 33775, 33875, 33880, 34042, 34165-7, 34183, 34209,
    34222, 34272-3, 34317, 34422, 34610, 34616, 34815, 34837-8, 34889, 34953,
    34972, 35060, 35093, 35247, 35321, 35473, 35657, 35940, 35943, 35956, 36023,
    36215, 36278, 36390, 36456, 36520, 36537, 36574
    36614-5, 36642, 36774, 36803, 36922, 36995, 37017-8, 37022, 37056, 37118,
    37157, 37170, 37246-7, 37264, 37267, 37276, 37349, 37502, 37511-2, 37520,
    37620-1, 37665, 37733, 37871, 37979, B38118, 38193, 38363, 38443, 38451,
    38651, 38852, 38861, 38863, 38977, 39029-30, 39114, 39287, 39325, 39350,
    39355, 39408-10, 39456, 39515, 39609, 39706, 39745, 39779, 39881, 39995,
    40259, 40289, 40303, 40350

Mineral elements (other than N, P, K) and chloroplast (chromatophore)
    22056, 22062, 22565, 22615, 22671, 22967, 22975, 23075, 23281, 23549, 24030,
    24075, 24386, 24615, 24985, 25029, 25084, 25096
    25697, 25700, 26447-9, 27261, 27487-9, 27492, 27715, 27923, 28084, 28104,
    28198, 28252, 28263
    28518, 28584-6, 29578, 29758, 29814, 29830, 29932, 30428, 30856, 31056,
    31209, 31264, 31375, 31697, 32042, 32079, 32319, 32535
    32659-61, 32769, 32964, 33648, 33791-2, 33900, 34033, 34127, 34155, 34165,
    34223, 34234, 34317, 34616, 35383, 35473, B35862, 35956, 35959, 35987,
    36323, 36520
    36773, 36829, 37055, 37270, 37365, 37665, 37667, 37730, 37843, 37852, 37932,
    37976, 38468, 38812, 38861, 38889, 38977, 38999, 39040, 39287, 39550, 39622,
    39745, 39881, 39887, 39999, 40322, 40367

Mineral elements (other than N,P,K) and ecosystem and plant productivity
    22035, 22638, 22908-9, 23271, 24075, 24080, 24331, 24748, 25125; 25435,
    25781, 26093, 26204, 26216, 26263, 26684, 26852, 26926-8, 27366, 27413,
    27489, 27707, 28113, 28312; 28711, 29042, 29157, 29163, 29233, 29448, 29705,
    29874, 29921, 30095, 31107, 31116, 31209, 31535, 31940, 32079; 33242, 33398,
    33668, 34507, 34635, 34916, 34972, 35256, 35321, 35473, 35691, 36021; 36641,
    36922, 37017-8, 37022, 37168, 37170, 37349, 37871, 38363, 38883, 38949,
    39031, 39109, 39325, 39350, 39355, 39409, 39417, 39750, 39785, 39788, 39995,
    40020, 40220-3

Mineral elements (other than N, P, K) and electron transport chain
    21646, 21688, 21694, 21746, 21764, 21818, 21820, 21824, 21888, 21935, 22104-5,
    22233, 22491, 22515, 22566, 23394-6, 23428, 23477, 23842, 23941-2, 24034,
    24077, 24102, 24133, 24238, 24258, 24494, 24667, 24730, 24815-6, 24856,
    24874, 25006, 25144-5
    25242, 25245, 25262, 25315-6, 25330, 25417, 25420, 25432, 25591, 25636,
    25657, 25671, 25697, 25700, 25830, 25914, 26193, 26210, 26224, 26275, 26335,
    26374, 26388, 26488, 26726, 26730, 26852, 26925, 26927, 26955, 26967, 27005,
    27037, 27039, 27104, 27131, 27244, 27487, 27489, 27495, 27562, 27747, 27776,
    27907, 27923, 28027, 28111-2, 28198, 28420, 28449
                                                              (continued)

Model of canopy photosynthesis ... (continued)
        23871, 23894, 23910-1, 23943, 23976, 24003-4, 24105, 24186, 24207, 24214-5,
        24401, 24473-4, 24610, 24689, 24740, 24773, 24798, 24835, 24839, 24863,
        24880, 24887, 24947, 24975, 25020, 25024
        25167, 25195, 25198, 25224, 25231, 25253, 25279, 25291, 25451, 25472, 25593,
        25660, 25681, 25683, 26026, 26047, 26145, 26296, 26331, 26364, 26409, 26419,
        26530, 26548, 26569, 26840, B26846, 26877, 26879, 26883, 26919-20, 26954,
        27030, 27095, 27139, 27209, 27361, 27437, 27601, 27630, 27656, 27658, 27682,
        27735, 27782, 27820, 27830, 27888, 27919, 28070, 28089, 28097, 28109-10,
        28195, 28231, 28241, 28413, 28431, 28459-60, 28462
        28689, 28706, 28711, 28865, 28899, 28972, 29045, 29086-8, 29103, 29123, 29129,
        29168, 29222, 29290, 29515, 29574, 29605, 29663, B29707, 29886-7, 29967,
        29975, 30011, 30029, 30048-9, 30084, B30281, 30349, 30387, B30511, 30515,
        30532, 30555, 30593, 30608, 30916, 30918, 30976, 31113-5, 31198, 31279, 31332,
        31517, 31527, 31537, 31539, 31546, 31760, 31849, 31859, 31935, 31962, 31967,
        32046, B32150, 32152, 32164, 32187, 32210, 32277, 32292, 32328, 32436
        32568-9, 32611, 32663, B32852, 32882, 32919, 33023, 33050, 33104, 33175,
        33199, 33223, 33245, 33260, 33321, 33342, B33343, 33358-9, 33395, B33510,
        33528, 33546-7, 33611, 33741, 33808, 33848, 33897, 33928, 34012, 34036-7,
        34190, 34195, 34228, 34288, 34291, 34323, 34479, 34536, 34611, 34613, 34622,
        34625, 34650, 34677, 34713, 34785, 34896, 34940, 34970, 35037, 35055, 35082-
        -3, 35092, 35118, 35150-1, 35169, 35171-2, 35243, 35272, 35608, 35610,
        35667, 35808, 35845, 35883, 36014, 36068, 36092, 36132-3, B36191, 36203,
        36242, 36565
        36602, 37063-4, 37073, 37123, 37128, 37190, 37192, 37220, 37289, 37292,
        B37299, 37334, 37352, 37373, 37389, 37394, 37407, 37588, 37696, 37709,
        37719, 37751, 37895, 37898, 37908, 37915, 37974, 38028, 38034, 38120,
        38124, 38174, 38209, 38485, 38516, 38520, 38563-4, 38764, 38776, 38858,
        38894, 38988, 39158, B39523, 39536, 39565, 39567, 39636, 39639, 39667,
        39692, B39711, 39748, 39799, 39807, 39824, 39905, 39980, 40043, 40118,
        40133, 40138, 40145, 40191, 40194

Model of carbon fixation pathways
        22794; 26315; 30631; 35686, 36091; 36917, 37471, 38168, 38235, 38673,
        38758

Model of chlorophyll energetics (cf. also Chlorophyll in model systems)
        21662, 21962, 22048, 22751, 22794, 23102, 24811, 24813; 27844, 28155, 28345;
        29519, 29521, 30189, 30347, 30359, 30679, 30759, 31556, 31649, 32254; 33141,
        33382, 33618, 34210, 34756-7, 34760, 34797, 36349, 36352, 36567; 36970,
        37598, 37759, 38537, 38571, 38653, 39149, 39373, 40201

Model of electron transport chain
        21752, 22794, 23241, 23520, 23633, 24811, 24886; 25719, 25984-5, 26156,
        27309, 27324, 27663, 27953, 28155; 28729, 29001, 29025, 29359, 29519, 29521,
        29654, 29720, 29830, 30000, 30158, 30479, 30626, 31346, 31368, 31410, 31635,
        31702, 31906, 32432; 34148, 34389, 34460, 34601, 35209, 35218, 35231, 35530-
        -2, 35649, 36106, 36567; 37370, 37561, 37762, 38105, 38319, 38652, 38673,
        38957, 39268, 39491, 39519, 39550, 39648, 40057, 40201, 40229, 40302, 40332

Model of leaf gas exchange, ideotype of leaf photosynthesis
        21620, 22038-40, 22479, 22686, 22749, 22875, 22993, 23170, 23380, 23467,
        23481, 23514, 23612, 23679, 23994, 24426, 24473, 24507, 24601, 24624, 24863,
        24947
        25167-8, 25231, 25300, 25476, 25609-11, 25613, 25684, 25722, 26047, 26088,
        26154, 26169, 26200, 26235, 26238, 26307, 26329, 26364, 26541, 26567, 26572,
        26690, 26692, 26843, 26883, 27182, 27359, 27424-5, 27437, 27524, 27754,
        27830, 27865, 27873, 27998, 28106-7, 28109, 28125, 28182, 28275, 28369-70
        28976, 29088, 29090, 29130, 29188, 29399, 29776, B30281, 30283, 30510, B30511,
        30515, 30583, 31214, 31256, 31258, 31261, 31372, 31432, 31470, 31484, 31592,
        31702, 31836, 31848, 31867, 32073, 32110, 32118-9, 32188, 32369
        32753, 32826, 33039, 33193, 33330, 33433, 33465, 33500, 33504-5, 33695,
        33845, 34218-9, 34289-90, 34511, 34550, 34625, 34881, 35285-6, 35450, 35842,
        36198, 36403, 36472, 36522-3, 36545
                                                    (continued)

Model of leaf gas exchange, ideotype of leaf photosynthesis (continued)
     37122-3, 37176, 37224, 37292, 37397, 37575, 37617, 37736, 37774, 38099,
     38168, 38522, 39016, 39093, 39309, 39360, 39536, 39650, 39667, 39867, 40080,
     40154, 40297, 40332

Model of photophosphorylation
     21752, 22794; 25522; 33142, 33819, 34025, 34033; 37704

Model of plant photosynthesis, ideotype
     22479, 22749, 22993, 23068, 23603, 25139; 25238, 25722, 25771, 25955, 26169,
     26364, 26541, 26690, 26692, 26843, 27097, 27361, 27424-5, 27437, 27524,
     27768, 27830, 28041, 28369-70; 29088, 29129, 29657, 29822, 29984, B30281,
     31148, 32171, 32210-1, 32525; B33343, 33368, 33465, 33507, 33595, 33845,
     34036, 34218-9, 34923, 34925, 34946, 35001-2, 35004, 35255, 35341, 35449,
     35534, 36344; 36695, 37334, 37587, 37751, 37825, 27898-9, 37909, 37953,
     38132, 38431, 38590, B38739, 38839, 39309, 40060-1, 40291, 40341, 40360

Morphogenesis see Leaf and plant development and ageing, morphology

Mutagens and further organic substances and algae productivity
     26287, 26825, 28016; 32669, 32694; 37478, 40364

Mutagens and further organic substances and carbon fixation pathways
     21784, 22346, 23211, 24101, 24139, 24352, 24383; 25529, B26551, 25714; 31154,
     31602, 32406, 32529; 33155, 33157, 33770-1, 33795, 34534, 35142, 35794,
     36413, 36415; 38247, 38289, 38955, 39167, 39505, 40197, 40204, 40206, 40213

Mutagens and further organic substances and carotenoids
     21904, 24203, 24731; 28239; 28948-9, 31851; 33177, 33476, 34583, 34791,
     36232; 37481, 38980, 39758

Mutagens and further organic substances and chlorophyll
     21596, 21729, 21992, 22204, 22337, 22506, 22674-5, 22827, 22864, 22911,
     22959, 23048, 23118, 23272, 23528, 23562, 23576, 23833, 23892, 24021, 24101,
     24383, 24388, 24539, 24653, 24728, 24731, 24847, 24916, 25129;
     25459, 25864, 26076, 26099, 27339, 27406, 28239, 28270; 28710, 28938,
     28961, 29177, 29403, 29499, 30202, 30857, 31602, 31744, 31850-1, 32511;
     33177, 33418, 33578, 33601, 33955, 33980, 34145-6, 34245, 34769, 34791,
     35019, 35232, 35898, 36083-4; 36867- 37362, 37401, 37481, 37519, 38203,
     38980, 39439, 39505, 39758, 39833, 40102, 40372

Mutagens and further organic substances and chloroplast (chromatophore)
     21700, 21992, 24365, 24847; 25529, 26008, 26260, 26342, 26496, 27738, 28263,
     28278; 29532, 30110, 30984, 31744, 32222; 33417-8, 33980, 35071, 35800,
     36302; 36913, 37362, 37916, 37921, 38732, 39505, 40100

Mutagens and further organic substances and ecosystem and plant productivity
     23817, 24363; 29411, 31509; 33680, 36371; 37194, 37401, 38943, 38979, 39856,
     39979

Mutagens and further organic substances and electron transport chain
     21698, 22244, 22345, 22513, 22666, 22801, 22891, 22894, 23077, 23435, 23541,
     23906, 24104, B24176, 24226, 24362; 25840, 26008, 26496, 28211, 28263;
     28523-4, 28875, 29184, 29416, 29532, 30110, 30115, 30984-5, 31668; 32649,
     32669, 33882, 33980, 34769, 35264, 35410, 35545, 36223; 36749, 36797, 37178,
     37362, 37644, 38552, 39020, 39075, 39397, 40120, 40183, 40185, 40206

Mutagens and further organic substances and gas exchange
     21696, 21817, 22418, 24021, 24225, 24557, 24580, 24884, 25058, 25122; 25458,
     25949, 26008, 26246, 26497, 27062, 27239, 27406, 27774, 27935, 28263, 28278,
     28480, 28484; 28898, 28961, 29177, 29298, 29416, 30490, 31292, 31602, 31969,
     32529; 32625, 32694, 33153, 33218, 33622, 34484, 34799, 35019, 35061, 35939,
     36371, 36557; 37481, 37950, 38145, 38532, 38943, 39064, 39856, 39982, 40209,
     40324

Mutagens and further organic substances and photorespiration
        23150; 28480; 28898, 30468, 32529; 32841, 33153, 35061; 38552, 38955, 40204,
        40324

Mutagens and further organic substances and resistances to CO_2 and water vapour trans-
        fer 28898; 38532, 39134

Mutagens and further organic substances and respiration
        23079, 24021, 24338, 24557; 28898; 35019, 35545

Mutants and algae productivity 32346

Mutants, biliproteins in
        28008; 28598, 28603; 34478, 35706; 36654, 37315, 37651, 39344, 39484, 39574,
        39966

Mutants, carbon fixation pathways in
        25497, 26597, 26809-10, 27246-7, 27272, 27668, 27698; 29040, 29098, 29322,
        29777, 29807, 30232, 30235, 30859, 30936, 30962, 31117, 31223, 32097; 32623,
        32958, 33625, 33817, 33827, 33842, 34379, 34456, 34692, 34905, 34958, 35035,
        35152, 35490, 35703, 35794, 35877, 35970, 36409, 36411; 36767, 37132, 37201,
        37650, 37756, 38033, 38625, 39571, 39713, 39733

Mutants, carotenoids in
        21613, 21870, 22393-5, 22616, 23021, 23272, 23404, 23746, 23810-1, 23825,
        24283, 24357, 24398, 24621; 25335, 25838, 26232, 26597, 26756, 27211, 27834,
        27884, 28261; 28551, 28816, 28944-7, 29025, 29240, 29268, 29661, 29712,
        29807, 29995-6, 30027, 30302, 30487, 30496, 30730, 30948, 31119, 31139, 31223,
        31349, 31504, 31619, 31623, 31832, 31834, 32014, 32018, 32190, 32195, 32481;
        33010, 33472, 33579, 33817, 33827, 33842, 33901, 34435, 34490-1, 34608,
        34642, 35596, 35617, 35763, 35780, B35862, 35901, 36411; 36654, 36894, 37155-
        -6, 37359, 37361, 37522, 37613, 37685, 38110, 38305, 38315, 38332, 38365,
        38380-1, 38992, 39285, 39474, 39568, 39606, 39766, 40244-5, 40355

Mutants, chlorophyll in
        21603, 21678, 21729, 21749, 21771, 21860, 21870, 22079, 22117, 22204-5,
        22317, 22393, 22395, 22432, 22596, 22630, 22674-5, 22683, 22799, 22800,
        22953, 22987, 23021, 23159, 23266, 23272, 23278, 23312, 23404, 23718, B23767,
        23768, 23810-2, 23823, 23833, 23954, 24041, 24109, 24141, 24190, 24235,
        24280, 24283, 24299, 24357, 24388, 24398, 24466, 24584, 24603, 24621, 24746,
        24788, 24962, 25088, 25133
        25184-5, 25378, 25410, 25427, 25456, 25504, 25690, 25715, 25734, 25749, 25791,
        25838, 25864, 25900, 26053, 26094, 26362, 26395, 26456, 26562, 26597, 26632-3,
        26700, 26755-6, 26768, 26774, 26829-30, 26899, 26990, 27137, 27211, 27246,
        27272, 27346, 27505, 27690, 27718, 27751, 27809, 27834-5, 27879, 27884,
        27902-3, 27924, 27973-4, 28008, 28148, 28216, 28218, 28255, 28260-1
        28597, 28639, 28816, 28863, 28892, 28942, 28946, 29012, 29025, 29040, 29067,
        29094, 29098, 29118, 29240, 29265, 29268, 29283, 29447, 29459, 29538, 29602,
        29661, 29706, 29807, 29818, 29875, 29878, 30006-7, 30027, 30050, 30228, 30302,
        30318, 30487, 30495-7, 30569-70, 30687-8, 30730, 30859, 30935, 30948, 30962,
        31044, 31071, 31118-9, 31139, 31223, 31298, 31349, 31355, 31431, 31575, 31588,
        31619, 31623, 31640, 31655, 31666, 31722, 31756, 31772, 31878, 31917, 32034,
        32144, 32190, 32222, 32228, 32272, 32326, 32360-1, 32419, 32432, 32481-2
        32651, 32707, 32788, 32861, 32889, 32892, 32917, 32975, 32977, 33025-6,
        33064, 33109, 33303, 33472, 33516, 33579, 33600-1, 33784, 33816, 33827,
        33842, 33855, 33880, 33901, 34050, 34098, 34236, 34261, 34305, 34327, 34340-
        -2, 34418, 34435, 34478, 34490-1, 34529, 34572, 34607-8, 34610, 34761,
        34770-1, 34779-80, 34858, 34883, 34959, 35035, 35191, 35376, 35502, 35560,
        35596, 35617-8, 35633, 35687, 35706, 35763, 35835, B35862, 35887, 35901,
        35922, 35970, 36029, 36119, 36156-7, 36187, 36230, 36258, 36266, 36268,
        36409, 36411
        36598, 36654, 36697, 36894-5, 37056, 37155-6, 37315-6, 37359, 37361, 37366,
        37461, 37558, 37607, 37613, 37703, 37738, 37756-7, 37835, 37918, 37957, 37963,
        38033, 38039, 38110, 38119, 38208, 38213, 38225, 38315, 38365, 38376, 38379-
                                                                        (continued)

Mutants, chlorophyll in (continued)
        -82, 38469, 38557, 38701, 38723-4, 38896, 39042, 39180, 39229, 39285, 39292,
        39310, 39318, 39344, 39452, 39474, 39484, 39517, 39568, 39571, 39574, 39606,
        39621, 39646, 39738-9, 39766, 39835, 39966, 40014, 40069, 40122, 40136,
        40179, 40314, 40355

Mutants, chloroplast (chromatophore) in
        21613, 21825, 21860, 22078, 22117, 22393, 22395, 22643, 22737, 22800, 23031,
        23129, 23265-6, 23279-80, 23512, 23636, 23746-7, 23820-2, 23849, 23885,
        23919, 24067, 24357, 24369, 24381-2, 24398, 24402, 24603, 24696, 24788,
        24847, 24855, 24959
        25184, 25378, 25427, 25463, 25497, 25504, 25726, 25749, 26094, 26144, 26260,
        26362, 26597, 26649, 26699-700, 26768, 26774, 26831, 26990, 26995, 27211,
        27246, 27268, 27351-2, 27467, 27576, 27614, 27668, 27694, 27884, 27924, 27962,
        27974, 28373
        28554, 28759, 28804-5, 28816, 28850, 28863, 28897, 28909, 28942, 28946, 29538,
        29706, 29777, 29807, 29824, 29839, 30006, 30027, 30345, 30569, 30673, 30688,
        30730, 31104-5, 31139, 31223, 31336, 31588, 31619, 31721-2, 31820, 31832,
        31834, 31876-8, 31917, 31975, 32030, 32221-2, 32276, 32350, 32397
        32651, 32811, 32958, 32975, 33598, 33625, 33816-7, 33827, 33861, 33942,
        34026, 34491, 34642, 34692, 34762, 34771, 34883, 35026, 35035, 35067, 35071,
        35128, 35700, 35703, 35873-4, 35922, 35967, 35988, 36257
        36590, 36793, 36894-5, 36957, 37056, 37201, 37756-7, 37770-1, 38208, 38382,
        38397, 38731-2, 38806, 38896, 38950, 38963, 39040, 39229, 39258, 39506,
        39571, 39606, 39646, 39922, 40051-2, 40165, 40251, 40255

Mutants, ecosystem and plant productivity of
        25827, 27991; 29067, 29565; 33899, 34959; 38315, 38809, 40069

Mutants, electron transport chain in
        21575, 21807, 21888, 21935, 22078-9, 22117, 22145, 22325, 22498, 22501,
        22842, 22957, 23404, B23407, 23548, 23822, 24041, 24190, 24280, 24283, 24329,
        24351, 24466-7, 24545, 24603, 24858, 24955, 25088
        25184, 25335, 25370, 25456, 26146, 26362, 26456, 26557, 26632-3, 26773-4,
        26793, 26830, 26887, 26990, 27086, 27132, 27137, 27246, 27668, 27681, 27809,
        27834, 27924, 28218, 28469
        28641, 28759, 28815-6, 28904, 29012, 29025, 29192, 29264, 29272, 29282, 29602,
        29664, 29807, 29818, 29840, 29960, 29995, 30006, 30027, 30569-70, 30599-600,
        30762, 30803, 30859, 30962, 31044, 31118-9, 31298, 31478, 31489, 31564, 31575,
        31588, 31665, 31668, 31722, 31877-8, 31907, 32029, 32097, 32228, 32417, 32419,
        32464, 32481-2
        32811, 32860, 32889, 33186, 33190, 33300, 33311, 33597-8, 33842, 33853,
        33901, 34026, 34098, 34185, 34214, 34303, 34478, 34490, 34529, 34607, 34726-
        -7, 34762, 34801, 34858, 34882-3, 35376, 35617-8, 35700, 35763, 35780,
        35922, 36257-8, 36409, 36411
        36731, 36957, 37315-6, 37350, 37558, 37607, 37613, 37664, 37708, 37737,
        37937, 37955-7, 38063, 38208, 38291, 38315, 38376, 38381-2, B38415, 38469,
        38619-20, 38724, 38731-2, 38963, 39176, 39258, 39358, 39473-4, 39574, 40098,
        40255

Mutants, gas exchange in
        21677, 21771, 22265, 22393, 22842, 23278, 23718, 23845, 23849, 24280, 24283,
        24466, 24483, 24848
        25184-5, 25335, 26362, 26597, 26632-3, 26700, 26829, 27132, 27187, 27210,
        27247, 27272, 27357, 27924
        29040, 29091, 29118, 29268, 29629, 29661, 29706, 29807, 29848, 30228, 30495,
        B30511, 30859, 30935-6, 30948, 30962, 31118-9, 31619, 31772, 32221, 32228
        32892, 33816, 33842, 33899, 33901, 34404, 34456, 34779, 34801, 34858, 34905,
        34958-9, 35035, 35121, 35191, 35856, 36411
        37201, 37315, 37613, 37737, 37757, 38663, 38723-4, 39712-3, 40009, B40331

Mutants, photorespiration in   28477; 30886, 31118-9; 34379, 34905, 35152; 38625,
        38663, 39474

Mutants, photosynthetic, isolation and selection   23513; 25378; 31668; 32623, 35780

Mutants, resistances to CO_2 and water vapour transfer in   27187; 30936, 30991; 34959

Mutants, respiration in   34490, 34905

N

N_2, anaerobic atmosphere, and biliproteins   37562

N_2, anaerobic atmosphere, and carbon fixation pathways 26866; 36658, 38908, 39024

N_2, anaerobic atmosphere, and chlorophyll   26736; 36292; 38908

N_2, anaerobic atmosphere, and chloroplast (chromatophore) 36645

N_2, anaerobic atmosphere, and electron transport chain 31919, 32050; 38482

N_2, anaerobic atmosphere, and gas exchange 21572, 23764; 28994; 38908, 39694

N_2, anaerobic atmosphere, and photorespiration 36687

N_2 and photosynthates 39130

NAD see NADP ...

NADP, NAD 21530, 21548, 21611, 21638, 21673, 21737, 21743, 21837, 22051, 22134,
        22164, 22347, 22601, 22659, 22706, 22786, 22821, 23078, 23146, 23170, 23201,
        23231, 23261, 23351, 23370, 23591, 23724, 23726, 23798, 23934, 24172, 24187,
        24242, 24258, 24384, 24415, 24468, 24489, 24491, 24651, 24665, 24790, 24804,
        24875, 25006, 25030, 25124
        25184, 25211, 25320, 25506, 25564, 25675, 25862, 25892, 25940, 25942, 25983-4,
        26291-2, 26380, 26486, 26574, 26677, 26718, 26722, 26765, 26767, 26872, 26900,
        26902, 26921, 27020, 27064, 27104, 27203, 27297, 27345, 27550, 27580-1, 27603,
        27610, 27628, 27681, 27772, 27776, 27841, 27856-7, 27901, 27979, 28105, 28184,
        28218, 28286, 28297, 28304, 28306-7, 28410
        28524-5, 28593, 28739, 28751, 28846, 28851, 28862, 29037, 29051, 29062,
        29070, 29397, 29435, 29639, 29792, 29885, 29949-50, 30034, 30093, 30118,
        30340, 30370, 30415, 30417, 30442, 30493, 30549, 30551, 30777, 30841, 30845,
        30868, 30914, 30963-4, 30984-5, 31016-7, 31025, 31027, 31151, 31352, 31389,
        31485, 31503, 31519, 31610, 31872, 31874, 31907, 31918, 32242, 32278, 32290-1,
        32315-6, 32329, 32333
        32656, 32674, 32848, 33402, 33462, 33710, 33811, 33911, 34013, 34080,
        34133, 34166-7, 34250, 34255, 34281, 34299, 34321, 34356, 34400, 34446,
        34573-5, 34746, 34767, 34866, 34952, 34982, 35012, 35036, 35075, 35077,
        35131, 35136, 35305, 35387, 35427, 35434, 35566, 35581, 35591, 35676, 35818,
        35969, 36043, 36240, 36301
        36632, 36664, 36849, 37098, 37200, 37410, 37467, 37561, 37642, 37787, 37813,
        37841, 37863, 37884, 38029-30, 38080, B38139, 38159, 38268, 38298, 38338,
        B38415, 38487, 38552, 38630, 38733, 38735, 38757, 38939-40, 39103, 39105,
        39214, 39247, 39250, 39288, 39331, 39336, 39356-7, 39466, 39502, 39602, 39612,
        39627, 39733, 39861, 39957, 40080, 40099, B40105, 40206, 40270

NADP, NAD, methods
        22921; 27841; 30551, 30914, 32278; 33456, 35978; 37098

Net assimilation rate see Growth analysis, net assimilation rate ...

Net photosynthetic rate see Gas exchange ...

Nitrogen   see N_2 ...; Mineral elements (N,P,K) ...

NMR see EPR, NMR

Nuclear magnetic resonance see EPR, NMR

Nucleic acids see Proteins, amino acids, nucleic acids ...

O

O_2 and algae productivity
      29346; 35965, 35995, 36339; 37058

O_2 and biliproteins
      24405; 35308; 37562, 39964

O_2 and carbon fixation pathways
      21592, 21618, 21652, 21883, 22397, 22530, 23452, 23841, 23877, 23926-7,
      24222, 24408; 25216, 25708, 25711, 25787, 25868, 25912, 26251, 26619, 26668,
      26959, 27026, 27182, 27247, 27494, 27639, 27760, 27928, 28268-9; 28931,
      29534-5, 29571, 29841, 29987, 30462, 30544, 30928, 31117, 31256, 31275, 32007;
      32585, 32685, 33213, 33699, 34003, 34278, 34861, 35142, 35468, 35535, 36561;
      36624, 36658, 36736, 37020, 37149, 37212-3, 37398, 37471, 37817, 37904,
      B38139, 38586, 38757, 39275, 39963, 40204

O_2 and carotenoids 25798, 27711, 28455; 33010, 33610, 34791

O_2 and chlorophyll
      22068, 22093, 22230, 22259, 23613, 23825, 23993, 24405, 24930; 25621, 25798,
      26727, 26737-8, 27494, 28240, 28379, 28455, 28476; 28892, 29338, 29433,
      31354, 31577, 32008, 32314; 33639, 33784, 34237, 34397, 34424, 34791, 35308,
      35725, 36119; 37767, 39443, 39708, 40152, 40190

O_2 and chloroplast (chromatophore)
      25621; 34237, 35333; 37427, 38366, 40152

O_2 and ecosystem and plant productivity
      22479, 24123-5; 25568, 26155, 26403, 26648, 26784, 27290, 27413, 27494,
      27502, 27539-40; 29987, B30405, 31400, 32010; 33871-2, 34719, 36070; 36708,
      37012, 37083, 37982, 38489, 38586, 39214-5

O_2 and electron transport chain
      21611, 22104, 22236, 22345, 22721, 22934-5, 23261, 23632, 24099, 24329,
      24859; 25242, 25401, 26146, 26333, 27064; 28594, 29051, 29282, 29383, 30309,
      30777, 31145, 31275, 32007; 32673, 32834, 35076, 35285, 35413; 36937, 37427,
      37644, 37841, 37950, 38483, 38580, 39025, 39104, 39458, 39602, 39647

O_2 and gas exchange
      21711, 21717, 21800, 21812, B21914, 21926, 22031, 22066, 22070-1, 22086,
      22131, 22152, 22251-2, 22305, 22521, 22542, 22544, 22608, 22649, 22796,
      22833, 22935, 23218, 23329, 23444, 23493, B23540, 23764, 23796, 23814, 23841,
      23926, 23928, 23994, 24082-3, 24296, 24405, 24447, 24593, 24611-2, 24804,
      24946, 24998
      25249, 25346, 25395, 25448, 25457, 25568, 25592, 25684, 25708-12, 25767,
      25776, 25883, 25913, 25923, 26091, 26110, 26155, 26188, 26214, 26403, 26566-
      -7, 26619, 26775, 26785, 26835, 26858, 26863, 26880, 27156, 27182, 27289,
      27370, 27424-5, 27494, 27502, 27548, 27760, 27951, 28092, 28205, 28229-30,
      28268-9, 28379, 28393, 28477

(continued)

O_2 and gas exchange (continued)

28573, 28664, 28742, 28772, 28795, 28901, 28993-4, 29050, 29156, 29328, 29346,
29489, 29536, 29601, 29766, 29942, 29989, 30097, 30099, 30224, 30227, 30273,
30289, 30453-8, 30462, 30469, 30499, 30543, 30622-4, 30684, 30777, 30927,
31066, 31102, 31145, 31231, 31256, 31266, 31278, 31314, 31372, 31400, 31457,
31466, 31519-20, 31636, 31733-4, 31873, 31944-5, 32082, 32219, 32297
32653-4, 32804, 32826, 33002, 33039, 33068, 33153, 33212, 33231, 33363,
33385, 33543, 33696, 33699, 33759, 34017, 34053, 34103, 34123, 34213, 34343,
34417, 34423, 34447, 34451-2, 34535, 34549, 34718-9, 34798, 34863, 35113,
35140, 35162-4, 35261, 35286, 35304, 35315, 35341-2, 35374, 35418, 35423,
35430, 35471, 35535, 35540, 35765, 35895, 36162, 36170, 36406, 36425
36630, 36687, 36693, 36897, 37025, 37099, 37140, 37212-3, 37216, 37325,
37387, 37452, 37505-6, 37516, 37744, 37754, 37766, 37774, 37881, 37950,
38103, 38122, 38366, 38429, 38492, 38518, 38586, 38793, 38908, 38926, 39057,
39073, 39094, 39105, 39149, 391-76, 39178-9, 39208, 39263-4, 39277, 39391,
39441, 39626, 39648, 39694, 39712, 39716, 39725, 39839, 39867-9, 40080,
40127, 40193, 40285, B40331, 40344

O_2 and photorespiration

21800, 21828, 21926, 22066, 22071, 22131, 22226, 22479, 22544, 23511, 23714,
23796, 23877, 23994, 24082, 24593, 24950; 25448, 25776, 26427-8, 26626,
26629, 26880, 27493, 27754, 27760, 27929, 28269, 28479; 29032, 29156, 29169,
29328, 30097, 30288, 30499, 30637-8, 30798, 31256, 32073, 32087-8, 32430;
32684, 32822, 32826, 32828, 33363, 33699, 33887, 34052-3, 34123, 34256,
34423, 34452, 34718, 35164, 35418, 35423, 35535, 35765, 36280, 36406; 36824,
37084, 37234, 37516, 37881, 38122, 38966, 39234, 39263-4, 39391, 39716,
39868, 39939, 40204

O_2 and resistances to CO_2 and water vapour transfer

25604, 26091, 26403, 27187; 33385, 34535, 35098, 35261; 39094, 39725

O_2 and respiration

22226, 22252, 23764, 23928, 24083, 24405, 24946; 25710, 25883, B28482; 28795,
30218, 30804; 33385, 34213, 36406; 37140, 37325, 40193, 40285

O_2 determination in water reservoirs   see   Algae, primary productivity, methods

O_2 determination (other than O_2 electrode)

22548, 23494, 24719, 25135; 25480, 25823, 26064, 27123, 28205, 28415, 28475;
30723, 32063; 36897, 36934, 37348, 38398, 39304

O_2 electrode

21609, 21693, 22386, 22548, 22655, 23049, 23442, 23487, 23553, 24055, 24220,
24514, 24637, 24774; 26009, 26381, 26544, 26577, 26661, 26924, 27117, 27146,
27773, 28205, 28349, 28475; 28507, 28579, 28615, 28664, 28728, 29248, 29842,
30100, 30255, 30490, 30578, 30600, 30955; 32700, 33385, 33430, 33813, 33828,
34208, 34568, 34617, 35494, 35715, 35796, 35879, 35995, 36509; 36708, 36892,
37758, 37790, 38019, 38462, B38477, 38622, 38924, 39407, 39427, 39473

O_2 evolution mechanism and kinetic

21528, 21617, 21620, 21631, 21644, 21750, 21754, 21806, 21817-8, 21912,
B21979, 21980, 22525, 22527, 22682, 22856-7, 22881-4, 22957, 23219, 23365,
23390, 23394, B23407, 23434-5, 23473, 23632-3, 24055, 24133, 24136, 24192,
24281, 24338, 24481, 24497, 24607, 24611-2, 24630, 24684, 24874, 24901,
24904, 24914, 25050, 25085, 25152
25180, 25189, 25242-4, 25276, 25322, 25368, 25379, 25384, 25432, 25466,
25616, 25626-7, 25695, 25836, 25840, 26056, 26267, 26325, 26480, 26487,
26659, 26718, 26720, 26765, 26767, 26859-61, 26942-3, 27122, 27332, 27375,
27512, 27548, 27556, 27603, 27617, 27655, 27763, 27876, 28027, 28029, 28083,
28213, 28330, 28334, 28437-9, 28468

(continued)

O$_2$ evolution mechanism and kinetic (continued)
        28519, 28539, 28548, 28550, 28581, 28599, 28621, 28655, 28679, 28745, 28747,
        28759, 28789, 28903, 29097, 29143, 29160, 29186, 29279, 29304-5, 29307-9,
        29412, 29521, 29525, 29539, 29570, 29634, 29666, 29688, 29714-5, 29733-4,
        29740, 29773, 29831, 29991, 30014, 30034, 30085-6, 30094, 30155, 30197, 30365,
        30399, 30442, 30539, 30569, 30708, 30718, 30774, 30779, 30892, 31128, 31151,
        31280, 31410, 31412, 31437, 31473, 31494-6, 31571, 31612, 31617, 31635, 31664,
        31666, 31747, 31776, 31809, 31858, 31949, 31960, 31971, 32070, 32072, 32270,
        32293, 32379, 32432, 32474-6, 32523-4, 32532-3
        32590, 32682, 32685, 32697, 32744-5, 32800, 32880, 32954, 33036, 33137,
        33250, 33253, 33315-6, 33354, 33356, 33458, 33460-1, 33563, 33585, 33605,
        33626, 33687, 33746-7, 33751, 33919, 33933, 33982, 34043, 34049, 34051,
        34106, 34179, 34186, 34202, 34205, 34208, 34210-1, 34223, 34286, 34357,
        34417, 34419, 34529, 34531, 34601, 34661, 34703, 34753, 34773, 34777, 34801,
        34858-60, 34893, 34901, 35157, 35327-8, 35366, 35571, 35615, 35649, 35653,
        35699, 35720, 35879, 35944-5, 35955, 36043, 36094, 36105-6, 36174, 36251,
        36256, 36263-4, 36439, 36507, 36511, 36541, 36554-5, 36568
        36622, 36710, 36928, 36930, 36937, 36969, 37178, 37245, 37254, 37315, 37373,
        37425, 37593-4, 37633, 37671, 37705-6, 37763, 37806, 37813, 37842, 37884,
        37920, 37965, 37996, 38049, 38095, 38179, 38291, 38342, 38349, 38507, 38526,
        B38528, 38620, 38708, 38762, 38932, 39078, 39103-4, 39176, 39301, 39458,
        39648, 39774, 39887, 39915, 40005, 40030-1, 40152, 40164, 40246-7, 40298,
        40330, 40340

O$_2$ exchange see Gas exchange ...

O$_2$ isotopes, use in photosynthesis measurement
        22252, 23225, 23632, 24946, 24976; 26135, 27122, 27991; 29488-9, 30767,
        31953, 32213, 32516; 32828, 33009, 34658, 36169, 36490; 36671-2, 36871-2,
        37286, 37635, 37653, 37763, 38708

Ontogeny of algae and algae productivity
        22258; 27652; 30836; 33613; 39883, 39959

Ontogeny of algae and biliproteins
        22490; 27177, 27257; 28569, 28631, 29680, 31757, 32053; 39484, 39804, 40036

Ontogeny of algae and carbon fixation pathways
        22899, 24963, 25034; 25280; 31275; 33198, 33625, 34473, 34978, 35241, 36409,
        36411; 38327, 40210

Ontogeny of algae and carotenoids
        21769, 21894, 22465, 24347; 25648, 26855; 28569, 28631, 30629, 31171, 31329,
        31502, 31757; 33001, 34845; 38484, 38850, 40088

Ontogeny of algae and chlorophyll
        21582, 21769, 21894, 21976, 21992, 22248, 22264, 22465, 22490, 22899, 23751,
        23794, 23867, 23915, 24339, 24347, 24397, 24401, 24713, 24768, 25001, 25121;
        25621, 25648, 25853, 26097, 26794-5, 26807, 26855, 26957, 27077, 27177,
        27257, 27652, 27796; 28569, 28631, 28723, 28980, 29017, 29096, 29680, 30440,
        30613, 30629, 30836, 30987, 31171, 31275, 31329, 31502, 31705, 31757, 32053,
        32433; 33001, 33047, 33118, 33131, 33423, 34070, 34560, 34691, 34845, 35836,
        36145, 36409, 36411; 36598, 37129-30, 37444, 37558, 38200, 38426, 38453,
        38484, 38850, 39171, 39321, 39484, 39598, 39883, 39944, 39960, 40036, 40298

Ontogeny of algae and chloroplast (chromatophore)
        21992, 22362, 23354, 23867, 24402, 24867, 24963; 25624, 25853, 26774, 26795;
        28723, 28924, 29017, 31725, 31757, 32030; 32854, 33225, 33810, 35241

Ontogeny of algae and electron transport chain
        21770, 22264, 22304, 23915, 24401, 24403, 24652, 24768, 25001; 25706, 25853,
        26367, 26774, 26794, 26807, 27257, 27652; 29017, 31275, 31723-5, 31738,
        32433; 33001, 33047, 33810, 34589, 34845, 35241, 36214, 36409, 36411; 37129,
        37316, 37558, 38453, 38515, 40298

Ontogeny of algae and gas exchange
        21770, 21966, 22089, 22258, 22304, 22407, 22490, B23407, 23751, 23766, 23827,
        23867, 23915, 24053, 24172, 24347, 24397, 24401-3, 24867, 25001, 25034,
        25121; 25215, 25395, 26367, 26584, 26642-4, 26794-6, 26807, 26957, 27257;
        28795, 28921, 29096, 29196, 31362, 31651, 31724, 31757, 32180; 32622, 32787,
        32813, 33001, 33423, 34070, 34473, 34589, 35241, 35712, 36195-6, 36214;
        37162, 37316, 37429, 37444, 38327, 38426, 38515, 39171, 39538, 40210, 40298

Ontogeny of algae and photosynthates
        27752; 34638, 35109; 39555

Ontogeny of algae and respiration
        23867, 24053; 25215, 26795; 28795; 36998

Ontogeny of canopy and ecosystem and plant productivity
        22254, 22447, B22822, 22823, 23232, 23740, 24007, 24110, 24605; 27175, 27293-
        -4, 27735-6, 28093, 28212, 28431; 29009, 29334, 30070, 30412, 31242, 31376,
        31868; 33073, 33199, 34365, 35787, 36138, 36281; 36593, 36734, 37168, 37412,
        39567

Ontogeny of canopy and gas exchange   23171-2, 24474; 27139, 28282, 28413; 28781,
        31759, 32449; 37987-8

Ontogeny of canopy and photorespiration   23427

Ontogeny of canopy and respiration   37987-8

Ontogeny of chloroplast and carbon fixation pathways
        24347; 25528; 28909, 31578; 35241; 37457, 37725, 39062

Ontogeny of chloroplast and carotenoids
        21872, 24340, 24347; 27443; 28651, 28944, 29846, 30673, 31504; 35071, B35862;
        37925, 38504, 39044, 40164

Ontogeny of chloroplast and chlorophyll
        21872, 22068, 22800, 23095, 23109, 23581, 24177, 24340, 24347, 24414, 24443,
        25075; 25501, 25748-9, 26278, 26842, 26928, 27407, 27443, 27831-2, 27922,
        28117, 28217; 28519-20, 28595, 28651, 28736, 28836, 30424, 30673, 31104,
        31578, 31719, 31779, 32085, 32419; 32778, 32890, 33108, 33389-90, 33598,
        33781, 34922, 35071, 35602, B35862; 36611, 36702, 36905, 37925, 37978, 38504,
        38547, 38562, 39044-5, 39062, B39191, 39300, 39814, 40164

Ontogeny of chloroplast and electron transport chain
        22193, 22463, 22800, 24545; 25616; 28518, 28595, 28836, 28896, 32385-6,
        32419; 32594, 32799, 33390, 34042, 35071, 35241, 35731, 36177, 36487; 37372,
        37558, 38055, 39045, 39357, 40164

Ontogeny of chloroplast and gas exchange
        22056, 24347; 25501, 25528, 25902, 26537, 27827; 28836, 29700, 32011; 32799,
        33390, 33916

Ontogeny of leaf, insertion level, and carbon fixation pathways
        21661, 21702, 22032, 22080, 22226-7, 22697, 22725, 22727, 22836, 23696,
        23835, 24336, 24523, 24976, 25097
        25221, 26068, 26626, 26940, 26991, 27431, 28351, 28389-90, 28393, 28430
        28915-6, 29132, 29273, 29316, 29351, 29641, 29700, 29800, 29951, 29983, 30088,
        30431, 30896-7, 30907, 31164, 31226, 31578, 32097, 32109, 32425, 32429-30,
        32440
        33273, 33654, 33731, 33835, 33893, 34683, 34907, 34937, 34949, 34972, 35023,
        35051, 35087, 35247-8, 35294, 35877, 35970, 36111, 36114, 36132, 36327,
        36426, 36440, 36442, 36450-1, 36527
        36600, 36724, 37215, 37274, 37862-3, B38139, 38192, 38428, 38542, 38612,
        38798, 39201, 39236, 39723, 39747, 39761, 39972, 40214, 40238, 40248, 40351,
        40360

(continued)

Ontogeny of leaf, insertion level, and electron transport chain (continued)
        26653, 27032, 27465, 27804, 28062, 28180, 28416; 28599, 28995, 29070, 29107,
        29700, 30080, 30579, 30594, 30881, 31082, 31736-7, 32097, 32310; 32706,
        32739, 34233-4, 34319, 34462, 34588, 35764, 35767, 35774, 36262, 36300,
        36581; 38053, 38644, 39243, 39543, 40360

Ontogeny of leaf, insertion level, and gas exchange
        21517, 21538, 21590, 21629, 21633, 21661-2, 21786, 21802, 21842, 21869,
        21970, 22004, 22014, 22025, 22031, 22226-8, 22678, 22725, 22776, 22786-7,
        22796, 22836, 22840, 22897, 22960, 22962, 23045, 23049, 23134, 23257, 23431,
        23465, 23493, 23539, B23540, 23643; 23694, 23706, 23756-7, 23779, 23783-4,
        23835, 23896, 23922, 24138, 24183, 24323, 24336, 24411, 24428, 24523, 24532,
        24592, 24596, 24601, 24692, 24759, 24787, 24821, 24830, 24853, 24880, 24915,
        24919, 24931, 24985, 25018, 25034, 25061, 25073, 25097, 25155
        25221, 25224, 25294, 25312, 25453, 25642, 25654, 25666, 25774, B25775, 25801,
        25829, 25873, 25928, 26093, 26123, 26188, 26353, 26400, 26471, 26512, 26537,
        26674, 26749, 26799, 26856, 26884, 26905, 26961, 26965, 26991, 27011, 27035,
        27207, 27216, 27327, 27348, 27380, 27389, 27460, 27595, 27688, 27725, 27813,
        27920, 27991, 27995, 28050, 28058, 28062, 28086, 28090, 28092, 28134, 28180,
        28379, 28408, 28413, 28426, 28428, 28430, 28461, 28474-5
        28501, 28504, 28571, 28605, 28649, 28685, 28782, 28810, 28866, 29090, 29132,
        29179, 29297, 29490-1, 29563-4, 29572, 29587, 29594, 29675, 29700, 29764,
        29795, 29800, 29951, 29954, 30011, 30057, 30088, 30353, 30413, 30476, 30513,
        30656, 30691, 30706, 30743, 30896-7, 30920, 31057, 31156, 31164, 31200,
        31245, 31525, 31584-5, 31627, 31695, 31853, 31866, 31896, 32011, 32020,
        32181, 32220, 32294, 32297, 32430, 32440, 32449-50
        32613, 32639, 32654, 32751, 32777, 32895, 32923, 33002, 33015, 33029, 33050,
        33052, 33068, 33162, 33290, 33409, 33489, 33523, B33628, 33630, 33671,
        33697, 33703, 33731, 33835, 33893, 33899, 33916, 33950, 33976, 34036, 34061,
        34284, 34295, 34348, 34463, 34471, 34485, 34523, 34639, 34683, 34707, 34712,
        34730-1, 34835, 34933, 34937, 34940, 34956, 34961-2, 34997, 35022-3, 35030,
        35035, 35051, 35087, 35092, 35106, 35201, 35224, 35233, 35247-8, 35255,
        35283, 35293, 35297, 35373-4, 35423, 35527, 35559, 35664, 35674-5, 35789,
        35807, 35866-7, 36077, 36090, 36114, 36132, 36199, 36270, 36300, 36418,
        36425, 36509, 36551-2
        36698, 36726, 36738, 36785, 36804, 36817, 36865, 36878, 36985, 37105, 37112,
        37122-3, 37215, 37265, 37287, 37289, 37347, 37381, 37395, 37434, 37472,
        37678, 37695, 37699, 37857, 37888, 37893, 37895, 37981, 37999, 38059, 38098,
        38231, 38312, 38353, 38420, 38516, 38542-3, 38710, B38739, 38798, 38891,
        38962, 38968, 39034, 39076, 39081-2, 39137, 39178-9, 39238, 39367, 39421-2,
        39543, 39649, 39732, 39756, 39808, 39856, 39888-9, 39968, 39972, 40075,
        40154, 40166, 40211, 40248, 40253-4, 40291, 40293

Ontogeny of leaf, insertion level, and photorespiration
        22226, 22228, 22542, 22796, 23171, 24915; 25654, 26188, 26626, 28474, 28478;
        29700, 31018, 32430; 33731, 34683, 35081, 35248, 35423, 36114; 37105, 39238,
        39710

Ontogeny of leaf, insertion level, and resistances to CO_2 and water vapour transfer
        21538, 22226, 22787, 22962, 22964, 23134, 23431, 23757, 24915; 25654, 25843,
        25947, 25998, 26164, 27035, 27133, 27139, 27595, 28125, 28426, 28428; 28504,
        29162, 29203, 29247, 29795, 29881, 30069, 30162, B30511, 30547, 30738,
        30743, 31164, 31627, 31659, 32020, 32051, 32126, 32194, 32369, 32450; 32777,
        33290, 33396, 33916, 33976, 33985, 34633, 34690, 35283, 35563, 35712, 36418;
        36609, 37265, 37347, 37413, 37415, 37434, 37699, 38016, 38542, 39421-2,
        39659, 39661, 39813, 39890, 40020, 40154, 40360

Ontogeny of leaf, insertion level, and respiration
        21590, 21842, 22226-8, 22786, 22796, 23091, 23257, 24821, 24830, 24919,
        25047, 25073; 25654, 26400, 26626, 26856, 26884, 26991, 26997, 27035, 27207,
        27327, 27380, 27725, 27935, 28050, 28086, 28408, 28430, B28482; 28685,
        29563, 29587, 29675, 29700, 30691, 31018, 31896, 32020, 32183; 32654, 33409,
                                                                        (continued)

Ontogeny of plant and gas exchange
21542, 21878, 21967, 22014, 22075, 22284, 22776, 22836, 22960, 23171-2,
23297, 23330, 23376, 23431, 23443, 23493, 23784, 23845, 23907, 24084, 24272,
24280, 24759, 25111, 25155
25164, 25312, 25349, 26065, 26392, 26537, 26642-3, 26812, 26905, 27011, 27062,
27216, 27348, 27389, 27409, 27560, 27594, 27837, 27930, 28018, 28050, 28133,
28169, 28205, 28282, 28380, 28475
28527, 28606, 28674, 28687, 28930, 29223, 29385, 29430, 29509, 29572, 29816,
29931, 29975, 29993, 30058, 30253, 30537, 30556, 30694, 30712, 30912, 30932,
30936, 31086, 31156, 31164, 31200, 31227, 31587, 31607, 31695, 31823, 31866,
32023, 32163, 32226, 32294, 32449, 32545
32639, 32689, 32831, 32931, 33013, 33242, 33278, 33290, 33335, 33790, 33859,
33898, 34295, 34317, 34370, 34456, 34649, 34835, 35057, 35074, 35106, 35161,
35201, 35222, 35255-6, 35349, 35476, 35611, 35892, 36049, 36135, 36137,
36207, 36234, 36371, 36418, 36457, 36478, 36497, 36527
36670-1, 36701, 36804, 37624, 37887, 37988, 37999, 38074, 38077, 38103, 38194,
38493, 38558, B38739, 38769, 38798, 38814, 38897, 38942, 39063, 39076, 39137-
-8, 39223, 39395, 39599, 39741, 39751, 39755, 39808, 39815, 39968, 40077,
40098, 40162, 40212, 40249, 40291

Ontogeny of plant and photorespiration
23973; 27837; 28929, 30164; 34379; 37624, 38769, 39599

Ontogeny of plant and resistances to CO_2 and water vapour transfer
23696, 23757; 25604, 25742, 26065, 26392, 27575, 27594, 27897, 28144, 28427;
28606, 28630, 29247, 29816, 30936, 31156, 31164, 31250, 31607, 31759, 31823;
33290, 33527, 33571, 33985, 36207, 36372, 36418, 36457; 37112, 38016, 38075,
38077, 38465, 38814, 39328, 39599, 40141, 40169

Ontogeny of plant and respiration
22563, 23443, 23587, 23973, 24084; 25164, 25435, 27389, 28050, 28167, 28408;
28527, 29223, 29509, 29819, 29993, 31587, 32023, 32163; 33242, 33859, 33898,
35074, 35106, 35161, 35255, 36135, 36178, 36299, 36497; 36671, 36804, 37624,
37887, 37953, 37988, 38198, 38431, 38558, 38769, 38814, 38897, 39755, 40291

Optical properties, leaf see Leaf optical properties

Oscillations, short-term fluctuations, steady and non-steady state in carbon fixation
26770

Oscillations, short-term fluctuations, steady and non-steady state in electron transport chain 30845, 32476; 33316, 33819, 34367; 38027

Oscillations, short-term fluctuations, steady and non-steady state in gas exchange
22582, 22935, 22999, 23609, 23649, 23840, 24449; 26583, 26835, 26909, 27461;
28930, 29613, 29766, 30596, 30708, 30723, 30845; 33002, 33315, 35140; 36701,
37915, 38027, 38745, 38996, 39552

Oscillations, short-term fluctuations, steady and non-steady state in resistances to
CO_2 and water vapour transfer
22403, 23609; 26397, 26909; 28930, 29766, 30738; 37260, 37414, 37915, 39552

Oscillations, short-term fluctuations, steady and non-steady state in respiration
33891

Osmotically active substances and algae productivity 23099; 26344

Osmotically active substances and carbon fixation pathways
22412, 22609, 23480-1, 25035; 26760, 27639, 28405; 28841, 29092, 29817, 30550,
31319, 32477; 33198, 36364; 36916, 38408, 39156

Osmotically active substances and carotenoids 25906, 27946; 33001, 35962

Osmotically active substances and chlorophyll
        23523, 24713, 24727-8; 25460, 25906, 26658, 27036, 27671, 28111; 28584, 30215
        30550, 30693; 33001, 33370, 33405, 34267, 35962; 36958, 38227, 38975-6,
        39312

Osmotically active substances and chloroplast (chromatophore)
        21755, 22062, 22536, 23995, 24036, 24311; 25814, 26103, 26276, 26487, 27355,
        27454; 28584, 29596, 30885, 31207, 31569, 32553; 32795, 33405, 34764, 34986,
        35325, 35501, 35603, 36486; 37336, 37811, 38733-4, 39392-3

Osmotically active substances and ecosystem and plant productivity
        21909, 22085, 23952; 25605, 26345, 27036, 27882, 28138; 28510, 29823; 32986,
        35512; 36958, 37482, 37897, 38975

Osmotically active substances and electron transport chain
        21755, 23415, 23842, B24176, 24720; 25242, 26523, 27064, 27354, 27492, 28111;
        28747, 29730, 30094, 30693, 30773, 30884, 31293, 31651; 32608, 32795, 32801,
        33001, 33160-1, 33207, 33370, 33687, 34769, 34799, 35320, 35688, 35935,
        36094; 36762, 37811, 38094, 38510, 38672, 39332, 39393

Osmotically active substances and gas exchange
        21755, 21908-10, 22321, 22938, 23099, 23523; 26227, 26344-5, 26524, 26669,
        27036; 28538, 29092, 30139, 30220, 30797, 30968, 31293, 31651, 31732, 32206;
        33001, 33219, 33687, 34267, 34799, 35320, 35325, 35688; 37189, 37614, 37897,
        38074, 38218, 38655, 39990

Osmotically active substances and photorespiration 22609; 26227; 30797, 31293;
        35325

Osmotically active substances and resistances to CO_2 and water vapour transfer
        26524; 33689; 37614, 37897

Osmotically active substances and respiration 22484; 31651

Osmotically active substances, use for water stress induction 21908-10

Oxygen see O_2 ...

Ozone see Pollution of air ...

P

P680   21643, 21645-6, 21662, 22243-4, 22580, 23518, 23954, 23957, 24041, 24133,
          24307, 24804, 24917, 25057, 25062, 25151; 25276, 26156, 26422, 26559, 26694,
          27100, 27571, 27863, 28334; 28547, 28550, 28744, 28823, 28884, 28968, 28998,
          29341, 29689, 29713, 29867, 30184, 30337-8, 30782, 30812, 31390, 31410,
          31486-7, 31511, 31666, 32239, 32249, 32301, 32370, 32393, 32456; 32620,
          32861, 33253, 33316, 33485, 33745, 33814, 34357, 34600, 34605, 34759, 35181,
          35366, 35572, 35844, 35897, 36246; 36741, 36969, 37193, 37254, 37282, 37315-
          -6, 37668-9, 37705-6, 37717, 37967, 37979-80, 37996, 38069, 38154, 38159,
          38232-3, 38381, 38557, 38698, 39222, 39446, 39451, 39721, 39901, 39920,
          40247

P700, P750, P890, etc.
          21528, 21534, 21536, 21567-8, 21638, 21644, 21662, 21687-9, 21743, 21773,
          21810, 21816, 21825, 21843, 21866, 21922-3, 21925, 21959, 22079, 22090,
          22134, 22141, 22219, 22288-9, 22354, 22374, 22390, 22393, 22498, 22501,
          22561, 22580, 22588, 22794, 22816, 22862, 23020, 23022, 23041, 23043, 23046-
          --7, 23056-7, 23076, 23084, 23102, 23111, 23120, 23140, 23184, 23187, 23239,
                                                                    (continued)

P700, P750, P879, etc. (continued)
        23241, 23347, 23420, 23470, 23515-6, 23518, 23552, 23573, 23586, 23600-1,
        23638, 23724, 23729-30, B23767, 23798, 23825, 23847, 23876, 23916, 23954,
        23966, 24096, 24163, 24189, 24208, 24238, 24241, 24258, 24327, 24478-80,
        24487, 24496, 24631, 24671, 24691, 24756, 24760-2, 24789, 24804, 24909,
        25002, 25050, 25057, 25062, 25070
        25166, 25180, 25184, 25186, 25202, 25217, 25281, 25293, 25295, 25315, 25323,
        25347-8, 25379, 25385-6, 25470, 25488, 25517, 25570, 25606, 25629, 25668-9,
        25773, 25819, 25846, 25884, 25888, 25918, 25957, 25963, 25968, 26041-2,
        26048, 26084, 26172-3, 26223, 26257-8, 26325, 26380, 26423, 26457, 26468,
        26610, 26616, 26632-3, 26687, 26713, 26718, 26720, 26731, 26734, 26761,
        26794, 26830, 26942-3, 26974, 26990, 27031, 27033, 27050, 27072, 27086,
        27099, 27173, 27203, 27222, 27254, 27273-4, 27340, 27449, 27518, 27525,
        27536, 27581, 27653, 27668, 27681, 27724, 27804, 27843, 27856, 27863,
        27866-71,   27917, 27979, 28021, 28028, 28036, 28105, 28127, 28131, 28135-6,
        28237, 28258-9, 28332-5, 28409, 28420
        28515, 28528-9, 28535, 28547, 28550, 28596, 28667-8, 28698-701, 28720, 28744,
        28746, 28751, 28769, 28798, 28814, 28816,' 28822, 28824, 28833, 28836, 28851,
        28859, 28878, 28893-4, 28932, 28964-9, 28995, 29073, 29079, 29124, 29126,
        29134, 29148-9, 29161, 29173, 29192, 29263, 29266, 29268, 29272, 29275, 29284,
        29310, 29406, 29438, 29440-2, 29453, 29474, 29497, 29505, 29518-20, 29525,
        29534, 29540, 29666, 29686-9, 29738, 29754-5, 29784-6, 29807, 29818, 29831,
        29867, 29872, 29878-80, 29939, 29949-50, 29955-6, 29958, 29970-1, 30050-1,
        30123, 30135, 30138, 30188-9, 30246, 30254-5, 30321, 30323, 30328, 30385-6,
        30436, 30586-7, 30594, 30626, 30718, 30742, 30780, 30799, 30915, 31009, 31025,
        31027, 31030, 31033, 31046, 31073-4, 31078, 31128, 31138, 31143, 31219, 31225,
        31230, 31297-9, 31370-1, 31374, 31391, 31410, 31426, 31478-81, 31485, 31503,
        31534, 31564-5, 31567, 31598, 31621, 31629-30, 31642, 31725, 31785-6, 31893,
        31900, 31979, 32005, 32045, 32072, 32112-4, 32127, 32134, 32168-9,
        32198-9, 32225, 32239, 32279-80, 32282-3, 32370, 32417, 32419, 32432, 32441,
        32467-8, 32550
        32587-8, 32600, 32619, B32630, 32665-6, 32801, 32848, 32860, 32889-90,
        32899, 32906, 32908, 32915, 32924-5, 32947, 32992-4, 33016, 33100, 33186-7,
        33203, 33253, 33314, 33355-6, 33402-3, 33469, 33486-7, 33515, 33551, 33564,
        33626, 33688, 33712, 33716-8, 33720, 33745-7, 33788, 33814, 33821, 33869,
        33882, 33901, 33919, 33930-1, 33956, 33958, 33983, 34020, 34205, 34219,
        34237, 34244, 34277, 34311, 34344, 34347, 34350, 34372, 34393, 34431-2,
        34460, 34490, 34566, 34662, 34687-8, 34726, 34758, 34766-7,   34775-6,
        34779, 34845, 34893, 34990, 34998, 35058, 35086, 35096, 35107, 35181, 35220,
        35238-41, 35246, 35327, 35333, 35380, 35387, 35409, 35427, 35502, 35530,
        35572, 35600, 35665-6, 35670, 35736, 35739, 35758, 35804, 35820, 35844,
        35847, 35899, 35975, 36095, 36118-9, 36129, 36139-40, 36239, 36247, 36272,
        36389, 36409, 36411, 36430, 36447, 36482, 36539
        36626-7, 36651, 36660, 36668, 36673, 36689, 36699, 36741, 36751, 36771-2,
        36786-7, 36823, 36860-1, 36911-2, 36944, 36952, 36955, 36971, 37045-6,
        37061-2, 37065, 37115, 37165-7, 37180-1, 37251-3, 37255, 37350, 37362,
        37370, 37446,   37454, 37494, 37545, 37547, 37565, 37584, 37598, 37611,
        37668, 37675, 37730-1, 37749, 37820, 37877, 37900, 37914, 37954-7, 37980,
        38004, 38015, 38023-5, 38063, 38069, 38095, 38136, 38153-5, 38159-60,
        38165, 38208, 38240, 38254, 38291, 38349, 38381, 38447, 38481, 38510,
        38526, 38538, 38549, 38557, 38571, 38593, 38613, 38615, 38620, 38638-9,
        38714, 38723-4, 38795-6, 38833, 38862, 38906, 38918, 38957, 38970-2, 39048,
        39110, 39203, 39222, 39230-1, 39292, 39295-6, 39303, 39318, 39320, 39379-80,
        39384-5, 39444, 39451-3, 39502, 39588, 39612-7, 39624-5, 39666, 39681,
        39700, 39822, 39901, 39920, 40011, 40029, 40107, 40129, 40149, 40182, 40186,
        40192, 40201, 40246-7, 40278, 40359

Paramagnetic oxygen analyser   see O_2 determination ...

Paramagnetic resonance   see   EPR, NMR ...

PEP carboxylase (PEPC)   see   Phosphoenolpyruvate carboxylase

Peroxisome, glyoxysome, microbody
        21535, 21957, 22009, 22071, 22111, 22410, 22906, 22980-2, 23011, 23195,
        23310, 23427, 23714, 24036, 24247, 24319, 24426, 24499, 24734, 25000, 25007;
        25443, 25512, 25542, 25878, 25940, 26218, 26574, 27121, 27754, 27759, 27975-
        -6, 28154, 28267, 28422; 28706, 28882, 29002, 29111, 29181, 29746-7, 29759,
        30374, 30526, 30559, 30801, 31677-8, 31685, 31768, 32337-8, 32547; 32759,
        32783, 32952, 33493, 35519, 33593, B33662, 33731, 33785, 33960, 34266,
        B35862, 36158, 36168-9, 36179, 36328, 36475, 36562-3; 36832, 37038, 37336,
        37729, 37772, 37858, 37981, B38037, 38517, 39401, 39482, 39860, 39939

Pesticides   see also   Inhibitors of electron transport chain

Pesticides, herbicides and algae productivity 22550; 28996; 39256

Pesticides, herbicides and carbon fixation pathways
        22183, 22192, 22398; 26513, 26889; 29321, 29471, 29669, 29888, 29910, 30695,
        30841, 30887, 30962, 31275; 33519, 33603, 33651, 34560-1, 34735, 35024,
        36192, 36366, 36527; 37479, 37792, 37972, 38589, 38908, 39368

Pesticides, herbicides and carotenoids
        22240, 22636, 23248-9, 23964, 24446, 24678, 24921-2; 25852, 26601, 26962,
        27184, 27387, 27677, 28012, 28217, 28240; 29573, 30230, 30594, 30597,
        31503-4, 32215, 32424; 32763, 33265, 33338-9, 33463, 33519, 33607, 34468,
        34995, 35519, 35964, 36142, 36580; 37000, 37294-5, 37388, 37480, 37574,
        37740, 37856, 38228-30, 38354-5, 38478-9, 38505, 38581, 39426, 40041, 40281,
        40319

Pesticides, herbicides and chlorophyll
        21679, 21732, 21781, 21842, 22266, 22339, 22555, 22670, 22804-5, 22911,
        23165, 23226, 23248-9, 23576, 23955, 23957, 23964, 23967, 24048, 24432,
        24446, 24549, 24785, 24882, 24902, 24920-2, 24987, 25000, 25146
        25162, 25852, 26123, 26167, 26601, 26962, 27159, 27387, 27397, 27497, 27637,
        27677, 27762, 28012, 28159, 28217-8, 28240, 28326
        28653, 28709, 28846, 29321, 29323, 29449, 29473, 29573, 29669, 29946, 30104-5,
        30222, 30230, 30500, 30594, 30597, 30740, 30794, 30867, 31015, 31285, 31395,
        31503, 31582, 31621, 31648, 31855, 32000, 32255, 32323, 32424, 32433, 32451,
        32482
        32708, 32728, 32763, 32780, 32899, 33159, 33338-9, 33391, 33463, 33519,
        33603, 33607, 33641, 33651, 33670, 33867, 33869, 34140, 34238, 34468, 34897,
        34995, 35343, 35947, 35957, 36142, 36250, 36527
        36711, 36921, 26993, 370/6-7, 37226, 37294-5, 37371, 37388, 37426, 37479-80,
        37486, 37574, 37692, 37723, 37738, 37740, 37856, 37861, 37901, 37972, 37997,
        38158, 38228-30, 38354-5, 38394, 38461, B38477, 38478-9, 38505, 38681,
        38688, 38735, 38836, 38867, 38946, 39042, 39047, 39123, 39307, 39311, 39370,
        39376, 39415, 39426, 39826, 39853, 40019, 40041, 40152, 40281, 40292, 40319

Pesticides, herbicides and chloroplast (chromatophore)
        21527, 22240, 22339, 22550, 23867, 24920-2, 25000; 25545, 25852, 26601,
        26613, 27184, 27497, 27690; 28709, 29024, 31504, 31621, 32535; 32763, 32883,
        32899, 33463, 33603, 33917-8, 35555, 35948, 36488; 36711, 36989, 37075-7,
        37480, 37850, 38305, 38445, 38688, 39762, 40100, 40152, 40319-20, 40323

Pesticides, herbicides and ecosystem and plant productivity
        21779, 21781, 22185, 22331, 22356, 23236, 23907, 24432; 25190, 25489, 26123,
        27145, 27338, 27397, 27918, 28254; 28780, 29323, 29334, 29577, 30132, 31015,
        31750, 32306; 33007, 33641, 33867, 34468; 36945, 37041, 37294, 37479, 37901,
        37972, 38478, 38581, 40292

Pesticides, herbicides and electron transport chain
        21527, 21559, 22202, 23165, 23242, 23486, 23657, 23837, 23941-2, 23944,
        24032, 24193, 24208, 24241, 24490, 24663, 24665, 24791, 24881, 24921
        25459, 26137, 26350, 26889, 27132, 27184, 27338, 27387, 27397, 27549, 27996,
        28254

                                                                      (continued)

(continued)

pH, effect on carbon fixation pathways (continued)
        36663, 36719, 36723, 36849, 37024, 37542, 37841, 37844, 37852, 37947, 38126,
        38247, 38297, 38356, 38487, 38789, 39098, 39339, 39689, 39837, 39876, 40203,
        40213, 40218

pH, effect on carotenoids 22644, 24489; 25714, 28239; 31810

pH, effect on chlorophyll
        22068, 22269, 22426, 22647, 22808, 22843, 23618, 23656, 23892, 24497, 24747,
        24757, 24860, 24902, 24911; 25550, 25714, 26011, 26192, 26602, 27111, 27346,
        28122, 28239, 28320; 28582, 28737, 28967, 29599-600, 30494, 30620, 30827,
        31110, 31129, 31254, 31323, 31488, 31491, 31782; 32992, 33268, 33577, 33782,
        33875, 34188, 34838, 35093, 35284, 35809, 35858, 36062, 36160, 36278; 36644,
        36699, 36774, 36786-7, 36995, 37280, 37572, 38187, 38274, 38556, 39045,
        39180, 39307, 39350, 39706, 39849, 39978

pH, effect on chloroplast (chromatophore)
        21631, 23995, 24355, 25094; 26011, 28084; 28936, 29932, 30268, 30856, 31056,
        31356, 31491; 33341, 33791, 34033, 34563, 34780, 35026, 35111, 35391, 35414,
        35606, 35651, 35853; 36625, 37047, 37626, 37841, 38971, 39999, 40290, 40366

pH, effect on electron transport chain
        21588, 21631, 21691, 21745-6, 21888, 21890, 21997, 22063, 22164, 22202,
        22233, 22325, 22349-50, 22456, 22525, 22536, 22567, 22575-6, 22586, 22594,
        22708, 22774, 22801, 22811, 22824, 22900, 22913, 22972, 22996, 23123, 23183-
        -4, 23192, 23362-3, 23449, 23555, 23591, 23615, 23721, 23798, 23851, 24032,
        24042, 24081, 24098, 24136, B24176, 24187, 24213, 24257, 24281, 24316, 24367,
        24374, 24417-8, 24497, 24637-8, 24757, 24790, 24869-70, 24886, 24911, 24938,
        25006, 25094, 25110, 25143
        25202, 25214, 25242, 25304, 25375, 25417, 25494, 25671, 25719, 25804, 25830,
        25861, 25980-1, 25984, 26056, 26198, 26231, 26258-9, 26325, 26333, 26367,
        26411, 26500, 26523, 26615, 26671, 26695, 26703, 26718, 26751, 26767, 26793,
        26967, 26989, 27005-6, 27031, 27075, 27125-6, 27194, 27328, 27355, 27447-8,
        27454-6, 27486, 27491-2, 27517-8, 27528, 27581, 27628, 27640, 27680, 27876,
        27878, 27916-7, 27923, 27955, 28083, 28165, 28176, 28235-6, 28307, 28420,
        28445, 28447-8, 28451-2
        28562, 28581, 28727, 28744, 28747, 28753-4, 28766, 28903, 28920, 28935-6,
        28967, 29107, 29114, 29192, 29216-7, 29245, 29263, 29416-7, 29539-41, 29583,
        29599-600, 29607, 29696, 29740, 29749, 29831, 29844, 29849, 29933, 29961,
        29973, 30002, 30108, 30114, 30126, 30180, 30245, 30291, 30335, 30370, 30402,
        30561, 30653, 30718, 30772-4, 30782, 30832, 30909, 31124, 31169, 31179, 31191,
        31291, 31459, 31462, 31469, 31483, 31487, 31520, 31549, 31560, 31562, 31617,
        31630, 31664, 31683-4, 31693, 31716, 31746-7, 31808, 31813-4, 31816, 31950,
        32042, 32135, 32245, 32332, 32362, 32480, 32513
        32631, 32657, 32697, 32702-3, 32801, 32807, 32886, 32935, 32954, 33016,
        33117, 33142, 33160, 33253, 33289, 33299, 33356, 33448, 33552-3, 33647,
        33691, 33718, 33778, 33787, 33864, 33929, 33945, 33989, 34025, 34032, 34126-
        -7, 34138, 34203, 34277, 34286, 34300, 34355, 34424, 34723, 34734, 34764,
        34803, 34805, 34868, 34918, 35068, 35076, 35125, 35177, 35209, 35252, 35322,
        35433, 35568, 35591, 35649, 35685, 35696-7, 35774, 35819, 36130, 36282,
        36397, 36482, 36484, 36506, 36538, 36558
        36606, 36699, 36829, 36937, 36968-9, 37047, 37218, 37245, 37280, 37285, 37507,
        37553, 37615, 37705, 37728, 37806, 37822, 37852, 37891, 37917, 37919, 37944,
        38002, 38023, 38030, 38055, 38094, 38109, 38250, 38541, 38630, 38647, 38773,
        38953, 38957, 38971, 38994, 39041, 39295, 39315, 39397, 39498, 39590, 39730,
        39913, 40109, 40120, 40124, 40247, 40278

    pH, effect on gas exchange
        22054, 22452, 22511, 22536, 22578, 22665, 22732, 22828, 22836, 22848, 22935,
        23058, 23462-3, 23596, 23840, 23843, 24834, 24919, 24979, 25069; 25163,
        25395, 25503, 25548, 26011, 26024, 26089, 26333, 26641, 27583-4, 27994,
        28000, 28088, 28205, 28229-30; 28579, 28728, 28900, 29570, 29629, 29877,
        30381, 30561, 31136, 31145, 31488, 31514, 31520, 31732, 31734; 32570, 32830,
                                                                          (continued)

pH, effect on gas exchange (continued)
        32953, 33146, 33384, 33426, 33759, 33929, 34162, 34188, 34558, 34671, 35402,
        35802, 36362, 36455, 36509; 36688, 36810, 36831, 36889, 36897, 37024, 37189,
        37351, 37497, 37927-9, 37932, 38250, 38274, 38293, B38477, 38843, 38845,
        39189, 39252, 39307, 39838, 40156

pH, effect on photorespiration
        25721, 26218, 27800; 30406, 31451, 31734; 32841; 37234, 38413

pH, effect on resistances to CO_2 and water vapour transfer 26748

pH-stat, buffers 25548, 28331; 32013

PhAR, PAR see Irradiance...; Canopy, radiation ...

Phosphoenolpyruvate carboxylase
        21722, 21736, 21739, 21782-3, 21814, 21874, 21932, 21941, 22054, 22077,
        22119-20, 22150, 22203, 22239, 22251-2, 22381, 22397, 22406, 22556, 22609,
        22635, 22697-8, 22727, 22835, 22837-8, 22905, 22951, 22958, 22997, 23030-1,
        23060, 23072, 23195, 23231, 23274, 23289, 23343, 23374, 23382, 23397, 23481,
        23610, 23632, 23754, B23767, 23772, 23797, 23814, 23877, 23926, 23973, 24008,
        24074, 24100, 24122, 24126, 24161-2, 24164-5, 24201, 24255, 24270, 24309,
        24344, 24469, 24523, 24525, 24615, 24657, 24797, 24804, 24933, 24984, 24986,
        25035-6, 25065-6, 25138
        25193, 25216, 25221, 25338, 25346, 25374, 25395, 25408-9, 25431, 25437-8,
        25443, 25445, 25582, 25759, 25767-9, 25787, 25797, 25831, 25835, 25845,
        25862, 25867, 25912, 25944-5, 26057, 26078, 26165, 26181, 26273, 26280,
        26303, 26314-7, 26428, 26444, 26446, 26485, 26539, 26549-50, B26551, 26574,
        26595, 26618-9, 26623, 26626, 26641, 26650, 26686, 26692, 26714, 26786,
        26806, 26872, 26908, 27007-8, 27052-3, 27106, 27147, 27183, 27198, 27315,
        27358-9, 27364, 27506, 27541-2, 27551, 27582, 27596-8, 27605, 27621, 27634,
        27709, 27753, 27896, 27906, 27948, 28040, 28083, 28142, 28146, 28183, 28220,
        28264, 28351, 28357, 28388-90, 28399
        28501, 28613, 28652, 28748, 28818, 28841, 28869, 28901, 28928, 29115, 29142,
        29165, 29231, 29262, 29273, 29363, 29463, 29534-5, 29669, 29709, 29777, 29850,
        29860, 29863, 29910, 29990, 29992, 30009, 30037, 30098, 30128, 30174, 30232,
        30234, 30332-3, 30435, 30537-8, 30695, 30749, 30754, 30778, 30784, 30823,
        30886, 30897, 30907, 30928, 30951-2, 30959-61, 30990, 31166, 31223, 31240, 31300,
        31419, 31441, 31447-9, 31465, 31472, 31474, 31576, 31611, 31637, 31977,
        32036, 32262, 32317, 32425-6, 32428-9, 32440, 32457, 32502
        32691, 32826, 32912, 32931, 32953, 32996, 33148-9, 33344, 33474, 33555,
        33571, 33612, 33654, 33777, 33843, 33913-4, 33981, 34017, 34052, 34196,
        34253, 34256, 34312, 34429, 34452, 34456, 34496, 34525, 34549, 34556, 34716,
        34738, 34800, 34834, 34905-6, 34949-50, 34952, 34978-9, 35029, 35034-5,
        35051, 35087, 35191-2, 35197-8, 35248-9, 35266-7, 35302, 35425, 35427,
        35431, 35466-9, 35471-2, 35489, 35593, 35631, 35701, 35892-3, 35924-5,
        35970, 35986, 36019, 36121, 36241, 36366, 36426, 36440, 36442, 36491, 36533
        36603, 36664, 36716, 36723, 36767, 36814, 36889, 36939, 37028, 37185, 37197,
        37213, 37215, 37249, 37274, 37303, 37317-9, 37387, 37391-3, 37457-8, 37479,
        37542, 37627-8, 37659, 37713-4, 37755-6, 37791, 37817, 37826-7, 37858, 37905,
        37933-4, 37962, 37998, 38036, 38087, 38103, 38125, B38139, 38235-7, 38323,
        38327, 38358, 38420, 38428, 38758, 38789, 39009, 39064, 39093, 39105, 39132,
        39153, 39167, 39217, 39235, 39237, 39239, 39263-5, 39284, 39335, 39366,
        39416, 39431, 39460, 39479, 39579, 39583, 39604, 39688-9, 39723-4, 39747,
        39876, 39954, 40046, B40105, 40119, 40210, 40212-4, 40238, 40264, 40307,
        40360

Phosphoenolpyruvate carboxylase, methods
        22381, 23343, 24309, 24387, 25065-6; 25408, 25797, 28201; 29165, 29990,
        30749, 30784, 30907, 30959, 31419; 33914, 34196, 34979, 36442; 37028, 38036,
        39579

Phosphorus see Mineral elements (N, P, K) ...

Photoperiod and algae productivity   26058

Photoperiod and carbon fixation pathways
     21932, 22366, 22725, 23539, 24126; 25515, 27073, 27541, 28388; 29273, 29641;
     35192; 37028, 37274, B37531, 37564, 38643, 38788-9, 39132, 39217, 39889,
     40010

Photoperiod and carotenoids 23417; 31237; B38021

Photoperiod and chlorophyll
     21734, 23035, 23417, 23528, 23564, 23861; 25596, 26377, 27810; 29150, 29354,
     31237, 31454; 32594, 33481, 34607, 35035; B37531, 37707, 37893, 37943, 38292,
     39203

Photoperiod and chloroplast (chromatophore)
     21851-2, 22978; 27714; 28871, 28896, 29765, 29871, 30896; 33481; B37531,39539

Photoperiod and ecosystem and plant productivity
     21882, 22725, 23712, 23815, 24625; 25231, 25596, 26426, 27283, 27428, 27673,
     27719, 27810, 28041, 28145, 28352; 30680, 30896, 31237, 32436, 32504; 33019,
     34545, 35052, 35458, 36219, 36348; 37124-5, 37298, 37564, 37898, 37974,
     38157, 38170, 38813, 39272, 39442, 40235

Photoperiod and electron transport chain
     23005; 31727; 32591, 32594; B37531

Photoperiod and gas exchange
     21696, 22366, 22726, 23371, 23712, 23764, 24556, 24625, 25118; 25332, 26386,
     26426; 28900, 29093, 29820, 30905, 30947, 32449; 32613, 32953, 34368, 35115;
     36604, 37125, B37531, 37656, 37857, 37893, 37943, 40254

Photoperiod and resistances to CO_2 and water vapour transfer   30903, 30905; 36984,
     40254

Photoperiod and respiration 38813

Photophosphorylation, cyclic
     21528, 21530, 21559, 21563, 21588, 21611, 21613, 21688, 21698, 21744, 21888,
     21913, B21914, 22017, 22051, 22134, 22233, 22392, 22527, 22540-1, 22566,
     22576, 22669, 22706-7, 22711, 22812, 22834, 22842, 22894, 22922-3, 23057,
     23209, 23242, 23273, 23351, 23415, 23624, 23647, 23660, 23668, 23837, 23866,
     23888, 23920, 23940, 23944, B24024, 24032, 24048, 24099, 24136, 24159,
     B24176, 24226, 24286, 24311, 24355, 24416, 24458, 24502, 24669, 24721,
     24730, 24789, 24791, 24856, 24869, 24875, 25006, 25030, 25143
     25222, 25240, 25365, 25399, 25425, 25493, 25508, 25520, 25846, 25891, 26148,
     26193, B26220, 26268, 26335, 26442-3, 26472, 26556, 26566, 26711, 26718,
     26767, 26852, 26872, 26889, 26967, 27060, 27114, 27125, 27365, 27403, 27485,
     27535, 27550, 27552-3, 27580, 27588-91, 27603, 27637, 27668, 27744, 27809,
     27856, 27901, 28002, 28112, 28164, 28203, 28225, 28235, 28416, 28447, 28452,
     28457-8
     28541, 28592-5, 28640, 28708, 28742, 28789, 28798, 28851, 28859, 28887,
     28889, 28894, 29051, 29070, 29180, 29282, 29353, 29376, 29449, 29519, 29525,
     29596-7, 29603, 29664-5, 29714, 29731, 29810, 29844, 39885, 29942, 29959,
     30035, 30050, 30103, 30110, 30114-5, 30135, 30223, 30336, -30361, 30436, 30449,
     30558, 30607, 30659, 30661, 30839, 30845, 30859, 30867, 30892, 31045, 31096,
     31291, 31379, 31416, 31420, 31455, 31458, 31461, 31503, 31522, 31533, 31551,
     31665, 31715, 31727-8, 31736, 31828, 31872, 31998, 32157, 32161, 32178,
     32198, 32204, 32216-7, 32235-6, 32247-8, 32384-5, B32410, 32496-9, 32521
     32636, 32673, 32706, 32769, 32792, 32834, 32848, 32853, 32862, 32908, 33207,
     33253, 33323, 33425, 33687, 33716, 33853, 33932, 33986, 33994-5, 34026-7,
     34059, 34093, 34372, 34431, 34441, 34447, 34598, 34805, 34866, 34891, 35009,
     35124, 35167, 35177, 35396, 35404, 35427-.35432, 35434, 35440, 35539, 35698,
     35752, 35764, 35892-3, 35917, 35926, 36002, 36175, 36210, 36316, 36382
                                                                      (continued)

Photophosphorylation, cyclic (continued)
        36701, 36710, B36768, 36783, 36935, 36975, 36982, 36993, 37005, 37042,
        37186, 37222, 37469, 37642, 37708, 37841-2, 37875, 37907, 37920, 37937,
        38030, 38080, 38191, 38394, 38453, 38723, 38733, 38735, 38769, 38926, 38978,
        39194, 39311, 39329, 39352, 39356, 39379, 39404, 39408-9, 39432, 39521,
        39539, 39557, 39584, 39648, 39674, 39707, 39997, 40019, 40038, 40040, 40058,
        40097, 40124, 40325

Photophosphorylation in photosynthetic bacteria   see   Photosynthetic bacteria, photo-
        phosphorylation

Photophosphorylation, methods
        21858, 21982, 23183, 23186, 24149, 24351, 24869; 27093, 27125, 27454-5, 28002,
        28108, 28452; 29665, 30658, 31818, 32245; 32697, 34680, 34918, 35650, 35889,
        36559; 37800, 37913, B38037, B38477, 38938, 39629

Photophosphorylation, model   see   Model ...

Photophosphorylation, non-cyclic
        21514-5, 21528, 21530, 21548, 21559, 21561, 21563, 21588, 21607-8, 21611,
        21613, 21687-8, 21727, 21744-5, 21753, 21766, 21859, 21888, 21912-3, B21914,
        21996, 22017, 22025, 22038, 22134, 22216-7, 22286, 22426-7, 22498, 22513,
        22540-1, 22566, 22568, 22574-5, 22580, 22586, 22625, 22657, 22659, 22669,
        22689-90, 22711, 22720, 22722, 22724, 22766, 22801, 22834, 22870, 22891-2,
        22894, 22900, 22906, 22913, 22946, 23005, 23008, 23066, 23079, 23118, 23163,
        23173, 23201, 23209, 23215, 23242, 23273, 23351, 23434, 23486, 23506-7,
        23541, 23555, 23624, 23647, 23660, 23718, 23723, 23762, 23847, 23888-9,
        23898, 23920, B23933, 23944, 23997, 24011, 24024, 24032, 24035, 24091, 24095,
        24104, 24107, 24136, 24159-60, B24176, 24188, 24224, 24242, B24250, 24257,
        24281, 24298, 24311-2, 24338, 24351, 24367, 24374, 24411, 24416, 24457,
        24466, 24471, 24502, 24568, 24721-2, 24730, 24737, 24789, 24840, 24846,
        24856, 24859-60, 24870, 24875, 24941, 24951, 24954, 24999, 25029-30, B25064,
        25090-1, 25094, 25100, 25102, 25124
        25177, 25182-3, 25197, 25362, 25365-6, 25375, 25377, 25387, 25399, 25401,
        25417, 25425, 25487, 25493-4, 25508, 25520, 25706, 25789, 25829, 25846,
        25891, 25984-5, 26148, 26152, 26175, 26193, 26210, B26220, 26230, 26268,
        26366, 26418, 26423, 26473, 26496, 26500, 26507, 26566, 26615, 26711, 26718,
        26720, 26726, 26730, 26767, 26780, 26790, 26852, 26862, 26872, 26902, 26967-8,
        26980, 26984, 27010, 27017-8, 27047, 27125, 27131, 27184, 27194, 27204,
        27243-4, 27335, 27354-5, 27442, 27471, 27485-6, 27491, 27523, 27532, 27550,
        27552, 27559, 27562, 27580, 27590-1, 27640, 27661, 27676, 27744-5, 27747,
        27786-7, 27796, 27815-6, 27856-7, 27901, 27910, 27934, 27992, 27996, 28105,
        28112, 28164-5, 28211, 28227, 28235, 28246, 28306-7, 28376, 28410-1, 28416,
        28445, 28452, 28458
        28523-4, 28592-3, 28679, 28739, 28755-6, 28766, 28777, 28779, 28784, 28789,
        28798, 28833, 28839, 28859, 28879, 28887, 28896, 28906-7, 29044, 29184,
        29285, 29353, 29376, 29382, 29449, 29464, 29525, 29599, 29639, 29665, 29695,
        29730-1, 29826, 29838, 29847, 29885, 29932, 29959, 30034, 30062, 30093,
        30103, 30110-1, 30115, 30119, 30126-7, 30132, 30136, 30178, 30197, 30211,
        30291, 30297, 30329, 30361, 30429, 30442, 30528, 30540, 30548, 30558, 30661,
        30726, 30805, 30859, 30867, 30892-3, 30909, 30962-4, 30966, 30969, 30985-6,
        31045, 31082, 31096, 31119, 31128, 31191, 31238, 31275, 31322, 31356, 31379,
        31418, 31456, 31458, 31461, 31522-3, 31541, 31554, 31626, 31664-5, 31667,
        31703, 31715, 31717, 31727-8, 31736-7, 31754-5, 31818, 31970, 31987, 31998,
        32072, 32135, 32145, 32161, 32178, 32198, 32204, 32216, 32223, 32233, 32246-7,
        32309, 32311, 32343, 32385, B32410, 32437-8, 32444, 32480, 32484, 32490,
        B32515, 32536
        32636, 32664, 32673, 32709, 32795, 32819, 32826, 32834, 32839, 32848, 32862,
        32866, 32886, 32908, 33008, 33032, 33041, 33108, 33207, 33323, 33425, 33530,
        33647, 33687, 33714-5, 33743, 33748-9, 33853, 33882, 33886, 33932, 33939-40,
        34025, 34059, 34205, 34256, 34372, 34422, 34441, 34475, 34521, 34575-6,
        34589, 34661, 34674, 34769, 34803-4, 34850, 34866, 34918, 34922, 34963,
        34984, 35009, 35012, 35054, 35131, 35134, 35177, 35187-8, 35209, 35306,
                                                                             (continued)

Photophosphorylation, non-cyclic (continued)
       35396, 35404, 35427, 35432, 35434, 35487, 35618, 35676, 35681, 35698, 35700,
       35716-7, 35764, 35843, 35868, 35893, 35915, 35926, 35981, B36081, 36173-4,
       36200, 36223-4, 36236, 36265, 36283, 36316, 36363, 36382, 36398, 36558,
       36581
       36611, 36701, B36768, 36833, 36842, 36884, 36975, 36993, 37005, 37026,
       37146, 37186, 37282, 37469, 37566, 37615, 37666, 37682, 37711, 37737, 37830,
       37841, 37843, 37850-1, 37875, 37920, 37937, 38022, 38026-7, 38030, 38042, 3837
       38377, 38394, B38477, 38482, 38487, 38527, 38531, 38580, 38623, 38658,
       38661, 38683, 38733, 38751, 38769, 38797, 38808, 38812, 38820, 38882, 38888,
       38901, 38926, 38936, 38940, 38978, 38994, 39041, 39048, 39152, 39194, B39220,
       39250, 39298, 39311, 39331, 39352, 39397, 39404, 39427, 39433, 39471, 39539,
       39543, 39557, 39584, 39608, 39629, 39673, 39730, 39795, 39810, 39912, 39997,
       40019, 40038-9, 40055, 40058, 40097, 40100, 40206, 40281, 40307, 40321

Photophosphorylation, pseudo-cyclic
       22548, 22707, 22833, 23273, 23847, 23888, 23944, 24170, 24351; 25493, B26220,
       26268, 26443, 26566, 26780; 28742, 28879, 29376, 29665, 30292, 30765, 32216-
       -7; 33425, 34250, 34255, 34447, 34694, 34764, 36175; B36768, 37026, 38029,
       39602, 39730, 40097, 40301

Photoreduction see H$_2$ evolution ...

Photorespiration enzymes
       21711, 21957, 22071, 22410, 22542, 22940, 22980-2, 23427, 24122, 24247,
       24933
       25542, 26004, 26162, 26218-9, 26402, 26478, 26784, 26786, 26858, 27359,
       27364, 27416, 27569, 27754, 27778, 28151-4, 28183, 28220, 28229, 28267,
       28342, 28478
       28513, 28928-9, 28931, 29109, 29231, 29512, 29536, 29604, 29637, 29700,
       29747, 29841, 30122, 30374, 30568, 30640, 30886, 30951, 31008, 31088, 31164,
       31182, 31251, 31300, 31315, 31433, 31457, 31914, 31928, 32422, 32461-2, 32529,
       32547
       32585, 32841-2, 32952, 33125, 33156, 33235, 33705, 33837, 33887, 33960,
       34073, 34254, 34256, 34424, 34509, 34658-9, 34683, 34905, 35061, 35152,
       35213, 35247-9, 35324, 35431, 35444, 35627, 35694, 35766, 36104, 36162,
       36410, 36415, 36561
       36624, 36814, 36824, 36832, 36844, 36871, 36889, 36978, 37216, 37234, 37451,
       37471, 37653, 37783, 37824, 37858, 37946, 38061-2, 38081, 38103, 38106,
       38413, 38517, 38625, 38868, 38955, 39205, 39235, 39338, 39460, 39839, 39939,
       40050, B40105, 40204-5, 40345

Photorespiration metabolic cycles
       21543, 21587, 21618-9, 21711, 21717, 21728, 21785, 21800, 22070, 22110-1,
       22192, 22310, 22320, 22347, 22442, 22542, 22548, 22609, 22649, 22942, 23072-
       -3, 23198, 23309, 23329, 23455, 23467, 23509, 23511, 23714, 23877, 23973,
       23981-2, 24064, 24113, 24222, 24293, 24408, 24779, 24804, 24949-50, 24969,
       25071, 25091, 25097, 25136, 25138-9, 25160
       25249, 25262, 25340, 25542, 25573, 25602, 25619, 25640, 25711-2, 25767-8,
       25940, 26038, 26162, 26218, 26272, 26314, 26402, 26574, 26595, 26620, 26623,
       26659, 26668, 26689, 26691, 26722, 26767, 26810, 26858, 26872, 26959, 26996,
       27247, 27416, 27462, 27493-4, 27524, 27754, 27778, 27797, 27989, 28154,
       28230, 28422, 28477-80
       28600, 28772, 28774, 28928, 29002, 29032, 29109, 29159, 29176, 29286, 29294,
       29328, 29412, 29504, 29571, 29586, 29637, 29746-7, 29759, 30098, 30106, 30122,
       30202, 30288, 30294, 30369, 30374, 30429-30, 30468-9, 30499, 30543, 30546,
       30631, 30635, 30637-40, 30686, 30798, 30951, 31034, 31134-5, 31164, 31227,
       31313-4, 31352, 31451, 31520, 31677, 31731, 31733, 31802, 31904, 31908, 31928,
       32036, 32337-8, B32410, 32460-2, 32529
       32577, 32585, 32638, 32683-5, 32822-3, 32826, 32828, 32855, 32953, 33069,
       33125, 33217, 33235, 33324, 33543, 33561, 33631, 33633, 33696, 33785, 33843-
       -4, 33960, 34103, 34251, 34254-6, 34379, 34383, 34410, 34424, 34524-5,
       34539, 34652, 34658-9, 34695, 34701, 34718, 34905, 35061, 35162-4, 35167,
       35186, 35265, 35324, 35375, 35468, 35535, 35601, 35675, 35765-6, 35944,

                                                              (continued)

Photorespiration metabolic cycles (continued)
        36116, 36162, 36169, 36185, 36280, 36330, 36410, 36432, 36475, 36480, 36561
        36679, 36714, B36768, 36810, 36824, 36889, 37012, 37084, 37116, 37425,
        37431, 37506, 37631, 37653, 37729, 37784, 37786-7, 37876, 37881, 37950,
        38114, 38176, 38429-30, 38433-4, 38445, B38477, 38496, 38552, 38608, 38868,
        38903, 38964-6, 39033, 39065, 39150, 39178, 39263, 39265, 39401, 39482,
        39710, 39939, B40015, 40080-1, B40105, 40113-4, 40188, 40268, 40310, 40345

Photorespiration metabolic cycles enzymes, methods see Enzymes of glycollate cycle,
        methods

Photorespiration rate
        21535, 21659, 21897, 21926, 21967, 21994, 22005, 22038, 22066, 22152, 22228,
        22521, 22649, 23073, 23171, 23329, 23493, 23634, 23796, 23994, 24072, 24122,
        24264, 24336, 25137, 25160
        25167-8, 25366, 25448, 25476, 25609, 25640, 25858, 25883, 26055, 26155,
        26214, 26317, 26626, 26988, 27106, 27313, 27414, 27424, 27469, 27521, 27524,
        27760, 27837, 27989, 28112, 28153, 28268, 28386-7, 28393, 28477
        28772, 28785, 28898, 28901, 28993-4, 29043, 29054, 29087, 29090, 29108-9,
        29156, 29169, 29178, 29191, 29328, 29514, 29604, 29795, 29989, 30047, 30457,
        30542, 30546, 30622-4, 30640, 30886, 31102, 31118, 31134, 31266, 31293, 31470,
        31636-7, 31836, 31913, 32073, 32104, 32219, 32428, 32461, B32528
        32652, 32773, 32804, 32823, 32965, 32967, 32996, 33062, 33123-4, 33153,
        33231, 33363, B33510, 33543, 33561; 33631, 33633, 33785, 34052-3, 34097,
        34122, 34154, 34379, 34524, 34535, 34549, 34689, 35044, 35081, 35138, 35140,
        35161-2, 35164, 35222, 35286, 35324-5, 35334, 35341-2, 35368, 35370, 35423,
        35465, 35471, 35535, 35574, 35678, 35714, 35765, 35842, 36114-6, 36162,
        36280, 36344, 36406, 36417
        36602, 36671, 36736, 36748, 36833, 36871, 36876, 36967, 36983, 37020, 37066-
        -7, 37105, 37291, 37312-3, 37431, 37502, 37515-6, 37576, 37612, 37624,
        37630, 37712, 37754, 37817, 37924, 37981, 38122, 38287, 38384, 38429, 38436,
        38464, 38755, 38769, 38903, 39033, 39057, 39065, 39073, 39099, 39179, 39235,
        39266, 39277, 39391, 39459, 39716, 39868, B40015, 40080-1, 40127, 40324

Photosynthate translocation and distribution
        21511-2, 21535, 21552-4, 21557, 21560, 21593, 21630, 21634, 21684, 21692,
        21740, 21762, 21786, 21847, 21853, 21878, 21883, 21951, 21966, 21999, 22004,
        22037-8, 22064, 22096, 22123, 22137, 22196, 22212, 22229, 22351, 22365,
        22367-9, 22428-9, 22470, 22489, 22547, 22638, 22663, 22678, 22687, 22714,
        22723, 22732, 22754, 22770-2, 22777, 22826, 22845, 22872, 22878-9, 22910,
        22932, 22965, 22974, 22985, 23032, 23045, 23073, 23085, 23175, 23212, 23220,
        23236-7, 23243, 23251, 23332, 23372, 23385, 23431, 23587, 23603, 23649, 23694,
        23696, 23713, 23756, 23773-4, 23803, 23817, 23902, 23908, 23912, 23937, 23958,
        23975, 23990, 24000-1, 24023, 24088, 24105, 24114, 24120, 24124, 24142, 24150,
        24204, 24246, 24252, 24259-60, 24275, 24301, 24326, 24332, 24337, 24359,
        24363, 24432, 24439, 24444, 24448, 24508, 24526, 24556, 24558, 24598, 24626,
        24628, 24632, 24639, 24656, 24708, 24715, 24765, 24771, 24799, 24805, 24830,
        24839, 24851-2, 24872-3, 24965, 24967, 24969, 24974, 24976-7, 25059-61,
        25072, 25089, 25109, 25125, B25153
        25170-1, 25190, 25193, 25206-7, 25226, 25229, 25256, 25299, 25324, 25328,
        25397, 25405, 25411, 25489, 25519, 25527, 25568, 25593, 25642, 25666, 25682,
        25686, 25722-3, 25727, 25740, 25765, 25774, B25775, 25799, 25801, 25806,
        25856, 25901, 25929, 25946, 25965-7, 26006, 26010, 26032, 26039, 26093,
        26109, 26121-2, 26135-6, 26150, 26161, 26169, 26190, 26221, 26320, 26349,
        26369, 26384-6, 26389-90, 26429, 26433-4, 26451-2, 26489, 26493, 26498,
        26522, 26569, 26596, 26607, 26623, 26742, 26778, 26784, B26815, 26816-7,
        26845, 26847, 26856, 26876, 26883, 26926, 26954, 26965, 26979, 26983-4,
        27011, 27016, 27055, 27098, 27199, 27210, 27215, 27240, 27255, 27290, 27331,
        27338, 27343-4, 27358, 27361, 27384, 27386, 27389, 27393, 27412, 27430,
        27438, 27445, 27477-8, 27502, 27514-5, 27540, 27573, 27585, 27593, 27600,
        27624, 27687-8, 27722, 27725, 27752-3, 27812, 27814, 27821, 27887, 27918,
        27930, 27932, 27942, 27990-1, 27995, 27999, 28005-6, B28015, 28018, 28020,
        28023, 28053, 28062, 28130, 28177, 28246, 28254, 28269, 28296, 28301-2,
        28304, 28322, 28336-9, B28340, 28343, 28374, 28382, 28394, 28425, 28435,
                                                         (continued)

Photosynthate translocation ... (continued)
        B28482, 28495
        28613, 28618, 28620, 28624, 28749, 28761, 28796, 28855, 28872-4, 28913, 28926,
        28957, 28975, 29009, 29046, 29052-3, 29102, 29119, 29132, 29157, 29221-2,
        29277, 29301-2, 29313, 29325, 29331, 29338, 29351, 29372, 29381, 29443, 29446,
        29460, 29482, 29485, 29487, 29492, 29500, 29506, 29509, 29577, 29642-6, 29648,
        29699, 29705, 29726, 29771, 29783, 29822, 29832, 29848, 29860, 29895, 29902,
        29921, 29923, 29951-4, 29963-4, 29966, 29992, 30010-1, 30021-2, 30206, 30236,
        30351-3, 30523, 30530, 30532, 30543-4, 30627, 30712-3, 30734, 30754, 30798,
        30811, 30820-1, 30840, 30858, 30896, 30920, 30929, 30937, 31229, 31245, 31262,
        31277, 31310, 31376-8, 31399, 31408, 31417, 31477, 31494, 31641, 31675, 31732,
        31759, 31763-4, 31790, 31792, 31823, 31849, 31882, 31889, 31894, 31916, 31940-
        -1, 31947, 31953-4, 31974, 31978, B32004, 32041, B32056, 32057, 32066, 32119,
        32122, 32171-2, 32182, 32200-1, 32218, 32306, 32339, 32343, 32347, 32363,
        32372, 32389, 32415, 32434, 32436, 32491, 32503, 32509, B32515, 32519-20,
        32542
        32576, 32582, 32612, 32686, B32723, 32742, 32808, 32850, 32882, 32971, 33002,
        33011, 33021, 33040, 33043, 33111, 33159, 33162, 33168-9, 33181, 33226,
        33232, 33260-1, 33270, 33279, 33318, 33324, 33358-9, 33369, 33396-7, 33415,
        33467, 33500, 33507, 33521, 33525, 33534, 33537, 33547-9, 33599, 33614,
        33624, 33658, 33671, 33813, 33831-2, 33846, 33859, 33867, 33871-2, 33893,
        33928, 33950, 33970, 33974-5, 34005-6, 34040, 34156, 34258, 34264, 34304,
        34312, B34354, 34408, 34429-30, 34437, 34458, 34510, 34525, 34536, 34555,
        34614, 34622, 34637-40, 34712, 34719, 34721, 34754, 34831, 34834, 34873,
        34896, 34923, 34932, 34937, 35033, 35110-1, 35169, 35180, 35205, 35225-6,
        35243, 35255-6, 35279, 35295, 35330-2, 35373, 35394, 35416, 35451, 35481,
        35504, 35524, 35656, 35712, 35721, 35744, 35759, 35766, 35790, 35824, 35866-
        -7, 35888, 35938, 35952-4, 35966, 35972, 35974, 35979-80, 35997-8, 36009,
        36014, 36021, 36044, 36116, 36126, 36136, 36144, 36172, 36194, 36204-5,
        36227, 36267, 36288, 36308, 36329, 36333, 36343-4, 36381, 36422-3, 36531
        36623, 36641, 36681, 36708, 36716, 36728, 36742, B36768, 36777, 36780, 36785,
        36788-9, 36818, 36840, 36851, 36883, 36925, 36950, 36986, 37012, 37024, 37043-
        -4, 37071-2, 37074, 37086, 37122, 37124-5, 37139, 37141, 37152, 37159, 37164,
        37185, 37191, 37207-8, 37277, 37289, 37297, 37327, 37383, 37390, 37464-5,
        37474, 37482, 37487, 37491-2, 37510, 37521, 37617-8, 37629, 37637, 37681,
        37687, 37709, 37787, 37801, 37803-4, 37807, 37847, 37863, 37867, 37876,
        37886-7, 37893, 37896-7, 37926, 37941, 37986, 37989, 38117, 38128, 38148,
        38156-7, 38168, 38170, 38174, 38195, 38206, 38218, 38246, 38256, 38317,
        38323, 38421, 38427, 38450, 38488-9, 38498, 38512, 38520, B38544, 38551,
        38583, 38586-7, 38607, 38627, 38635, 38645, 38658, 38664, 38673, 38676-7,
        38715, 38742, 38749, 38755, 38770-1, 38787, 38812, 38834, 38873, 38880,
        38893, 38943, 38993, 39015, 39019, 39031-2, 39065-7, 39070, 39076, 39084,
        39088, 39109, 39117, 39130, 39142, 39190, 39214-5, 39219, 39233, 39238,
        39313, 39337, 39360, 39389, 39398, 39417, 39428, 39433, 39464, 39480-1,
        39493, B39513, 39565, 39570, 39595-6, 39627, 39637-8, 39680, 39692, 39697,
        39750, 39761, 39793, 39815, 39852, 39858, 39864, 39900, 39905, 39933, B39952,
        39989-90, 39993, 40008, 40074, 40103, 40138, 40143-4, 40146, 40154, 40176,
        40211, 40216, 40242-3, 40263, 40271, 40275, 40296, B40331

Photosynthates and intermediates of carbon fixation pathways
        21573, 21618, 21673, 21699, 21784-5, 21799-800, 21847, 21859, 21883, 21966,
        22018, 22051, 22064, 22080, 22137, 22239, 22251-2, 22308, 22399, 22530,
        22552, 22608, 22697, 22808, 22828, 22838-9, 22859, 23001, 23030-1, 23033,
        23060, 23072, 23198, 23210, 23213-4, 23309-10, 23370, 23382, 23386, 23467,
        23508, B23767, 23772, 23776, 23785, 23835, 23841, 23843, 23973-4, 23989,
        24000, 24008, 24059, 24064, 24066, 24074, 24101, 24137, 24139, 24160-1,
        24173, B24176, 24344, 24442, 24452, 24615, 24656, 24708, 24800, 24850, 24982,
        24984, 25009, 25034, 25036, 25097, 25116, 25138
        25188, 25249, 25280, 25331, 25338, 25358, 25373-4, 25449, 25515, 25584, 25619,
        25642, 25678, 25682, 25766, 25768, 25796, 25831, 25863, 25866-9, 25912,
        25954, 26005-6, 26018, 26052, 26181, 26206, 26244, 26333, 26343, 26398,
        26402, 26492, 26504, 26549-50, B26551, 26574-5, 26595-7, 26623, 26627-8,
        26668, 26727, 26760, 26767, 26770-2, 26864, 26866, 26888, 26911, 26996,
                                                                        (continued)

Photosynthates and intermediates of carbon fixation pathways (continued)
        27007, 27010, 27021, 27026, 27236, 27247, 27255, 27279, 27292, 27361, 27384,
        27399, 27414, 27530, 27621, 27639, 27705, 27797, 27862, 27901, 27980, 28116,
        28130, 28269, 28303, 28351, 28357
        28567, 28841, 28855, 28940, 28957, 28975, 28981, 29019, 29040, 29098, 29142,
        29191, 29214, 29253, 29270, 29321, 29460, B29468, 29495, 29534, 29699, 29807,
        29814, 29825, 29891, 29911, 29983, 30022-3, 30038, 30088, 30098, 30109,
        30232-4, 30236, 30275, 30277, 30279, 30433, 30435, 30544, 30564, 30604,
        30810, 30817, 30841, 30887, 30897, 30936, 30968, 30988, 31040, 31048, 31109,
        31246, 31260, 31277, 31314-5, 31420, 31448, 31451, 31520, 31525, 31554,
        31599, 31628, 31676, 31765, 31889, 31908, B31990, 32036, 32075, 32092, 32121,
        32179, 32205, 32297, 32343, 32425, 32430, 32439-40, 32529
        32584, 32795, 32803, 32855, 32894, 33054-5, 33069, 33116, 33148, 33151,
        33154, 33198, 33491, 33553, 33696, 33777, 33842, 33887, 33943-4, 34199,
        34250, 34292, 34427-8, 34496, 34559-61, 34614, 34657, 34659, 34742, 34800,
        34813, 34824, 34906, 34978, 35005, 35064, 35087, 35192, 35368, 35378, 35390,
        35428, 35469, 35535, 35925, 35940, 35976, 36002, 36019, 36204, 36308, 36331,
        36365
        36800, 37033, 37143, 37311, 37876, 37903, 37962, 37975, 38001, 38088, B38139,
        38142, 38258, 38326-7, 38412, 38586, 38633, 38757, 38788, 39079, 39150,
        39167, 39217, 39313, 39390, 39546, 39713, 39832, 39834, 39876, B40105, 40119,
        40210, 40236, 40267, 40360

Photosynthates and intermediates of carbon fixation pathways and chloroplast (chroma-
        tophore) 33146, 35747

Photosynthates and intermediates of carbon fixation pathways and electron transport
        chain 23989

Photosynthates and intermediates of carbon fixation pathways and gas exchange 21573,
        21672, 21869, 21999, 22038, 22502, 22749, 22782, 23649, 23785; 25258, 26021,
        26121, 27171, 27215, 27908, 28003; 28975, 29133, 29202, 29363, 29564, 30904-
        -5, 31134; 34923, 35162-3, 35368, 36557; 36810, 39173, 39371

Photosynthates and intermediates of carbon fixation pathways and photorespiration
        22320, 22749, 23150, 24949

Photosynthates and intermediates of carbon fixation pathways and respiration 35471

Photosynthetic bacteria carbon fixation pathways
        23760, B24250, 24342, 24681, 25067; 25478, 25582; 29098, 30382; 32623,
        32683, 33401, 33612, 33707, 33807, 34658, 35323, 35693, 35937, 35975, 36189;
        37033, 37149, B38415, 38952, 39205, 39468, 39472, 39657, 40028

Photosynthetic bacteria carotenoids see Xanthophylls of photosynthetic bacteria

Photosynthetic bacteria chlorophylls see Bacteriochlorophylls; Chlorophylls,
        Chlorobium

Photosynthetic bacteria chromatophores see Chloroplast and chromatophore ...;
        Chromatophore ...

Photosynthetic bacteria electron transport chain
        21531, 21566, 21668-9, 21815, 21821, 21843, 21856, 21866, 21913, B21914,
        21993, 22105, 22154, 22253, 22259, 22289, 22390, 22499, 22580, 22612, 22648,
        22659, 22679, 22733, 22759-60, 22773, 22903, 23020, 23122, 23124, 23126-7,
        23207, 23292, 23391, B23407, 23420, 23501, 23516, 23563, 23615, 23886, 23965-
        -6, 24028, 24031, 24095-8, 24174, 24227, B24250, 24329, 24344, 24690, 24814,
        24859, 25051, 25076
        25302, 25656, 25669, 25746, 25773, 25855, 25884, 25953, 25964, 26423, 26456,
        26486, 26695-6, 26720, 26808, 26902, 26942-3, 27068, 27282, 27326, 27340,
        27447-9, 27516-9, 27536, 27615, 27653, 27731, 27841, 27945, 28103, 28136,
        28163, 28237, 28259, 28376, 28469

                                                                (continued)

Photosynthetic bacteria electron transport chain (continued)
          28530, 28548, 28551, 28559, 28580, 28610, 28641, 28671, 28692, 28701, 28769,
          28822, 28824, 28832, 28904, 29038, 29074, 29078, 29192, 29263-4, 29266, 29272,
          29359, 29439, 29452, 29664, 29679, 29748-9, 29818, 29827, 29870, 29959-61,
          29969-70, 30050, 30116, 30265, 30340, 30398-9, 30535, 30625-6, 30718, 30780,
          30832, 30939, 31059-60, 31288, 31370, 31455, 31479, 31534, 31598, 31624,
          32014, 32169, 32250, 32279, 32281, 32370, 32467-8, 32550
          32657, 32683, 32711, 32764, 32803, 32860, 32915, 32926, 32955, 33114, 33119,
          33188, 33190, 33250-1, 33253, 33300, 33325, 33400, 33402, 33406, 33494,
          33552, 33710, 33786-7, 33838, 34013, 34023-4, 34130, 34229-30, 34237, 34277,
          34281, 34356, 34393, 34460, 34584, 34726, 34762, 34772, 34789, 34862, 34904,
          34963, 34965, 35131, 35167, 35239-40, 35306, 35311, 35333, 35386-8, 35409,
          35531, 35724, 35900, 35927, 36129, 36188, 36376, 36482-5, 36513
          36626, 36649, 36651, 36699, 36771, 36798, 36860-1, 36911-2, 36929, 36944,
          37033, 37117, 37167, 37200, 37218, 37345, 37438, 37446, 37507, 37546, 37611,
          37660-1, 37759-60, 37773, 37800, 37822, 37900, 37914, 37954-7, 38041, 38095,
          38111, 38189, 38240-1, B38415, 38538, 38569-70, 38615, 38637, 38795, 38905-6,
          38957, 38972, 39078, 39121, 39231, 39268, 39293-7, 39386-8, 39451, 39491,
          39502, 39673, 39685, 39822, 39842, 39863, 39901, 39909, 39961, 40012-3,
          40279

Photosynthetic bacteria gas exchange
          25249, 26727, 28484; 28600, 28750, 29229, 29684, 29913, 29940-1, 30227,
          30698, 31274, 31730; 32684-5, 33734, 34131, 34863, 36359; 36866, 37149,
          37994, 38264, 38336, 39440, 39839

Photosynthetic bacteria photophosphorylation
          21649-50, 21668-9, 21780, 21797, 21913, B21914, 22127, 22179, 22230, 22259,
          22499, 22520, 22612-3, 22760, 22773-4, 22824, 22904, 22913, 23124, 23362,
          23484, 23615, 23825, 23839, 24034, 24174, 24315-6, 24671, 24858, 24867, 25110
          25645, 25721, 25804, 25914, 26146, 26222, 26233, 26266, 26439, 26486, 26631,
          26969, 27374, 27447-9, 27455, 27504, 27535, 27537, 27627, 27806, 27916,
          28163, 28451
          28640-3, 28671-2, 28992, 29630-1, 29748-50, 29959-61, 29980-1, 30050, 30280,
          30340, 30386, 30658, 30831-2, 31179, 31191, 31287, 31295, 31299, 31549, 31871,
          32375-6, 32438, 32459
          32670, 32685, 32710, 32732-4, 33086, 33114, 33251-2, 33406, 33448, 33661,
          33707, 33753, 33786-7, 34130, 34282, 34299-300, 34355, 34762, 34841, 35306,
          35568, 35651, 35934, 36382, 36473-4
          36764, 36796, 36852, 37218, 37642, 37800, 37815, 38172, 38241, 38243, B38415,
          38647, 38825, 38923, 39020, 39118-20, 39293, 39483, 39554, 39704, 39706,
          39842

Photosynthetic bacteria reaction centres   see P700 ...

Photosynthetic (chlorophyll) unit
          21600, 21607-8, 21613, 21620, 21662, 21690, 21871, B21914, 21959, 21964,
          22134, 22164, 22193, 22230, 22264-5, 22408, 22580, 22957, 23187, 23327,
          23434, 23520, 23566, 23751, B23933, 24051, 24193, B24250, 24300, 24315,
          24691, 24825
          25184-6, 25240, 25263, 25315, 25335, 25353, 25616, 25631, 25751, 25846,
          26060, B26220, 26257, 26353-4, 26655, 26718, 26943, 26990, 27128, 27378,
          27399, 27511, 27581, 27953, 28055-6, 28488
          28518-9, 28528-9, 28589-90, 28595, 28655-6, 28724, 28726, 28779, 28997, 29001,
          29012, 29025, 29040, 29083, 29161, 29268, 29341, 29505, 29525, 29609, 29735-6,
          29785, 29830, 29978, 29991, 30034, 30179, 30181, 30719, 30759, 30831, 30969,
          30993, 31030, 31044, 31078, 31105, 31118-9, 31232, 31354, 31485, 31489, 31692,
          31713, 31725, 32113, 32124, 32407, B32410, 32432, 32441
          32590, 32594-5, 32658, 32664-7, 32675, 32744, 32889-90, 33195, 33198, 33203,
          33311, 33380, 33382, 33389-90, 33499, 33551, 33579, 33833, 33869, 33932,
          33990-1, 34237, 34319, 34393, 34478, 34729, 34756-7, 34760, 34797, 34845,
          34893, 35054, 35096, 35219-20, 35376, 35380, 35725, 35945, 35956, 36279,
          36409, 36411, 36447
          36646, 36741, B36769, 37171, 37173, 37316, 37362, 37603, 37820, 37893,
                                                                    (continued)

Photosynthetic (chlorophyll) unit (continued)
     37958, 38042, 38154, 38160, 38245, 38367, 38479, 38596, 38729, 38745, 38843-
     -5, 38905, 39027, 39073, 39149, 39165, 39185, 39379, 39502, 39701, 39822,
     39901, 40122, 40338

Photosystem 1
     21548, 21566-8, 21575, 21589, 21598, 21600, 21607-8, 21613, 21638, 21644,
     21662, 21664, 21672, 21685, 21688, 21690, 21697, 21706, 21712, 21720, 21727,
     21738, 21743-6, 21763-4, 21773, 21790, 21807, 21810, 21816, 21825, 21839,
     21856, 21867-8, 21963-5, 21983, 22079, 22104-5, 22134, 22136, 22142, 22193,
     22195, 22214, 22265, 22326, 22330, 22338, 22373-6, 22392-3, 22478, 22501,
     22514, 22526, 22534, 22559, 22561, 22574-6, 22580, 22583, 22605, 22629,
     22637, 22640, 22657, 22682, 22704-7, 22711, 22720, 22748, 22800, 22821,
     22900, 22906, 22922-3, 22933, 22970, 22972-3, 23020, 23024, 23055-7, 23066,
     23102-3, 23110, 23129, 23139-41, 23184-5, 23187, 23199, 23227, 23238, 23241-
     -2, 23253, 23351, 23353, 23390, 23394, 23396, B23407, 23459, 23486, 23515-6,
     23518, 23600-1, 23624, 23631, 23655, 23704, 23724, 23742, B23767, 23794,
     23799, 23847, 23852, 23860, 23866, 23876, 23879, 23915-6, 23920, B23933,
     23934-5, 23940, 23954, 23957, 23970, 24041, 24102, 24133, 24159, 24163,
     24191, 24208, 24224-6, 24238, 24241, B24250, 24306, 24317, 24351, 24361,
     24373, 24379, 24401-2, 24415, 24462, 24478, 24487-8, 24490, 24493-4, 24545-6,
     24559, 24603, 24631, 24664, 24684, 24721, 24724, 24727, 24737, 24756, 24762,
     24768, 24789-91, 24804, 24808, 24864, 24875, 24879, 24881, 24908-11, 24917,
     24923, 24939, 24951, 25002, 25057, 25107, 25114, 25124, 25148
     25166, 25174, 25180, 25182-4, 25186, 25196, 25202, 25214, 25217, 25240-1,
     25244-6, 25293, 25295, 25304, 25315, 25320, 25339, 25344, 25347-8, 25362,
     25370, 25375, 25379, 25385-7, 25401, 25403, 25425, 25438, 25456, 25470,
     25477, 25490, 25504, 25589, 25635-6, 25762, 25782, 25819, 25853, 25879,
     25888, 25893, 25900, 25918, 25942, 25963, 25968, 26040, 26056, 26084,
     26148, 26172-4, B26220, 26223-4, 26253, 26255, 26257-8, 26268, 26288,
     26294, 26298, 26310, 26334, 26362, 26367, 26378, 26411-2, 26423-4, 26457,
     26544, 26556-7, 26575, 26616, 26632-3, 26675-7, 26682, 26700, 26704, 26709,
     26711, 26718, 26720, 26758, 26761, 26765, 26774, 26792-4, 26807, 26813,
     26822, 26830, 26872, 26890, 26899, 26910, 26914, 26917, 26921, 26938, 26946,
     26955, 26995, 27031, 27033, 27060-1, 27064, 27070, 27075, 27086, 27114-5,
     27131, 27203, 27222, 27284, 27312, 27324, 27387, 27394, 27478, 27485, 27519,
     27550, 27581, 27589, 27593, 27603, 27610, 27619, 27637, 27640, 27655, 27668,
     27699, 27724, 27764-5, 27784, 27786, 27809, 27816, 27832, 27863-4, 27864,
     27866, 27870, 27876, 27917, 27924, 27943, 27955, 27977, 27979, 28021, 28028,
     28045, 28080, 28105, 28165, 28221, 28234, 28258, 28260, 28276, 28328-9,
     28333, 28376, 28395, 28416, 28424, 28445, 28458
     28514, 28528, 28539, 28547-8, 28551, 28554, 28575, 28581, 28584-5, 28587-8,
     28590, 28595, 28599, 28655, 28667-8, 28676, 28690-1, 28693-4, 28696, 28699,
     28708, 28717-8, 28720, 28739, 28744-5, 28751, 28758, 28766, 28789, 28798,
     28815, 28833, 28836, 28845-6, 28859, 28878, 28884, 28886, 28918, 28920, 28932,
     28934, 28965-6, 28968-9, 28997-9001, 29079, 29081, 29107, 29114, 29138, 29143,
     29161, 29173, 29211, 29231, 29245, 29275, 29278, 29284, 29300, 29305, 29310,
     29353, 29360, 29384, 29406, 29412, 29416, 29438-42, 29449, 29458, 29518-20,
     29525, 29527, 29533, 29540-1, 29557, 29570, 29601, 29619, 29621, 29670, 29682,
     29684, 29686, 29688, 29698, 29719, 29731, 29734, 29754-5, 29768, 29773, 29807,
     29831, 29836, 29857, 29867, 29872, 29880, 29932-3, 29939, 29948-50, 29971,
     29973-4, 30002, 30006, 30089, 30093-4, 30112, 30114, 30123, 30156-7, 30172,
     30179, 30181, 30246, 30251, 30254-5, 30291, 30320-3, 30336, 30359-62, 30402,
     30415, 30436-7, 30442-3, 30449, 30472, 30493, 30518, 30525, 30570, 30579-80,
     30586-7, 30589, 30607, 30630, 30716-8, 30771, 30777, 30780, 30799, 30839,
     30859, 30878, 30892, 30955, 30982-3, 31007, 31009, 31025, 31027, 31030, 31053,
     31078-80, 31087, 31091, 31098, 31108, 31128, 31143, 31145-6, 31155, 31161,
     31169, 31194, 31205-7, 31264, 31291, 31298, 31322, 31336, 31374-5, 31426,
     31485, 31503, 31505-6, 31521-2, 31551, 31567, 31581, 31588, 31610, 31612,
     31625-6, 31629-30, 31666, 31668, 31707, 31713, 31722, 31724-6, 31736-7,
     31785-6, 31809, 31813-4, 31816, 31828, 31835, 31877, 31918-9, 31931, 31979-82,
     32005, 32029, 32071-2, 32130, 32157, 32161-2, 32199, 32216, 32239, 32242,
     32246-8, 32256, 32283, 32303, 32315, 32331-2, 32359, 32370, 32392, 32407,
     32419, 32432, 32471, 32496, 32498-9, 32534

                                                           (continued)

Photosystem 1 (continued)
        32595, 32607-8, 32620, 32628, B32630, 32631, 32667, 32675, 32682, 32701,
        32744, 32747, 32769, 32799, 32801, 32811, 32860-1, 32866, 32899, 32926,
        32932-3, 32947, 32993, 33016, 33024, 33026, 33035, 33100, 33142-3, 33170,
        33201, 33250, 33253, 33311, 33314, 33350, 33355-6, 33370, 33442, 33458,
        33460-2, 33487, 33502, 33551, 33649, 33687, 33712-3, 33745-7, 33750, 33788-9,
        33802, 33806, 33821-2, 33827, 33850, 33869, 33880-1, 33901, 33919, 33927,
        33930-2, 33983, 33994-6, 34027, 34093, 34138, 34165, 34167, 34185, 34204,
        34208, 34244, 34277, 34344, 34371, 34390, 34417, 34431-3, 34441, 34446,
        34462, 34529, 34569, 34572, 34588, 34607, 34628, 34694, 34703, B34706,
        34723, 34726, 34766-7, 34775, 34779, 34799, 34832, 34845, 34848, 34850,
        34882-3, 34891, 34919, 34990, 35058-9, 35076, 35114, 35124-5, 35143, 35177,
        35181, 35218-9, 35230, 35264, 35269, 35291, 35306-7, 35380, 35392, 35415,
        35427, 35434, 35530, 35539, 35545, 35548, 35572, 35576, 35613, 35644, 35666,
        35670, 35688, 35698, 35720, 35731, 35739, 35755, 35763, 35774, 35780, 35804,
        35818, 35844, 35878, 35894, 35915, 35944, 35969, 35987, 36024, 36043, 36048,
        36095, 36099, 36122, 36139-40, 36155, 36173-5, 36177, 36200, 36234-5, 36244,
        36257, 36277, 36307, 36310, 36355, 36398, 36409, 36411, 36430, 36439, 36469,
        36487, 36502-3, 36506, 36511, 36539-40, 36554
        36660, 36663, 36665, 36673-4, 36701, 36731, 36740-1, 36771, 36783,
        36803, 36808, 36828, 36862, 36892, 36899, 36932, 36953, 36993, 36995, 37026,
        37045-6, 37055-6, 37064, 37078, 37080-1, 37182, 37222, 37242, 37245, 37251-2,
        37262, 37315, 37427, 37452, 37454, B37493, 37494, 37547, 37565, 37607,
        37610, 37633, 37668, 37705, 37707, 37730, 37739, 37749-50, 37770-1, 37786-7,
        37794, 37830, 37840, 37842-3, 37870, 37878, 37882-3, 37901, 37907, 37920,
        37922, 37980, 38002, 38015, 38030, 38053, 38080, 38089, 38094-6, 38136-7,
        B38139, 38150, 38160, 38165, 38191, 38208, 38234, 38254-5, 38275, 38315-6,
        38318, 38342, 38349, 38373, 38376-7, 38379-82, 38453, 38458-9, B38477,
        38479, 38481, 38483, 38492, 38510, 38515, 38526, B38528, 38557, 38589,
        38593, 38620, 38634, 38638-9, 38656, 38667-9, 38698, B38706, 38723-4, 38730-
        -2, 38775, 38806, 38816, 38833, 38838, 38862, 38896, 38919, 38932, 38946,
        38953, 38972, 39000-1, 39047, 39075, 39099, 39110, 39152, 39168, 39176,
        39178, B39191, 39198, 39225, 39235, 39258, 39292, 39305-6, 39310-1, 39318,
        39329, 39379, 39384-5, 39427, 39441, 39444-5, 39448, 39451, 39453, 39458,
        39462, 39473-4, 39498, 39517, 39538, 39543, 39557, 39574, 39577, 39588,
        39597, 39614-7, 39624-5, 39645, 39676, 39679, 39745, 39763, 39774, 39835,
        39880, 39901, 39920, 39957, 40019, 40024, 40029-31, 40038-9, 40052, 40068,
        40084, 40107, 40122, 40124, 40164, 40185, 40255, 40260

Photosystem 1 activity measurement
        21547, 21685, 21688, 21727, 21766, 21825, 21959, 22078-9, 22164, 22309-10,
        22347, 22491, 22605, 22683, 22705, 22720, 22800, 22935, 22973, 23110, 23199,
        23227, 23541, 23647, 23655, 23724, B23767, 23822, 23917, 23940, 23942,
        24136, 24184, 24187, 24224, 24351, 24401, 24403, 24457, 24479, 24481, 24496,
        24545, 24630, 24664, 24668, 24879, 24881
        25166, 25240, 25320, 25362, 25375, 25386-7, 25425, 25459, 25636, 25853,
        25918, 26026, 26367, 26467, 26487, 26565, 26677, 26711, 26773, 26790, 26794,
        26830, 26910, 27114, 27552, 27580, 27603, 27640, 27668, 27681, 27724, 27786,
        27809, 27816, 27841, 27856, 27917, 27924, 27943, 28330, 28445, 28452
        28524, 28539, 28595, 28655, 28690, 28693, 28708, 28738-9, 28771, 28789,
        28845-7, 28887, 28918, 28995, 29138, 29143, 29384, 29533, 29557, 29637,
        29639, 29688, 29730, 29807, 29831, 29867, 29885, 29974, 30034, 30107, 30323,
        30630, 30881, 30955, 30984-5, 31087, 31091, 31206-7, 31522, 31643, 31663,
        31666-7, 31725-6, 31737, 31813, 31816, 31919, 31959, 32072, 32256, 32305,
        32315, 32418, 32499, 32512
        32595, 32607-8, 32610, 32664, 32667, 32726, 32769, 32861, 32866, 32899,
        33201, 33459, 33551, 33687, 33842, 33864, 33869, 33929, 33994-5, 34042,
        34093, 34138, 34165, 34316, 34441, 34446, 34576, 34606-7, 34703, B34706,
        34766, 34799, 34845, 34883, 35012, 35058, 35076, 35124-5, 35177, 35697,
        35699, 35739, 35763, 35773, 35780, 35804, 35818, 35894, 35918, 36175, 36355,
        36506, 36511, 36576
        36803, 36871, 36892, 36993, 37062, 37148, 37170, 37178, 37262, 37607, 37624,
        37843, 37878, 37922, 37937, 37967-8, 38055, 38208, 38316, 38321, 38458-9,
        38510, 38515, 38552, 38723, 38775, 38816, 38838, 38978, 39075, 39168, 39179,
        39311, 39379, 39441, 39590, 39843, 39880, 40038-40, 40124, 40164, 40185, 40307

Photosystem 1 activity measurement, methods
        21743, 22372, 23876, 24668; 27312; 28739, 28852, 29768, 29933, 32013; 35889,
        36082; 36892, 37980, B38037, 38634, 38862, 39625

Photosystem 1, primary acceptor
        21638, 22376, 22870, 23056, 23420, 23515-6, 23518, 23600, 23954, 23966, 24133,
        24688, 25076; 25477, 25942, 25963, 25968, 26468, 26480, 26942, 27099; 28548,
        28551, 28588, 28599, 28720, 28859, 29284, 29439, 29518, 29688, 29818, 30716,
        31027, 31030, 31629, 32432; 32682, 32915, 33016, 33253, 33356, 33494, 33930,
        34205, 34277, 34566, 34779, 35387, 36540; 37882-3, 39615

Photosystem 1 reaction centre see P700

Photosystem 2
        21528, 21548, 21551, 21566-8, 21575, 21598, 21600, 21606-8, 21612-3, 21638,
        21643-6, 21662, 21664, 21685, 21697, 21706, 21720, 21727, 21738, 21744,
        21746, 21750, 21773-4, 21790, 21804, 21806-7, 21810, 21817, 21819, 21825,
        21856, 21868, 21963-5, 22078, 22106-7, 22117, 22134, 22136, 22142, 22193,
        22195, 22202, 22214, 22224, 22233, 22236, 22243-4, 22264-5, 22291, 22325-6,
        22330, 22338, 22359, 22393, 22397, 22400, 22415, 22478, 22482, 22498, 22501,
        22514, 22525-8, 22534, 22540, 22574-5, 22580, 22582-3, 22629, 22637, 22657,
        22683, 22704, 22708-9, 22720, 22748, 22794, 22800, 22821, 22856, 22900,
        22933-4, 22936, 22957, 22972-3, 22995, 23020, 23024, 23043, 23066, 23102-3,
        23118, 23123, 23132, 23141, 23148, 23173, 23184, 23187, 23199, 23215, 23238-
        -42, 23323, 23327-8, 23351, 23353, 23365, 23390, 23394-6, 23400, B23407,
        23446, 23459, 23477, 23486, 23516, 23518-20, 23552, 23573-4, 23618, 23622,
        23631, 23638, 23659, 23724, B23767, 23794, 23847, 23852, 23860, 23866, 23876,
        23879, 23884, 23891, 23915-6, B23933, 23934-5, 23940, 23954, 23957, 23970,
        24032, 24099, 24102, 24133, 24136, 24159, 24169, 24187, 24190-1, 24193,
        24224-5, 24238, 24241, B24250, 24281, 24307, 24311, 24317, 24338, 24361,
        24371, 24373, 24401-2, 24466, 24478, 24483, 24490, 24494, 24545-6, 24559,
        24575-7, 24583, 24603, 24612, 24630-1, 24684, 24727, 24737, 24756, 24762,
        24768, 24789-91, 24804, 24850, 24864, 24874-5, 24876-7, 24881, 24908-9,
        24917, 24923, 24939, 24951, 25002, 25057, 25062, 25081-3, 25107, 25114,
        25131, 25151-2
        25166, 25174, 25180, 25182-3, 25186, 25202, 25217, 25240-1, 25244-6, 25276,
        25293, 25295, 25304, 25315, 25320-2, 25338-9, 25344, 25362-3, 25375, 25379,
        25385-7, 25396, 25401, 25425, 25432, 25438, 25456, 25477, 25491, 25504,
        25520-1, 25548, 25589, 25616, 25629, 25635-6, 25650, 25695, 25782, 25819,
        25836, 25840, 25853, 25861, 25879-80, 25888, 25893, 25900, 25932, 25941,
        25988, 26007, 26040, 26056, 26084, 26156, 26160, 26176, 26196, B26220,
        26224, 26230-1, 26252-4, 26257-8, 26268, 26288, 26294, 26325, 26362, 26367,
        26378, 26396, 26422, 26480-1, 26507, 26523, 26538, 26559, 26561, 26575, 26608,
        26615, 26632-3, 26640, 26659, 26664, 26675-6, 26694, 26703-4, 26711, 26716,
        26718, 26720, 26761, 26765, 26773-4, 26792-4, 26822, 26830, 26860, 26872,
        26890, 26899, 26910, 26914, 26917, 26921, 26938, 26955, 26995, 27034, 27060-
        -1, 27064, 27070-2, 27086, 27100, 27111, 27114, 27122, 27131-2, 27137, 27211,
        27222, 27264, 27275, 27312, 27319, 27324, 27332, 27345, 27353, 27378-9,
        27387, 27391, 27426, 27453, 27465, 27478, 27496, 27528-9, 27571, 27581,
        27589, 27593, 27603, 27617, 27619-20, 27637, 27640, 27655, 27659, 27668,
        27699, 27724, 27744, 27746, 27764-5, 27784, 27786, 27809, 27816, 27832, 27863-
        -4, 27870, 27878, 27921, 27924, 27943, 27953-4, 27977-8, 27996, 28001, 28021,
        28027-8, 28045, 28080, 28165, 28184, 28221, 28236, 28260, 28276, 28328-30,
        28333-4, 28376, 28395, 28424, 28445, 28449, 28457-8, 28468
        28514, 28518-9, 28539, 28547, 28549-51, 28554, 28575, 28581, 28584-5, 28587-8,
        28590, 28595, 28599, 28621, 28633, 28651, 28655-7, 28676, 28690-1, 28693-6,
        28699, 28718, 28738-9, 28741, 28744-5, 28754, 28759, 28766, 28777-8, 28789,
        28814-5, 28823, 28833, 28846, 28867, 28875, 28884-6, 28903, 28918, 28934-6,
        28968, 28991, 28997-9, 29001, 29081, 29096-7, 29107, 29114, 29138, 29143,
        29161, 29183-4, 29186, 29211, 29231, 29244, 29275, 29279, 29285, 29300, 29304-
        -5, 29307-9, 29341, 29353, 29360, 29384, 29406, 29416, 29436, 29439, 29442,
        29458, 29518-9, 29525, 29527, 29532, 29539-41, 29557, 29570, 29578, 29599-600,
        29602, 29619, 29634, 29670, 29684, 29688-9, 29698, 29713, 29731, 29740, 29754,
        29773, 29784-5, 29807, 29830-1, 29836, 29867, 29872, 29875, 29906, 29928,
                                                                           (continued)

Photosystem 2 activity measurement
        21527-8, 21531, 21544, 21559, 21563, 21600, 21606, 21675, 21685, 21687,
        21694, 21746, 21755-6, 21770, 21774, 21804, 21825, 21911, 21959, 21974,
        22025, 22028, 22051, 22056, 22078-9, 22103, 22136, 22142, 22164, 22215,
        22244, 22298, 22304, 22309, 22349-50, 22400, 22409, 22456, 22463, 22483,
        22491, 22514, 22528, 22536, 22607, 22644, 22664, 22683, 22720, 22773, 22780,
        22800, 22815-6, 22821, 22842, 22857, 22870, 22881-3, 22891, 22983, 22996,
        23008, 23041, 23043, 23079-80, 23086, 23110, 23119, 23147, 23183-4, 23199,
        23209, 23219, 23242, 23339, 23351, 23394-5, B23407, 23415, 23428, 23435,
        23464, 23477, 23486, 23548, 23585, 23632, 23638, 23647, 23655, 23657, 23672-
        -3, 23701, 23722-3, 23730, 23749, B23767, 23771, 23822, 23837, 23851, 23891,
        23917, 23920, 23934, 23940, 23942, 23944, 24011, B24024, 24032, 24077,
        24081, 24091, 24104, 24136, 24147-8, 24159-60, B24176, 24184, 24191, 24198-9,
        24224, 24280-1, 24283, 24292, 24311, 24336, 24338, 24367, 24401, 24403,
        24411, 24457, 24466, 24478-9, 24494, 24496, 24545-8, 24583, 24603, 24607,
        24630, 24637, 24663, 24665, 24667, 24684, 24722, 24730, 24850, 24860, 24866,
        24879, 24886, 24909, 24920; 24965, 24973, 25001, 25029, 25038, 25084, 25087-
        -8, 25096, 25102, 25134
        25166, 25214, 25244, 25262, 25304, 25320-2, 25330, 25335, 25338, 25375, 25385-
        -6, 25425, 25432-3, 25459, 25548, 25616, 25618, 25635-6, 25700, 25840, 25853,
        25913, 25941, 25988, 26007, 26107, 26137, 26152, 26175, 26224, 26268, 26275,
        27288, 26350, 26367, 26374, 26381, 26399, 26411, 26415, 26422, 26480-2,
        26496, 26507, 26711, 26716, 26720, 26753, 26773, 26786, 26790, 26792, 26794,
        26830, 26910, 26925-7, 27028, 27060, 27071, 27114, 27122, 27126-7, 27161,
        27163, 27204, 27218, 27257, 27276-7, 27279, 27284, 27332, 27345, 27354,
        27365, 27387, 27399, 27422, 27442, 27453, 27465, 27471, 27485, 27487, 27489,
        27491, 27496, 27549-50, 27552, 27558-9, 27580-1, 27583-4, 27640, 27655,
        27668, 27681, 27787, 27809, 27816, 27834, 27841, 27856, 27921-4,
        27943-4, 27996, 28062, 28083-4, 28111, 28180, 28198, 28203, 28263, 28306,
        28330, 28438, 28445, 28452, 28491
        28508, 28519, 28551, 28587, 28592, 28595, 28655, 28665, 28682, 28690, 28693-6,
        28717-8, 28743, 28753, 28766, 28771, 28777, 28789, 28817, 28835, 28846-7,
        28875, 28886-7, 28896, 28918, 28920, 28935-6, 28938, 28995, 29012, 29081-2,
        29138, 29143, 29184, 29238, 29244, 29285, 29384, 29416-7, 29449, 29534, 29556,
        29586, 29602-3, 29635, 29639, 29730, 29737, 29826, 29831, 29867, 29906, 29928,
        29991, 30062, 30085, 30089, 30103, 30110-2, 30115, 30119, 30132, 30291, 30295,
        30297, 30301, 30306, 30309, 30320, 30329, 30335, 30403, 30415, 30436, 30540,
        30548, 30558, 30579, 30589, 30594-5, 30600, 30639, 30681, 30701, 30826, 30841-
        -2, 30848, 30866-7, 30878, 30881, 30909, 30951, 30955, 30984-5, 30998, 31012,
        31038, 31050, 31075, 31087, 31091, 31116, 31119-20, 31128, 31195, 31206-7,
        31238, 31251, 31297, 31336, 31356, 31379, 31410, 31426, 31522, 31564, 31575,
        31621, 31625, 31643, 31651, 31663, 31666-7, 31715, 31723, 31725-8, 31737,
        31813, 31835, 31876-8, 31918-9, 31955, 31959, 32005, 32007, 32042, 32071,
        32079-80, 32084, 32086, 32161, 32228, 32236, 32255-6, 32267, 32310-1, 32362,
        32385, 32414, 32418-9, 32475, 32480, 32482-3, 32499, 32512-3
        32591, 32594-5, 32604, 32610, 32667, 32706, 32725-6, 32739, 32765, 32769,
        32786, 32819, 32834, 32861-2, 32864, 32867, 32889, 32934, 33001, 33030,
        33047, 33058, 33108, 33200-1, 33248, 33310, 33347, 33353, 33391, 33502,
        33513, 33530, 33539, 33551, 33582, 33663, 33665, 33687, 33719, 33748-9,
        33761, 33790, 33802-3, 33842, 33901, 33982, 33995, 34043, 34045, 34093,
        34100, 34102, 34126-7, 34165, 34180, 34231-4, 34286, 34316, 34319, 34372,
        34441, 34475, 34490, 34527, 34576, 34588, 34606-7, 34609, 34630, 34643,
        34663, 34693, 34703, B34706, 34766-7, 34799, 34845, 34848, 34866, 34883,
        34897, 34918, 35058, 35068, 35076, 35124-5, 35157, 35177, 35207, 35320,
        35328-9, 35358, 35427, 35432, 35434, 35440, 35507, 35547, 35555, 35637-8,
        35685, 35697-9, 35739, 35763-4, 35767, 35773, 35780, 35804, 35818-9, 35840,
        35894, 35921-2, 35935, 35947, 35987, 36003, 36101, 36112, 36171, 36175,
        36192, 36210, 36214, 36250-1, 36257, 36300-1, 36310, 36312, 36316, 36355,
        36409, 36411, 36506-7, 36511, 36541, 36556, 36576, 36581
        36711, 36762, 36766, 36782, 36803, 36807, 36877, 36892, 36914, 36931, 36968,
        36977, 36993, 37062, 37118, 37146, 37148, 37170, 37178, 37244, 37262, 37294,
        37371, 37388, 37393, 37469, 37524, 37553, 37574, 37607, 37668, 37711, 37792,
        37806, 37850-1, 37870, 37901, 37920, 37922, 37934, 37937, 37966-8, 38026-7,
                                                              (continued)

Photosystem 2 activity mesurement (continued)
        38054-5, 38089, 38104, 38109-10, 38129, 38133, 38150, 38159, 38228-9, 38291,
        38316, 38342, 38350, 38439, 38458-9, 38468, B38477, 38497, 38510, 38515,
        38588, 38617, 38644, 38717, 38723, 38775, 38838, 38862, 38865, 38877, 38896,
        38901, 38956, 38960, 38978, 38994, 39025, 39045, 39075, 39099, 39104, 39168,
        39179, 39194, 39226, 39243, 39247, 39253, 39288, 39306, 39311, 39316, 39393,
        39403, 39408-9, 39427, 39441, 39538-9, 39543, 39557, 39584, 39590, 39625,
        39653, 39679, 39702-3, 39728, 39730, 39767, 39885, 39955-6, 39973, 40014,
        B40015, 40020, 40024, 40038, 40040, 40049, 40072, 40093, 40098-100, 40109,
        40120, 40124, 40164, 40185, 40200, 40206, 40307, 40343, 40350, 40352

Photosystem 2 activity measurement, methods
        22371, 22656, 23749, 24198, 25135; 25548, 26288, 26381, 26753, 27126, 27312;
        28665, 29525; 33464, 34406, 34918; 36892, B38037, B38477, 40076

Photosystem 2, primary acceptor
        21638, 22525, 22527-8, 22709, 23122-3, 23516, 23552, 24133, 24864, 24876,
        25082, 25152; 25375, 25477, 25861, 25903, 26559, 26615, 26694, 27955;
        28549-50, 28588, 28633, 28741, 28754, 29307-9, 29439, 29541, 29688, 30108,
        30245, 30256, 30335, 30716, 30781, 30813, 31666, 31688, 32238, 32293,
        32432; 32755, 33253, 33720, 33748-9, 34127, 34205, 34357, 34601-2, 34724,
        34775, 35240, 35362, 35526, 35548, 35945, 36173, 36462; 36711, 36745,
        36993, 37475, 37705-6, 37786, 37968, 38023, 38698, 38733, 39301, 39451,
        40076

Photosystem 2 reaction centre see   P680

Photosystems stabilization, chloroplast immobilization, methods
        26487, 26664, 26677, 26728, 27375; 28765, 28839, 28934-6, 29756-7, 29831,
        30094, 30306, 30436, 30731, 30938, 31079, 31091, 31205, 31207, 31738, 31958,
        32247, 32479, 32513; 32934, 33789, 33995, 34217, 34956, 35124-5, 35230,
        35456, 35992, 36003, 36402; 37178, 37469, 37783, 37878, 37892, 38015,
        38089, 38142, 38155, 38282, 38393, 38397, 38487, 38932, 39048, 39252, 39590,
        39988, 40109, 40185

Phycobilins   see   Biliproteins

Phycobilisome
        22495-7, 22590, 24909, 24983; 25579, 25770, 26098, 26158, 26916, 27818,
        28008; 29589-92, 29670, 29723, 30372-3, 30587-8, 30940-1. 31108, 31709,
        31937, 32054, 32540; 32927, 33233, 33757-8, 33883, 34029, 35366, 35738,
        36179, 36504, 36575; 36836, 37030, 37310, 37314, 37537, 37599-601, 37652,
        38150, 38270, 38447, 38554, 38922, 39359, 39516, 39610, 39965, 40051,
        40161, 40255

Phycocyanins
        21725, 21836, 21920, 22047, 22350, 22490, 22495, 22497, 22538, 22545-6,
        22580, 22590, 22767, 23010, 23083, 23152, 23178-9, 23182, 23187, 23247,
        23383-4, 23532, 23843, 23884, 24496, 24697, 24810
        25361, 25393, 25402, 25439, 25579, 25867, 25869, 26061, 26098, 26148,
        26158-60, 26168, 26207, 26219, 26250, 26576, 26608, 26718, 26992, 27257,
        27259, 27319, 28008, 28184
        28569, 28598, 28603, 28631, 28675, 28737, 28760, 28959-60, 29152, 29552,
        29589-92, 29670-2, 29723, 29866, 30311, 30322, 30372-3, 30442, 30525, 30531,
        30552, 30588, 30592, 30678, 30872, 30877, 30940-1, 31011, 31108, 31142,
        31146, 31206, 31937, 32071, 32102, 32173, 32258, 32357, 32359, 32540
        32617, 32785, 32981, 33134, 33234, 33258, 33420, 33535, 33569, 33576, 33584,
        33657, 33757-8, 33776, 33811, 34324, 34339, 34394, 34629, 34696-8, 34846,
        34976, 35015, 35058, 35147, 35308, 35366-7, 35452, 35478, 35621, 35684,
        35738, 35804, 35935, 36180, 36186, 36294, 36429, 36504, 36575
        36608, 36654, 36706, 36836, 36902, 36965, 36982, 37022, 37030, 37134, 37315,
        37526, 37535, 37562, 37599, 37601, 37651, 38201, 38212, 38223, 38248, 38270,
                                                                           (continued)

Phycocyanins (continued)
        38290, 38295, 38347-8, 38386, 38554, 38899, 38917, 38929, 38945, 38948,
        39046, 39110, 39198, 39344, 39465, 39477, 39499, 39516, 39553, 39574,
        39610, 39744, 39804, 39829, 39901, 39965, 39994, 40034, 40036, 40153,
        40163

Phycoerythrins
        21836, 22047, 22092, 22495-7, 22538, 22546, 22580, 22590, 22767, 23083,
        23152, 23177-9, 23247, 23843, 24697-8
        25393, 25439, 25579, 25958, 26061, 26098, 26158-9, 26718, 26914, 26992,
        27177-8, 27615, 27818, 27901
        28598, 28603, 28631, 28675, 28737, 28760, 29152, 29545, 29589-91, 29670-2,
        29723, 29866, 30372-3, 30442, 30588, 30592, 30678, 30940-2, 31142, 31206,
        31423, 31499, 31757, 31937, 32053, 32102, 32258, 32288, 32368
        32557, 32617, 32981, 33420, 33535, 33576, 33613, 33657, 33757-8, 33811,
        34339, 34596, 34629, 34696-7, 34740, 34846, 34976, 35147, 35366-7, 35738,
        36071, 36294
        36836, 37014, 37030, 37448, 37495, 37535-6, 37599-601, 37622, 37651, 37784,
        38201, 38270, 38290, 38295, 38367, 38554, 38917, 38947, 39344, 39516,
        39610, 39744, 39804, 39901, 39994, 40153

Phylogeny of algae productivity 39061

Phylogeny of biliproteins
        22767; 26158; 34382, 35167; 38554

Phylogeny of carbon fixation pethways
        24130, 24554, 24804; 25437, 25692, 27928, 28061, 28385, 28493; 28853, 28951,
        29909, 30473, 30702, 31653; 32803, 32911, 34772, 35006, 35167, 36379;
        39468-9, 39687, 40202

Phylogeny of carotenoids
        21864, 21971, 22767; 25218, 25376; 39007, 39819

Phylogeny of chlorophyll
        22767, 24613, 24642; 25834, 26614, 27083, 27904, 28061; 34382, 34772, 34795,
        34917, 35167, 35868; 38266, 38900

Phylogeny of chloroplast (chromatophore)
        21835, 21846, B21914, 21978, 22803, 23769, 23788, 23870, 23914; 25495, 25587,
        25625, 25665, 25770, 26614, 27233; 28513, 28951, 29371, 29918, 30472, 30682,
        31092, 31650, 31937, 31951, 32420; 33417, 33674, 33726, 34669, 35729; 36621,
        B36768, 37549, 37769, 38474, B38475, 39136, 40178

Phylogeny of electron transport chain
        21668, 21912-3, B21914, 22658, 23020, B23407, 23474, 24944, 25119; 25511,
        25628, 26640, 26765, 27083, 27904, 28061, 28385; 28670, 28951, 29796, 29798,
        29853, 30208, 30421, 31652, 32295; 32614, 32679-81, 32732, 32911, 32948,
        33032, 33294, 34013, 34669, 34772, 34795, 35167, 35776, 35926; 36650,
        38716, 39468-9

Phylogeny of gas exchange
        21912, B21914; 25965, 26640, 26719, 27928, 28061; 28669, 28951, 29379, 29525,
        29601, 32317; 32803, 32983, 33294, 35167, 35686, 35868, 36549; 37778, 38266,
        38319, 38336, 39759, 40042

Phylogeny of photorespiration B21914

Phylogeny of respiration 28951; 35167

Phytochrome see Growth regulators ...

Phytoflavin see Ferredoxin ...

Phytopathological effects on biliproteins   38917

Phytopathological effects on carbon fixation pathways
        23508, 24452; 25799, 26777, 27010; 30713-4; 34716; 37972, 38104, 39366-7,
        39654

Phytopathological effects on carotenoids
        21563-4, 22696, 25142; 25705, 26493, 26777, 26999, 27490, 27794, 27895;
        31774, 32055; 32740, 34175, 34443, 34941, 35528, 35840, 36307; 36638,
        38383, 38511, 39426, 39758

Phytopathological effects on chlorophyll
        21563-4, 22085, 22298-9, 22696, 22715, 22804-5, 22851-2, 22911, 23246, 23687,
        23958, 24217, 24714, 24824, 25142, 25146; 25543, 25601, 25705, 26107, 26493,
        26777, 26999, 27042, 27059, 27490, 27751, 27794, 27895, 27941, 27971,
        27992, 28038, 28114, 28384; 28654, 28799, 29039, 29711, B29763, 29859, 30263,
        30649, 31221, 31229, 31774, 31784, 31800, 31885; 32055, 32402; 32740, 33129,
        33372, 33909, 34175, 34443, 34501-2, 34847, 34894, 34898, 34941, 35342,
        35528, 35720, 35840, B36081, 36184, 36307, 36335; 36638, 36905, 37504,
        37517, B37532, 37748, 37834, 37972, 38104, 38133, 38138, 38383, 38463,
        38511, 38777, 38865, 38917, 39082, 39149, 39366, 39377, 39653-4, 39758,
        39802, 39925, 40262

Phytopathological effects on chloroplast (chromatophore)
        21680, 21984, 22313, 22355, 22924-5, 23508, 23670, 23681, 23744, 24216,
        24559; 25959, 26170, 26234, 26339, 27010, 27134, 27675, 27877, 28384; B29763,
        30819, 31160, 31557, 32041, 32692, 33701, 34159, 34175, 34828, 34894, 36047,
        36167, 36307; 37548, 38101, 38383-4, 38390, 38741, 39082, 39653, 39983,
        40003

Phytopathological effects on ecosystem and plant productivity
        22004, 22331, 22470, 22917, 23385, 23424, 23582, 24585, 24632, 25041,
        25072; 25781, 25799, 26010, 26134, 26138, 26349, 26369, 27793, 27895, 27942,
        28296, 28374; 29487, 29885, 29966, 30445, 30932, 31800; 32850, 32869, 33950,
        34732, 35914, 36053, 36194, 36388, 36428; 37867, 37972, 38749, 38777, 39081,
        39464, 39653-4, B39903, 40074, 40263

Phytopathological effects on electron transport chain
        21563, 22106, 22298, 23687; 25799, 26148, 27010, 27675, 27898, 27992; 29711,
        30008; 33309, 35404, 35840, 36307; 36812, 38104, 38133, 38865, 39653, 40132

Phytopathological effects on gas exchange
        21563, 21783, 22004, 22053, 22085, 22298, 22661, 22752, 22851, 22872, 23508,
        24138, 24452, 24559; 25128; 25223, 25271, 25543, 25601, 25687, 25799,
        26010, 26013, 26107, 26138, 26148, 26170, 26269, 26369, 26493, 26776-7,
        26999, 27010, 27221, 27490, 27501, 27898, 27941, 28368; 28541, 28623, 28705,
        29487, 29711, B29763, 29895, 30130, 30167, 30263, 30704, 30932, 31181,
        31885, 32055; 33372, 34412-3, 34443, 34732, 34894, 35210, 35342, 35423,
        35527; 36638, 36748, 37291, B37532, 37746, 37761, 38312, 38718, 38856, 39081-
        -2, 39149, 39366-7, 39493, 39732, 40074, 40262-3, 40373

Phytopathological effects on photorespiration
        22752, 24113; 25271, 27898; 35423; 37291, 38718, 39149, 39234

Phytopathological effects on resistances to CO_2 and water vapour transfer
        22283, 22662, 23424; 26993, 27692; 31885, 32064, 32506; 34443, 34851; 36812,
        37300, 38718, 40262-3

Phytopathological effects on respiration
        21564, 22752, 25128; 25799, 26107, 26997; 30167, 30445, 31160, 31885; 35423-
        -4, 36638, 36748, 37291, 38718, 39234, 39653

Pigments determination, sampling and extraction
        21759, 21830, 22477, 24522, 24897; 26129, 27858; 28962, 30091, 31797; 32822,
        33535, 33733, 34021, 34532, 34624, 34634, 36211; B37528, 37880, 37939, 38196,
        B38477, 38675, 38904, 39148, 40283, 40365

Plastochron index see Leaf life span, plastochron index

Plastocyanin
        21516, 21519, 21537, 21638, 21743, 21763-4, 21804, 21915, 22243, 22640,
        22705-6, 22794, 22821, 22972, 23041, 23057, 23173, 23227, 23253, 23546,
        23668, 23748, 23750, B23767, 23798, 23851, 24182, 24226, 24238, 24241, 24334-
        -5, 24351, 24395-6, 24455, 24487, 24493, 24789, 25030, 25070, 25105, 25145
        25304, 25385, 25425, 25470, 25511, 25565, 25750, 26102, 26104, 26254, 26289,
        26309-10, 26442-3, 26653, 26793-4, 26807, 26872, 27086, 27275, 27310, 27566,
        27617, 27660,27668, 27737, 27875, 28314-5, 28333, 28376, 28424, 28458
        28525, 28599, 28744, 28746, 28751, 28766, 28778, 28812, 28816, 28845, 28852,
        28893-5, 29106, 29143, 29439, 29525, 29555, 29647, 29784-6, 29807, 29831,
        29857, 29939, 30118, 30123, 30244, 30246, 30444, 30449, 30717, 30913, 30983,
        31025, 31027, 31099, 31125, 31542, 31665, 31718, 31883, 32005, 32072, 32315,
        32464-5, 32496
        32769, 32801, 32898, 32905-6, 32908-9, 32948, 32972, 33142, 33220,
        33253, 33356, 33590-1, 33821, 33901, 33907-8, 33919, 34205, 34224, 34446,
        34605, 34717, 34725, 34938, 34983, 34990, 35104, 35420, 35566, 35644, 35679,
        35700, 35729, 35743, 35753, 36052, 36174, 36220, 36244, 36336, 36385, 36389,
        36476
        36612, 36619, 36963, 37045, 37115, 37134, 37251-3, B37458, 37885, 37920,
        38159, 38377, 38481, 38510, 38656, 38958, 39357, 39920, 40055, B40105,
        40183, 40247

Plastocyanin, methods
        21516, 23253, 23750, 24334, 24396, 24493, 25105; 26102, 27310, 27875; 32905-6,
        32948, 34983, 35753; 36619, 37253, B37458, 37885, B38037

Plastoquinones see Quinones ...

Pollution of air and algae productivity
        29346, 32107; 32984, 34058, 34885, 35127; 37008

Pollution of air and carbon fixation pathways
        23997, 24682; 26492, 27021, 27928; 28558, 30128, 30769, 30961, 31300; 34409,
        34786, 35023; 38748, 39218, 39213, 39438

Pollution of air and carotenoids
        22614; 26716, 27021; 30364, 31403; 33556, 35417, 35949; 37232-3, 40073

Pollution of air and chlorophyll
        22028, 22062, 22130, 22334, 22614, 22793, 23342, 23450, 23547, 23613, 23716,
        24716; 25364, 26716, 27021, B27046, 27824; 29717, 29753, B29763, 30087,
        30263, 30348, 30364, 30715, 31006, 31403-4, 31540; 33013, 34786, 35010,
        35022-3, 35127, 35417, 35720, 35949, 36087, 36124, 36217; 36825, 36845,
        37008, 37232-3, 39182, 39218, 40073

Pollution of air and chloroplast (chromatophore)
        22793, 23392, 23450, 24256, 24754; 26492, 27028; B29763, 30786, 30889; 36217;
        38748

Pollution of air and ecosystem and plant productivity
        21748, 22028, 22130, 22793, 23620, 25013-4; 25279, 25364, 25895, B27046;
        28602, 28757, 28927, 29036, 29806, 30262, B30281, 30787, 32066; 33013,
        33242, 33508, 34285, 34654-5, 34786-7, 35010, 35276, 36087; 36845, 36908,
        37036, 37807, 38357, 38616, 38622, 38678, 38856, 38920, 38995, 39182,
        39857, 39864, 39938

Pollution of air and electron transport chain
        22028, 22130, 22136, 23997, 24716; 26716; 30128, 31322, 32107; 33557, 35507;
        39584

Pollution of air and gas exchange
        21628, 21748, 22028, 22130, 22665, 22791, 22793, 22966, 23450, 23716-7,
        23832, 24501, 24592-3, 24833-4, 24861, 25054
        25255, 25641, 25646, 25724, 25982, 26621, 26895, 27028, 27221, 27334, 27824,
        27839-40, 28096
        28616-7, 29277, 29346, 29732, B29763, 29806, 30019, 30262-4, B30281, 30563,
        30691, 30715, 30769, 30787, 30961, 31090, 31388, 31853, 32022, 32066, 32146,
        32220, 32448
        32806, 32833, 32990, 33013, 33132, 33242, 33244, 33508, 33922, 34046, 34283-
        -4, 34408-9, 34484, 34505, 34654-5, 34787-8, 35022-3, 35099, 35727, 35949,
        36087
        36907-9, 37036, 37230, 37506, 37577, 37586, 37742, 38493, 38523, 38582, 38646,
        38678, 38828, 38856, 38866, 38916, 39113, 39199-200, 39313, 39503, 39584,
        39628, 39753, 39855, 39857, 39864

Pollution of air and photorespiration
        23427, 24593; 31300; 34408; 37506

Pollution of air and resistances to CO_2 and water vapour transfer
        23547, 23716; B27046, 27839, 28209; 28602; 33244, 33508, 34505, 35010, 35563,
        36087; 36609, 36826-7, 36907-9, 37418, 37514, 38828, 38920, 39857, 39938

Pollution of air and respiration
        23450, 24833, 25054; 27021, 27334; 30787; 33013, 33242, 34409, 34654-5;
        36907-8, 38523, 38622, 38916, 39113

Porometer 22598, 22653, B22718, 23295, 25046; 25327, 25604, 25971, 26282, 26739;
        28803, 29255, 30250, 30392, B30432, 30507, 30738, 31968, 32485; 33451, 33646,
        34954; 37224, 38066, 38721, 39926

Potassium see Mineral elements (N, P, K) ...

Potential photosynthetic rate, in artificial conditions
        21662, 22157, 22950, 23049, 23304, 24201, 24629, 24830; 25309, 25476, 25687,
        25828, 25954, 26300, 26370, 26646, 26835, 26905, 26932, 27873, 28134, 28230,
        28463, 28474; 28629, 28706, 28901, 29090, 29569, 29572, 30663, 31067, 31732,
        31760, 31862, 31865, 31885, B32150, 32175; 33513, 33611, 33759, 33879, 34137,
        34416, 34646-7, 35045, 35062, 36049, 36190, 36195-6; 36604, 36676, 37049,
        37114, 37296, 37702, 37744, 37895, 38180, 38472, 39073, 39137, 39402, 39420,
        39640, 39869, 40154

Potential photosynthetic rate, in nature
        21538, 21705, 22016, 22321, 22993, 23168, 23243, 23464, 23545, B23767,
        23776, 23814, 24172, 24296, 24324, 24588, 24608, 24692, 24758-9, 24935,
        24969, 25086, 25123; 25195, 25224, 25229, 25238, 26300, 26533, 26932, 27180,
        27190, 27813, 28109, 28167-9, 28372, 28474-5; 28629, 28706, 28866, 28988,
        28990, 29332, 29470, 30656, 30943, 30945, 31040, 31633, 31696, 31866, 31967,
        32026, 32031, 32047, B32318, B32515; 32751, 33262, 33436, 33620, 34227,
        34485, 34511, 34730, 34886, 34962, 34987, 35062, 35129, 35341, 35476,
        35561, 35791, 36010, 36050, 36132, 36530; 36758, 36839, 37049, 37296, 37338,
        37434, 37693, 37895, 37935, 38285, 39649, 39665, 40079, 40291

Precipitation, dew, and carbon fixation pathways 33130, 36443; 37331, 39605

Precipitation, dew, and chlorophyll   24512; 27386

Precipitation, dew, and ecosystem and plant productivity
        22343, 22380, 22593, 22676, 22701, 22932, 23038, 23063-4, 23235, 23733,
        24105, 24504, 25052; 25238, 25274-5, 25970, 26256, 26373, 27290, 27386,
                                                                        (continued)

Proteins, amino acids, nucleic acids and chlorophyll
        21870, 21886, 21976, 23095, 23356, 23947, 24381, 24615, 24693, 24984, 25029,
        25121; 25200, 27258, 27926, 28281; 28953, 29547, 29901, 29904, 29907, 30880,
        31924, 32302; 33094, 33600; 37468, 38917, 39862, 40190

Proteins, amino acids, nucleic acids, and chloroplast (chromatophore)
        21527, 21545, 21575, 21685, 21835, 21846, 21860, 21863, 21931, 22032, 22078-
        -9, 22117, 22135, 22159, 22338-9, 22422, 22424, 22469, 22641, 22729, 22737,
        22781, 22978, 23094, 23281, 23373, 23482-3, 23498, 23623, B23767, 23769,
        23788, 23822, 23825, 23836, 23860, 23870, 23880, B23933, 23969, 24049-50,
        24184, 24190, 24237, 24348, 24353, 24397, 24461, 24464, 24559, 24696, 24701,
        24738, 24750, 24793, 24832, 24892-4, 24997, 25002, 25107, 25120
        25174, 25177, 25212, 25263, 25354, 25377-8, 25427-8, 25463, 25497, 25504,
        25587, 25689, 25726, 25745, 25749, 25757, 25888, 25890-1, 26003, 26071,
        26094, 26144, 26153, 26158, 26252-3, 26327, 26340, 26342, 26352, 26414-6,
        26431-2, 26444, 26455, 26575, 26665, 26768, 26774, 26833, 26893, 26899,
        26942, 26994-5, 27025, 27141, 27167-70, 27186, 27214, 27249, 27286, 27407,
        27462, 27499, 27694-5, 27713, 27771, 27791, 27817, 27962, 27974, 28045,
        28127, 28132, 28146, 28248, 28250, 28281, 28417, 28476
        28499, 28542, 28574, 28577, 28608, 28690, 28697, 28709, 28755-6, 28759, 28794,
        28804-5, 28816, 28850, 28891, 28896, 28910, 28912, 28923-4, 28980, 29033,
        29186, 29300, 29315, 29339, 29374, 29404-6, 29431-2, 29502, 29560, 29627,
        29636, 29773, 29780-1, 29826, 29845, 29907, 29918, 29922, 29938, 30024, 30027-
        -8, 30038, 30082, 30242, 30287, 30295, 30320, 30334, 30361-2, 30472, 30534,
        30569, 30589, 30682, 30688, 30752, 30816, 30872, 31027-9, 31080, 31104-5,
        31137, 31143, 31155, 31323, 31345, 31365, 31414, 31434, 31545, 31575, 31590,
        31643, 31656, 31662, 31664, 31666, 31674, 31705, 31743-4, 31754, 31815, 31877-
        -8, 31892, 31900, 31923, 31937, 31973, 32048, 32060, 32125, 32149, 32276,
        32283, 32392, 32397, 32420, 32452
        32593, 32650-1, 32661, 32671, 32783, 32811, 32854, 32858, 32890, 32902,
        32920, 32932, 32944, 32975, 33053, 33090, 33098, 33170, 33172, 33180, 33202,
        33256, 33381, 33383, 33419, 33452-4, 33517, 33592, 33597, 33625, 33648,
        33678, 33686, 33766, 33773, 33801, 33810, 33820, 33827, 33838, 33841, 33942,
        33957, 34054, 34060, 34164, 34223, 34237, 34360, 34578, 34692, 34704, 34762,
        34816, 34871-2, 34892, 34908, 34927, 34930, 34956, 35034-5, 35054, 35066,
        35072, 35114, 35128, 35130, 35143, 35211, 35241-2, 35382-3, 35479, 35497-8,
        35515, 35544, 35590, 35655, 35689, 35698, 35730, 35755, 35804, 35859, 35873-
        -4, 35967, 35991, 36031, 36127-8, 36177, 36210, 36261, 36284, 36327, 36459,
        36461, 36529
        36645, 36667, 36690, 36711, 36841-2, 36854-5, 36894-5, 36933, 36961, 36979,
        36987, 36996, 37035, B37037, 37053, 37055, 37078, 37119, 37127, 37153-4,
        37175, 37182, 37223, 37228, 37275, 37281, 37326, 37336, 37366-7, 37408,
        B37458, 37468, 37640-1, 37716, 37725, 37737, 37747, 37756-7, 37858, 38146,
        38146, 38214, 38301, 38350, 38373, 38395, 38418, 38449, 38466-8, 38474,
        B38475, B38477, 38545-6, 38670, 38736, 38775, 38806, 38833, 38912, 38963,
        38991, 39001, 39044, 39062, 39145, 39149, 39168, 39258, 39273, 39282,
        39353, 39445, 39506, 39535, 39571, 39582, 39606, 39624, 39703, 39762-3,
        39797-8, 39929, 40063-4, B40105, 40121, 40161-2, 40314, 40359

Proteins, amino acids, nucleic acids and electron transport chain
        22243, 23495; 25177, 25585, 27019, 28450; 31025, 31027, 32260; 33170, 33906,
        35100, 36476; 38811, 39168, 39645, 39846, 39863, 40013

Proteins, amino acids, nucleic acids and gas exchange
        21699, 22228, 23766, 24138, 25121; 26589, 27141, 27926; 31519; 36407, 36557;
        38650, 39064

Proteins, amino acids, nucleic acids and photorespiration 28224; 30033; 32822,
        33593, 35601

Protochlorophyll(ide) see Chlorophyll biosynthesis ...

Proton transport in chloroplast see Chemiosmotic hypothesis ...

Protoplasts, isolated see Tissue culture ...

Pteridines see Ferredoxin-NADP reductase ...

Pyrenoid 22743, 23366, 24243, 24500, 25080; 25500, 26359, 27316; 29510, 29728,
     30758, 31589; 33211, 33765, 35626; 36645, 38178, 38382, 38614

Q

Quantum yield and requirement
     21585, 21638, 21662, 21690, 21727, 21750, 21834, 21891, 21893, B21914,
     21993, 22134, 22174, 22236, 22310, 22580, 22722, 22957, 23103, 23147-8,
     23209, 23365, B23407, 23673, 23965, 24133, 24315, 24401-2, 24433, 24576,
     24804, 24811, 24838, 24904, 25057
     25240, 25242, 25362, 25396, 25437, 25441, 25490, 25533-4, 25616, 25631-2,
     25741, 25751, 25836, 25923, 26001, 26040, 26211, B26220, 26224, 26267,
     26294, 26332, 26423, 26472, 26476, 26655, 26718, 26720, 26780, 26794,
     26859, 26861, 27045, 27075, 27163, 27173, 27537, 27617, 27663, 27796,
     27970, 27977, 28008, 28056, 28173, 28328, 28376, 28395, 28473
     28593, 28745, 28753, 28833, 28858, 28860, 28877, 28879, 28884, 28886, 28922,
     29000, 29012, 29018, 29097, 29244, 29383, B29468, 29518-9, 29530, 29539, 29614,
     29621, 29714, 29733, 29735, 29754-5, 29797, 29805, 29847, 29885, 29936, 30062,
     30080, 30220, 30347, 30454, 30580, 30591, 30607, 30637, 30640, 30656, 30742,
     30892, 30919, 31146, 31342, 31363, 31412, 31485, 31488-9, 31504, 31564, 31571,
     31690-1, 31709, 31980, 32031, 32071, 32162, 32251, 32359, 32433, 32534
     32587, 32615, 32664, 32744, 32792, 32812, 32826, 32866, 32897, 32914, 32916,
     33024, 33047, 33056, 33062-3, 33187, 33323, 33433, 33436, 33626, 33707,
     33714, 33758, 33779, 33788, 33799, 34023, 34099, 34208, 34277, 34393, 34447,
     34451, 34462, 34572, 34588, 34859, 34893, 35096, 35138, 35285, 35320, 35374,
     35762, 35819, 35945, 35955, 36150, 36250, 36256, 36569
     36749, 36772, 36911-2, 36955, 36991, 37315, 37613, 37660, 37673, 37706,
     37813, 37842, 37950, 38059, 38099, 38150, 38224, 38244, 38322, 38349,
     38376, B38477, 38638, 38654, B38892, 38923, 38945, 39027, 39276-7, 39372,
     39538, 39763, 39777, 39836, 39996, 40011, 40056, 40348, 40360

Quantum yield and requirement, methods see Quantum yield and requirement

Quinones in photosynthesis
     21568, 21631, 21638, 21646, 21697, 21721, 21752, 21804, 21810, 21867, 21954,
     21959, 22100, 22141, 22325, 22354, 22525-8, 22580, 22588, 22618-9, 22626,
     22637, 22667, 22673, 22706, 22759, 22794, 22812, 22934, 22969, 22972, 23076,
     23116, 23129, 23184-5, 23200, 23215, 23272, 23290, 23292, 23398, 23400-4,
     23552, 23566, 23574, 23655, 23687, 23730, 23886, 23921, B23933, 24028,
     24032, 24133, 24163, 24281, 24333, 24346, 24403, 24490, 24497, 24583, 24691,
     24784, 24789, 24864, 24876-7, 24921, 25006, 25029, 25057, 25076, 25099
     25293, 25321, 25339, 25362, 25370, 25387, 25392, 25401, 25416, 25470, 25616,
     25773, 25846, 25953, 25962, 25964, 26147, 26196, B26220, 26231, 26257-8,
     26262, 26376, 26410, 26442-3, 26459, 26556, 26566, 26610, 26615, 26751,
     26766, 26790, 26887, 26942-3, 27070, 27100, 27309, 27326, 27328, 27345,
     27447-9, 27529, 27580, 27617, 27653, 27668, 27809, 27825, 27868, 27878,
     28155, 28376, 28409, 28423
     28535, 28549-50, 28641, 28643, 28667, 28692, 28694-6, 28744, 28746, 28769,
     28778, 28794, 28814, 28822, 28833, 28904, 28995, 29192, 29263, 29266, 29304,
     29308, 29359, 29525, 29541, 29602, 29698, 29713, 29719, 29760, 29784-5, 29818,
     29868-70, 29948, 29969, 29973, 30000, 30118, 30182, 30245, 30291, 30337-8,
     30385, 30402, 30594-7, 30625-6, 30716, 30718, 30780-1, 30812, 30859, 30939,
     31009, 31056, 31061, 31095, 31288, 31291, 31368, 31371, 31410, 31414, 31479,
     31486, 31521, 31534, 31573, 31598, 31612, 31617, 31736, 31746-7, 31785-6,
     31819-20, 31959, 31970-1, 31998, 32072, 32127, 32130, 32148, 32249, 32255,
     32279, 32281, 32295, 32303, 32408, 32467-8, 32489
                                                          (continued)

Quinones in photosynthesis (continued)
        32703, 32755, 32769, 32834, 32845, 32860-1, 32915, 32955, 33030, 33142,
        33187, 33190, 33253, 33300, 33350, 33400, 33402, 33460, 33515, 33551, 33717-
        -8, 33748-9, 33802-3, 33864, 33901, 33919, 34032, 34126, 34138, 34202,
        34205, 34277, 34357, 34377, 34389, 34606-8, 34727, 34743, 34767, 34774-5,
        34789, 34845, 34965, 35066, 35167, 35239, 35282, 35292, 35366, 35615, 35649,
        35741, 35812, 35881, 35917, 35945, 36122, 36164, 36173, 36175, 36189, 36246,
        36250, 36256, 36262, 36265, 36389, 36449, 36483-4, 36502, 36511, 36540
        36779, 36782, 36798, 36861, 36888, 36911-2, 36937, 36993, 37042, 37062,
        37104, 37167, 37218, 37316, 37370, 37578-9, 37611, 37664, 37705, 37718,
        37741, 37745, 37773, 37842, 37870, 37878, 37900, 37914, 37961, 37964, 38039,
        38052, 38111, 38228-30, 38351, B38477, 38478-80, 38538, 38593, 38613, 38615,
        38619, 38698, 38707, 38735, 38773, 38795, 38887, 38905, 38957, 38969-70,
        38972, 39222, 39295-6, 39299, 39303, 39329, 39357, 39386-7, 39427, 39451,
        39491, 39674, 39681, 39842, 39913, 39915, 39920, 39941, 39955-7, 39970,
        40013, 40030, 40039, 40055, 40068, 40247, 40251, 40278-9

Quinones, methods
        25392; 30596, 30598, 30939; 35239, 35812; B38477, 39184

R

Radiation in canopy   see   Canopy, radiation ...

Radiation, light   see   Irradiance ...

Rain, precipitation, methods   see   Aerodynamic methods, bioclimatological methods ...

Reaction centres   see   P680; P700 ...

Recycling of CO_2 inside the cell and leaf   21897, 22066, 22397, 23171, 24264, 25136;
        25619, 26618, 27402, 28474; 29992, 30499, 31352, 31451, 32073, 32327; 36185;
        38868, 39263, 39716

Relative growth rate   see   Growth analysis, net assimilation rate ...

Relative water content   see   Water saturation deficit

Resistance, carboxylation and excitation
        22749, 23609, 23994, 24100, 24201, 24411, 24537, 24915; 25620, 25684, 27899;
        28785, 28896, 29090, 29156, 29319, 30221, B30281, 30456, 31067, 31232,
        31251, 31372, 31596, 31627, 31866, 32073, 32126; 32573, 33002, 33879, 34625,
        34648, 35634, 36190, B36191, 36521; 37025, 39420, 39971

Resistance, cuticular   23634, 23703, 24624; 25521, 25998, 26541, 28367; 29053,
        32020, 32485; 35355; 36682, 36821, 38075, 38847, 39005, 39492, 40141

Resistance, intracellular (mesophyll)
        22039, 22041, 22075, 22157, 22209, 22284, 22314, 22321, 22484, 22649, 22652,
        22686, 22787, 22875, 22937, 22938, 22950, 22960, B23006, 23430-1, 23609,
        23653, 23696, 23773, 23780, 23791, 23887, 23994, 24100, 24201, B24271,
        24411, 24537, 24592-3, 24675-6, 24730, 24776, 24837, 25106, 25111
        25419, 25620, 25654, 25873, 25881, 26065, 26091, 26154, 26163-4, 26199,
        26227, 26403, 26438, 26515, 26524, 26532, 26646, 26775-6, 26785, 26863,
        26971, B27046, 27171, 27215, 27270, 27363, 27422, 27595, 27700, 27749,
        28102, 28194, B28340, 28368, 28391, 28428, 28463
        28504, 28606, 28688, 28717-8, 28732, 28785, 28835, 28898, 28899, 28987, 29053-
        -4, 29090, 29156, 29319, 29331, 29486, 29676, 29795, 30078, 30221, 30270-1,
        B30281, 30457-8, 30608, 30633, 30904-5, 31066-7, 31164-5, 31214, 31232, 31251,
        31256, 31264, 31372, 31464, 31596, 31613, 31627, 31836,   31866-7, 32020,
        32126, 32252, 32450

                                                                      (continued)

Resistance, intracellular (mesophyll) (continued)
        32573, 32772, 32810, 32913, 32967, 32973, 32997, 33002, 33231, 33244,
        33304, 33306, 33358-9, 33506, 33703, 33879, 33976, 34038, 34097, 34170,
        34375, 34453, 34625, 34646-8, 34683, 34689, 34886, 35097-8, 35234, 35261,
        35370, 35610, 35634, 35712, 35910, 36190, B36191, 36358, 36424, 36457,
        36521-2
        36742, 36908, 36964, 37025, 37108, 37112, 37269, 37638, 37649, 37799, 37915,
        38018, 38285, 38436, 38472, 38518-9, 38542, 39073, 39094, 39099, 39147,
        39270, 39420-3, 39599, 39649, 39667, 39751, 39809, 39874, 39888, 39926,
        40154, 40254

Resistance, leaf boundary layer
        21538, 21748, 22075, 22209, 22284, 22316, 22584, 22589, 22649, 22652, 22749,
        22938, 23296, 23493, 23846, 24100, 24536, 24593, 24676, 24863, 24915
        25283, 25419, 25451, 25620, 26015, 26114, 26154, 26199-200, 26239, 26307,
        26397, 26515, 26532, 26541, 26843, 26865, 26947, B27046, 27269, 27401,
        27609, 27728, 28087, 28209, 28257, 28366, 28369-70, 28428, 28463
        28602, 28785, 28898, B29023, 29054, 29090, 29130, 29179, 29855, 30633,
        31164, 31613, 31793, 31803, 32120, 32277, 32296, 32485
        32810, 32973, 32997, 33223, 33358, 33385, 33436, 33451, B33510, 33547, 33985,
        34038, 34713, 34743, 34886, 35129, 35234, 35278, 35303, 35529, 35610, 35827,
        36013, 36545
        36907, 36964, 37021, 37025, 37224, 37260, 37819, 38529, 38858, 39147, 39259,
        39667, 39751, 39809, 39971

Resistance, stomatal (and intercellular) (*cf*. also Resistances for water vapour ...)
        21538, 21549-50, 21713, 21748, 21787-8, 21810, 21813, 21887, 21968, 22039,
        22041, 22075, 22116, 22157, 22186, 22189-91, 22209, 22263, 22283-4, 22314-5,
        22321, 22344, 22403, 22504, 22584, 22589, 22649, 22652-3, 22662, 22686,
        22719, 22749, 22761-2, 22787, 22875, 22920, 22938, 22950, 22960, 22964,
        B23006, 23007, 23134, 23218-9, 23295-6, 23329-30, 23380, 23424, 23430-1,
        23608-9, 23648, 23679, 23696, 23716, 23756-7, 23773, 23780, 23784, 23791,
        23801, 23813, 23848, 23887, 23945, 23978, 24063, 24100, 24155-6, 24214,
        24261, B24271, 24376-7, 24411, 24437, 24449, 24507, 24536-7, 24592-3, 24624,
        24649, 24675-6, 24688, 24742, 24776, 24795, 24804, 24826, 24837, 24842,
        24880, 24915, 24932, 25045-6, 25111
        25167, 25173, 25195, 25251, 25261, 25271, 25283, 25309, 25326-7, 25332,
        25364, 25407, 25419, 25430, 25454, 25514, 25521, 25537, 25571-2, 25592,
        25620, 25642, 25654, 25742, 25755, 25813, 25832, 25842-3, 25873, 25881,
        25947, 26015, 26065, 26091, 26116, 26124, 26154, 26164, 26199, 26204, 26227,
        26235, 26239, 26264, 26282, 26336, 26389, 26392, 26403, 26421, 26436, 26438,
        26514-5, 26532, 26535, 26579, 26612, 26646, 26684, 26739, 26775-6, 26785,
        26843, 26848, 26863, 26865, 26941, 26947, 26971, 26993, B27046, 27073,
        27109, 27133, 27148, 27171, 27180, 27187, 27269-70, 27278, 27280, 27356,
        27360, 27363, 27401, 27410, 27573-4, 27578, 27594-5, 27630, 27636, 27692,
        27749, 27839, 27846-7, 27865, 27882, 27897, 27899, 28064-5, 28067, 28089,
        28123, 28143-4, 28182, 28197, 28209, 28244, 28287, 28321, B28340, 28342,
        28350, 28368, 28374, 28391, 28426, 28428, 28463
        28504, 28509, 28606, 28688, 28718, 28731-2, 28785, 28803, 28810, 28835, 28896,
        28898-9, 28925, 28927, 28987-8, 29015, 29020, B29023, 29053-4, 29122, 29130,
        29162, 29179, 29232, 29243, 29247, 29255, 29331, 29373, 29395, 29425, 29467,
        29486, 29551, 29676, 29724, 29766, 29793, 29816, 30076, 30212, 30221, 30249,
        30271, 30455, 30457-8, 30508, B30511, 30547, 30633, 30743, 30818, 30861,
        30903-5, 30918-9, 30931, 31087, 31164, 31214, 31231, 31250-1, 31269, 31303,
        31321, 31432, 31470, 31516, 31536, 31596, 31613-4, 31627, 31633, 31759, 31762,
        31769, 31823, 31866-7, 31956, 31994, B32004, 32020, 32051, 32064-5, 32110,
        32226, 32252, 32296, 32352, 32369, 32450, 32458, 32485
        32562, 32573, 32690, 32698, 32713, 32772, 32777, 32810, 32866, 32913, 32967,
        32973, 32997, 33000, 33002, 33050, 33068, 33105-6, 33195, 33244, 33280,
        33285-6, 33358-9, 33388, 33396, 33435, 33451, 33506, 33527, 33547, 33558-9,
        33703, 33742, 33818, 33857-8, 33858, 33879, 33916, 33976, 33985, 34008,
        34038, 34097, 34132, 34135, 34170, 34181, 34193, 34267, 34295, 34332-4,
        34352, 34368, 34375, 34384, 34453, 34494, 34497, 34504-5, 34511, 34535,
                                                                      (continued)

Resistances to water vapour transfer ... (continued)
        -8, 35129, 35194, 35234, 35335, 35354, 35369, 35460, 35594-5, 35610, 35712,
        35773, 35815, 35827, 35905, 35931, 35939, 36013, 36069, 36075, 36132, 36132,
        36417-8, 36472
        36637, 36683, 36738, B36781, 36821, 37108, 37171, 37224, 37244, 37265,
        37268-9, 37325, 37360, 37413-5, 37441, 37470, 37544, 37614, 37649, 37702,
        37752, 37874, 37897, 38066, 38075, 38077, 38217, 38285, 38310, 38400, 38518,
        B38740, 38779, 38858, 38915, 38934, 38996-7, 39088,
        39143, 39259, 39259, 39424, 39511, 39709, 39737, 39813, 39828, 39891, 40020,
        40079, 40126, 40169

Respiration and photosynthesis
        21534-5, 21626, 21628, 21633, 21659, 21681, 21903, 21912, 21945-6, 22031,
        22231, 22245, 22292-4, 22353, 22368-9, 22406, 22488, 22494, 22548, 22623,
        22654, 22778, 22793, 22848, 22868, 22898, 22910, 22919, 23013, 23105, 23145,
        23170, 23188-9, 23304, 23369, 23444, 23464, 23478, 23561, 23695, 23714,
        23827, 23975, 23999, B24024, 24048, 24053, 24084, 24144, 24301, 24427,
        24535, 24562, 24657, 24862, 24880, 25005, 25053, 25086, 25128, 25136, 25160
        25195, 25224, 25252, 25291, 25307, 25312, 25365-6, 25406, 25452, 25457,
        25476, 25485, 25502-3, 25549, 25601, 25610-1, 25615, 25753, 25774, 25776,
        25810, 25829, 25848, 25860, 25883, 26004, 26009, 26045, 26081, 26116,
        26123, 26135, 26164-5, 26240, 26296, 26304, 26386-7, 26420, 26461, 26472-3,
        26516, 26529, 26578, 26620, 26651, 26659, 26669, 26719, 26744, 26829, 26856,
        26863, 26883-4, 26980, B27001, 27021, 27141, 27149, 27191, 27196,
        27227, 27257-8, 27317-8, 27337, 27348, 27389, 27396, 27416, 27429, 27470,
        27545, 27591-2, 27611, 27728, 27793, 27837, 27862, 27873, 27895, 27899,
        27914, 27919, 27958-9, 27972, 28083, 28086, 28088, 28206, 28297, 28349,
        28353, 28379, 28408, 28429, 28446, 28471-2, 28478
        28541, 28604, 28648, 28687, 28703, 28706, 28773, 28780, 28795, 28855, 28880,
        28896, 28901-2, 28918, 28979, 29008, 29101, 29191, 29201-2, 29206, 29231,
        29242, 29252-3, 29286, 29294, 29296-8, 29303, 29336-7, 29343, 29350, 29379,
        29451, 29495, 29514, 29563, 29700, 29721, 29819-20, 29842, 29893-4, 29914,
        29957, 30026, 30047, 30124, 30151, 30167, 30285, 30322, 30381, 30575, 30608,
        30622, 30624, 30750-1, 30762, 30768, 30873, 30904, 30922, 30936, 30947, 30968,
        31003, 31018, 31040, 31065, 31077, 31142, 31218, 31220, 31272, 31281, 31285,
        31327-8, 31346, 31362, 31402, 31429-30, 31460, 31468, 31493-4, 31527, 31555,
        31570, 31586-7, 31592, 31651, 31700, 31802, 31835, 31849, 31882, 31890, 31896,
        31937, 32032, 32046, 32055, 32079, 32088, 32093, 32115, 32118, 32121, 32137,
        B32150, 32174-5, 32181-2, 32312, 32327, 32329, B32410, 32432, 32484, B32515,
        32525
        32656, 32694, 32710-1, 32732, 32744, 32798, 32802, 32806, 32868, 32899,
        32941, 32953, 32967, 32969, 32974-5, 32980, 33001, 33070, 33124, 33162,
        33201, 33210, 33231, 33243, 33260, 33278, 33292, 33294, 33340, 33378-80,
        33384, 33409, 33496, 33514, 33652, 33675, 33687, 33731, 33759, 33764, 33832,
        33896, 33898, 33912, 33915, 33999, 34012, 34052, 34100, 34158, 34213, 34216,
        34252, 34281, 34293, 34370, 34373, 34404, 34409, 34437, 34490, 34518, 34520,
        34528, 34538, 34562, 34650, 34654-5, 34661, 34674, B34706, 34741, 34807,
        34809-10, 34853, 34867, 34905, 34912, 34915, 35138, 35140, 35167, 35171-2,
        35175, 35179, 35319, 35333, 35368, 35381, 35413, 35431, 35454, 35574, 35622,
        35643, 35676, 35688, 35762, 35802, 35868, 35872, 35885, 35920, 35927, 35957,
        36077, 36135, B36191, 36198-9, 36373, 36424, 36434, 36445, 36497, 36512,
        36547
        36613, 36638, 36657, 36662, 36670, 36712, 36805, 36818, 36865, 36871, 36876,
        36880, 36985, 37091, 37140, 37142, 37209-10, 37244, 37249, 37257, 37290,
        B37299, 37318, 37338, 37351, 37445, 37452, 37497, 37521, B37532, 37566,
        37678, 37696, 37712, 37818, 37893, 37975, 37981, 37989, 38020, 38145, 38166,
        38175, 38274, 38444, 38456-7, 38464, B38477, 38623, 38625, 38703, B38739,
        38839, 38893, 38916, 39075, 39078, 39114, 39181, B39220, 39257, 39297,
        39334, 39381, 39411, 39421, 39441, 39458, 39482, 39530, 39584, 39635, 39639-
        -40, 39653, 39672, 39697, B39711, 39728, B39759, 39852, 39885, 39924, 39931,
        39937, 39946, 39981, 40082, 40103, B40105, 40145, 40266, 40270, 40275,
        40332

Respiration, dark CO_2 efflux
        21526, 21535, 21590, 21640, 21794, 21803, 21828, 21845, 22005, 22031, 22038,
        22075, 22153, 22165, 22209-10, 22228, 22252, 22263, 22321, 22369, 22548,
        22563, 22649, 22768, 22819, 22830, 22832, 22937, 22950, 22952, 23077, 23145,
        23225, 23244, B23262, 23298, 23329, 23339, 23443, 23464, 23494, 23545, 23587,
        23632, 23634, 23652, 23714, 23796, 23828, 23863, 24015, 24017, 24055, 24105,
        24275, 24291, 24324, 24437-8, 24593, 24662, 24692, 24730, 24830, 24934,
        24989, 25015, 25032, 25047, 25118
        25167, 25271, 25283, 25480, 25520, 25592, 25613, 25666-7, 25674, 25710,
        25801, 25858, 25991, 25993, 26037, 26046, 26055, 26073, 26085, 26112,
        26135, 26166, 26241, 26296, 26300, 26370, 26400, 26461, 26506, 26578,
        26593, 26626, 26954, 26971, 26988, 27011, 27097, 27142, 27148, 27159,
        27187, 27196, 27207, 27209, 27212, 27267, 27302, 27327, 27348, 27398,
        27478, 27521, 27524, 27540, 27592, 27638, 27687, 27725, 27774, 27873,
        27930, 28003, 28050, 28086-7, 28092, 28102, 28167, 28244-5, 28300, 28355,
        28408, 28435, B28482
        28527, 28647, 28685, 28809, 28898, 28970-2, 28988-9, 28994, 29016, 29053,
        29065, 29068, 29104, 29109, 29207, 29212, 29222, 29319, 29409, 29421, 29425,
        29443, 29462, 29509, 29517, 29569, 29576, 29605, 29795, 29819, 29864, 29887,
        29923, 29952, 29967, 29992-3, 30011, 30139, 30218, 30271, 30448, 30476, 30513,
        30515-6, 30532, 30537, 30571, 30622-3, 30642-3, 30691, 30804, 30883, 30891,
        30947, 31077, 31107, 31200, 31232, 31251, 31293, 31492, 31592, 31633, 31759,
        31836, 31885, 31941, 32020, 32023, 32118, 32163, 32175, 32219, 32312, B32318,
        32339, 32341, 32364, 32412, 32491
        32562, 32568, 32573, 32633, 32654, 32742, 32773, 32804, 32837, 32882, 32965,
        32969, 33013, 33084, 33124, 33192, 33215, 33245, 33260, 33262, 33335, 33358,
        33385, 33465, B33510, 33523, 33566, 33660, 33703, 33725, 33740, 33859,
        33891, 33917-8, 33925, 34005-6, 34012, 34017, 34096, 34137, 34219, 34269,
        34302, 34326, 34330, 34471, 34485, 34536, 34613, 34677, 34689, 34754, 34792,
        34886, 34888, 34940, 34945, 34997, 35019, 35040, 35074, 35092, 35106, 35117,
        35129, 35154, 35161, 35169, 35173, 35233, 35243, 35255, 35261, 35286, 35370,
        35423-4, 35449, 35559, 35636, 35664, 35674-5, 35761, 35789, 35807, 35932,
        36037, 36049, 36113, 36116, 36132, 36135, 36203, 36213, 36299, 36303, 36329,
        36406, 36427, 36435, 36515-6
        36602, 36629, B36648, 36671-3, 36683, 36714, 36736, 36742, 36748, 36776,
        36804, 36817, 36880, 36908, 36940, 36967, 37002, 37049, 37066-7, 37112,
        37122, 37259, 37289, 37291, 37308, 37325, 37360, 37400, 37463, 37502, 37515-
        -6, 37576, 37624, 37638, 37649, 37654, 37687, 37702, 37754, 37768, 37836,
        37860, 37866, 37886-7, 37888, 37899, 37924, 37953, 37981, 37987-8, 38098,
        38141, 38198, 38287, 38335, 38358, 38392, 38431, 38436, 38464, 38472, 38558,
        38676, 38758, 38769, 38813, 38880, 38897, 39063, 39078, 39088, 39097, 39099,
        39163, 39211, 39232, 39325, 39343, 39421-2, 39665, 39716, 39748, 39755,
        39808, 39874, 39935, B39951, 40045, 40080, 40139, 40154, 40291, 40293

Respiration, "growth" and "maintenance"
        22910, 24001, 24172; 26572, 26867, 26883-4, 27097, 27171, 27212, 27591-2,
        27687, 28106-7; 29103, 29820, 29967, 30011, 30891, 30904, 31262, 32118,
        32341; 32753, 33260, 33808, 33859, 33999, 34325-6, 34650, 34809-10, 36344,
        36516; B37299, 37751, 37801, 37953, 38392, 38839, 39097, 39565, 39635,
        40138

Respiration of achlorophyllous tissues in light, light inhibition of respiration
        23652, 24730, 25047; 25226, 27478; 29016, 31077; 34301, 35974

Ribosome of chloroplast
        21545, 21931, 22117, 22338, 22423, 22539, 22743-4, 23265, 23543, 23623,
        23637, 23885, 23904, 23914, 24086, 24348, 24484, 25073, 25087, 25111
        25199, 25463, 25497, 25935, 26003-4, 26022, 26144, 26253, 26295, 26359,
        27056, 27076, 27246, 27300, 27451, 27669, 27739, 27962, 28007, 28266, 28330
        28542, 28574, 28736, 28849-50, 28897, 28912, 28915-6, 28980, 29033, 29371,
        29405, 29478, 29560, 29824, 29839, 29845, 30345, 30534, 30597, 30752, 30889,
        31092, 31126, 31139, 31553, 31705, 31874, 31917, 31973, 32390, 32397

                                                                    (continued)

Ribosome of chloroplast (continued)
    32616, 32671, 32763, 32902, 32949, 32958, 33170, 33282, 33518, 33520, 33686,
    33817, 33827, 33861, 34164, 34616, 34750, 34929, 34934, 35382-3, 35689,
    35755, 35936, 35967, 36532
    36763, 36793, 36957, 36987-8, B37037, 37154, 37281, 37336, 37388, 37650,
    37757, 38301, 38449, B38475, 39504, B39551, 39571, 40121, 40165

Ribulose 1,5-bisphosphate carboxylase
    21535, 21576-7, 21580, 21592, 21604, 21613, 21616, 21618, 21652, 21665,
    21702, 21711, 21724, 21736, 21773, 21782, 21814, 21875, 21933, 21939, 22019,
    22041, 22049-50, 22054-5, 22069-71, 22077, 22117, 22150, 22157, 22192,
    22203, 22226-7, 22239, 22251-2, 22303, 22339, 22381, 22393, 22406, 22469,
    22523, 22529, 22533, 22548, 22562, 22609, 22635, 22649, 22671, 22698, 22723,
    22725, 22727, 22730, 22833, 22899, 22905, 22940, 22951, 22958, 22980-1,
    22997, 23030-1, 23051, 23060, 23072, 23121, 23195, 23219, 23254-6, 23272,
    23274, 23282-3, 23286, 23309, 23374, 23397, 23416, 23451-4, 23504, 23511,
    23588, 23597-8, 23610, 23674, 23696, 23725, 23754, B23767, 23771-2, 23797,
    23814, 23858, 23877, 23903, 23926-7, 23940, 23973-4, 24008, 24016, 24100,
    24116, 24121-2, 24128, 24130, 24139, 24161-2, 24165, 24201, 24223, 24262-5,
    24292, 24309, 24336, 24346-8, 24357, 24368, 24397, 24484, 24492, 24523,
    24525, 24615, 24681-3, 24730, 24752, 24779, 24795, 24800, 24804, 24806, 24843,
    24860, 24933, 24984, 24986, 25044, 25063, 25073, 25097, 25106, 25138-9, 25148
    25179, 25192, 25199, 25216, 25221, 25249, 25261, 25284-5, 25290, 25325,
    25338, 25346, 25372-3, 25377, 25395, 25431, 25438, 25443, 25445, 25471,
    25496-7, 25504, 25528-9, 25574, 25584, 25650, 25671-2, 25692-3, 25711-3,
    25721, 25730, 25745, 25759, 25767-9, 25786-7, 25833, 25835, 25845, 25867,
    25891, 25912, 25923, 25934-5, 25944, 26003-4, 26052, 26057, 26068, 26077-8,
    26116, 26118, 26142, 26144, 26162, 26165, 26181, 26208-9, 26280, 26303,
    26314-5, 26317, 26352, 26354, 26361, 26402, 26405, 26430, 26444-6, 26485,
    26505, 26526, 26539, 26546, 26549-50, B26551, 26560, 26574, 26595, 26597,
    26618-9, 26623, 26626, 26650, 26665, 26727, 26776, 26784, 26786, 26806,
    26809-10, 26834-6, 26858, 26872, 26880, 26886, 26900-1, 26908, 26940, 26958-
    -9, 26991, 27007-8, 27052-3, 27063, 27074, 27116, 27138, 27156, 27180,
    27198, 27246-7, 27263, 27272, 27313, 27359, 27363-4, 27384, 27420, 27422,
    27424, 27503, 27506, 27531, 27548, 27597, 27621, 27634, 27639, 27651,
    27668-70, 27698, 27709, 27759-60, 27791, 27797, 27800, 27823, 27825,
    27896, 27906, 27911, 27928, 27948, 27966, 27995, 28009, 28032-3, 28071-2,
    28079, 28083, 28102, 28112, 28121, 28154, 28170, 28183, 28220, 28229, 28248,
    28268, 28285, 28303, 28305, 28357, 28386-7, 28398-9, 28430, 28477-8, 28480
    28501, 28513, 28613, 28645, 28652, 28658-9, 28697, 28717-8, 28748, 28775,
    28818-9, 28835, 28883, 28891, 28901, 28910-1, 28915-6, 28928-9, 28931, 29030,
    29098, 29107-8,   29132, 29144-7, 29231, 29262, 29273, 29315-7, 29322,
    29374, 29376, 29405-6, 29463, 29478, 29503, 29534, 29567-8, 29603-4, 29612,
    29624-6, 29633, 29641, 29650-1, 29700, 29709, 29728, 29777, 29781, 29795,
    29800-1, 29807, 29826, 29841, 29865, 29910, 29922, 29945, 30009, 30024-5,
    30038, 30097-8, 30128, 30143-4, 30170, 30174, 30232, 30234, 30332-3, 30435,
    30472-4, 30537-8, 30566, 30639-40, 30647, 30695, 30714, 30754, 30769, 30778,
    30783-4, 30809-10, 30823, 30844, 30893-5, 30897, 30912, 30928, 30933, 30969,
    30980, 30990, 31008, 31062, 31096, 31102, 31117, 31164-6, 31182, 31223, 31226-
    -7, 31240, 31250-1, 31256, 31263-4, 31300, 31319, 31419-20, 31433, 31447,
    31457, 31472, 31476, 31552-3, 31576, 31578, 31589, 31602, 31660-1, 31679,
    31734, 31806, 31844, 31923-4, 31930, 31963-4, 32035-6, 32097, 32125, 32236,
    32244, 32262, 32264, 32317, 32334-5, 32373, 32397, 32406, 32421-2, 32428-9,
    32440, 32457, 32459, 32488, 32502, 32529
    32583-5, 32623, 32661, 32685, 32691, 32717, 32719-20, 32759, 32767-8, 32783,
    32803, 32826, 32840-2, 32862, 32866, 32902, 32929-31, 32944, 32953, 32962-3,
    32996, 33002, 33042, 33076, 33090, 33110, 33135-6, 33148-9, 33154-7, 33171-2,
    33180, 33198, 33213, 33271, 33273, 33296, 33310, 33387, 33445, 33453-4,
    33473-4, 33513, 33518-20, 33571, 33597, 33612, 33625, 33633, 33642-5, 33648,
    33672-3, 33686, 33731, 33767-8, 33770-2, 33817, 33827, 33835-7, 33842,
    33849, 33873, 33887, 33892, 33895, 33913, 33944, 33979, 33992, 34003, 34017-
    -9, 34045, 34054-5, 34065, 34071-3, 34173, 34178, 34199, 34219, 34253,
    34270, 34276, 34312, 34315, 34320, 34379, 34388, 34429, 34437, 34452, 34456,
                                                              (continued)

Ribulose 1,5-bisphosphate carboxylase (continued)
        34470, 34473, 34475, 34509, 34525, 34534, 34549, 34556, 34574, 34618-20,
        34656-8, 34683, 34692, 34710, 34716, 34738, 34749, 34786, 34800, 34812-3,
        34817, 34834, 34911, 34922, 34937, 34949, 34972, 34978-9, 35006, 35023,
        35034-5, 35051, 35087-90, 35141-2, 35152, 35183-4, 35186, 35191-2, 35212-5,
        35226, 35241, 35247-9, 35294, 35372, 35382, 35411, 35420, 35425, 35427,
        35431, 35444, 35467-8, 35471, 35475, 35479, 35490, 35497-9, 35509, 35544,
        35569-70, 35596, 35626-7, 35646-7, 35692-4, 35701, 35703-4, 35755, 35762,
        35794, 35877, 35892-3, 35970, 35986, 36002, 36026, 36114, 36116, 36121,
        36132, 36159, 36162, 36234, 36327, 36366, 36379, 36395-6, 36410-2, 36414-6,
        36426, 36440, 36450-2, 36473, 36486, 36491, 36495, 36561, 36572
        36600, 36603, 36624, 36658, 36663, 36676, 36701, 36712, 36724, 36754, 36767,
        36784, 36800, 36807, 36814, 36889, 36901, 36961, 37015, B37037, 37093,
        37132-3, 37143-5, 37149, 37170, 37179, 37187, 37212-3, 37215-6, 37274,
        37311, 37318, 37322, 37360, 37391-3, 37400, 37421, 37457-8, 37471, 37479,
        37513, 37542, 37557, 37616, 37627-8, 37650, 37659, 37725, 37755-6, 37783,
        37793, 37807, 37817, 37824, 37858, 37862, 37902, 37933-4, 37946-9, 37962,
        37973, 37998, 38033, 38056, 38061-2, 38081, 38087, 38103, 38113, 38125,
        B38139, 38146, 38178, 38247, 38288, 38301, 38323, 38327, 38340, 38356,
        38420, 38428-9, 38487, 38517, 38525, 38542, 38598, 38612, 38625-6, 38696,
        38758, 38767, 38868, 38872, 38909, 38951, 38955, 39006, 39010, 39037, 39062,
        39065, 39093, 39098-9, 39105, 39127, 39135, 39154, 39174-5, 39204-5, 39235-6,
        39237, 39264, 39266, 39284, 39310, 39334-5, 39338-9, 39351, 39366, 39369,
        39410, 39430-1, 39441, 39460, 39472, 39478-9, 39482, 39526, 39571, 39594,
        39604, 39618, 39620, 39657, 39695, 39703, 39713, 39723, 39725, 39733,
        39741, 39747, 39761, 39830, 39834, 39837, 39839, 39869, 39876, 39926-7,
        39972, 40000-2, B40015, 40020, 40046, 40050, 40083, 40096, B40105, 40111,
        40159, 40161, 40181, 40201-2,   40204-5,   40210,   40218, 40248-9,
        40261, 40264, 40267, 40286, 40306, 40314, 40345, 40351, 40360

Ribulose 1,5-bisphosphate carboxylase, methods
        21577, 21616, 22050, 22055, 22381, 22523, 22562, 23256, 23453, 23598, 24263,
        24309, 24492, 24681, 24779
        25496, 25934, 26208-9, 26405, 26546, 26560, 26665, 27263, 27503, 27531,
        27669, 27966, 28009, 28032, 28071-2
        29144-5, 29374, 29626, 29650-1, 29841, 29865, 30025, 30275, 30279, 30636,
        30647, 30784, 31419, 31553, 32035, 32264, 32334
        32962-3, 33155, 33171, 33271, 33273, 33473-4, 33767, 33837, 33842, 33887,
        33895, 34178, 34388, 34812, 34925, 34979, 35006, 35510, 35569-70,
        36412, 36414, 36451, 36491
        36784, 36813, 37179, 37187, 37421, B37458, 37946, 38061, 38081, 38288,
        38340, 39127, 39135, 39369, 39430, 39472, 40001, 40096, 40202, 40267,
        40306

Ribulose 1,5-bisphosphate oxygenase see Ribulose 1,5-bisphosphate carboxylase ...;
        Photorespiration enzymes

Root removal see Defoliation ...

Root, underground part, and carbon fixation pathways 23333; 28855, 29016

Root, underground part, and chlorophyll 23771; 27162; 33034

Root, underground part, and chloroplast 32377; 32717

Root, underground part, and ecosystem and plant productivity
        21692, 22348, 22547, 22714, 23168, 23271, 23332, 24185, 24303; 26017, 26349,
        26498, 26723, 27707, 27814; 28521, 28937, 29016, 29022, 29492, 29726, 30317,
        30463, 30674, 31086, 31748, 32403; 32808, 33043, 33756, 34444, 34459, 34854,
        34875, 36172; 37250, 37464, 39035, 39596, 39638, B39951, 40219, 40292

Root, underground part, and electron transport chain 34049

Root, underground part, and gas exchange
21609, 21666, 21692, 21786, 22938, 23786, 24837; 26498; 32717, 32808, 33079, 35373, 35727

Root, underground part, and resistances to CO_2 and water vapour transfer   26421; 30249; 37250

Root, underground part, and respiration   25666, 27011; 30891, 32363; 36406; 36670, 40193

Rooted leaves, chlorophyll in   28791, 30683

Rooted leaves, gas exchange in   22782; 27331; 30683

Rooted leaves, productivity of   28791

RuBP carboxylase, RuBPC   see   Ribulose 1,5-bisphosphate carboxylase

Rubredoxin   see   Ferredoxin ...

S

Saccharides and algae productivity   25121

Saccharides and biliproteins   35935

Saccharides and carbon fixation pathways
21639, 21661, 21739, 21784, 21798, 21942, 22080, 22250, 22411, 22449, B22572, 22608, 22697, 22828, 22868-9, 22975, 23073, 23210-1, 23214, 23379, 23422, 23467, 24000, 24059, 24161, 24262, 24319, 24382, 24404, 24442, 24552, 24652, 24659, 24836
25188, 25374, 25449, 25512, 25620, 25685, 25727, 26206, 26244, 26251, 26343, 26360, 26513, 26525, 26770, 26777, 26907, 27091, 27101, 27284, 27509, 27589, 27790, 28116, 28477
28855, 29253, 29430, 29571, 29629, 29699, 29891, 30526, 30686, 30809, 30817, 30936, 30962, 31168, 31247-8, 31314-5, 31360, 31520, 31806, 31938, B31990, 32297, B32515
32792, 32841, 32855, 32894, 32927, 32970, 33378, 33491, 33671, 33842, 33975, 34250, 34276, 34409, 34477, 34749, 34905, 34988, 35024, 35084, 35592, 35848, 35976, 36289, 36364
36665, 36840, 37074, 37116, 37124, 37311, 37844, 37876, 37903, 37972, 38087, 38104, 38115-6, 38126, 38271, 38325, 38412, 38428, 38629, 38738, 38798, 38959, 39054, 39173, 39183, 39351, 40189

Saccharides and carotenoids   22673; 28240, 32819, 33952; 39400

Saccharides and chlorophyll
22337, 22673, 22735, 22789, 23951, 23959, 24397, 24985; 26292, 28240; 28762, 29220; 32819, 33129, 33596, 33952, 35145, 35232; 36701, 36884, 38946, 39400

Saccharides and chloroplast (chromatophore)
22176, 22673, 22783, 23681, 23924, 24381, 24476, 24865, 24985; 25621, 26095, 26141, 26276, 26795, 27268, 27742; 28896, 29728, 29871, 30016, 31616, 31677; 32659, 34685, 34907, 35749-50, 36488; 36684, 36712, 37293, 38546, 38624, 39921, 40157, 40189

Saccharides and ecosystem and plant productivity
21853, 22229, 23984; 27430, 27918; 31914, 31916, 32010, 32172, 32542; 32699, 33181, 33936, 35331, 35932, 35938, 36219; 37266, 39637

Saccharides and electron transport chain
        22673, 23726, 23917, 23989, 24568; 26292, 27284, 27589-90; 29270, 30551,
        31651; 35731, 35935; 38275, 38864, 39432

Saccharides and gas exchange
        22673, 23649, 24397, 24568; 26513, 26642, 27258, 27641, 27840; 29133, 29220,
        29500, 29921, 30546; 33153, 34409, 35164, 35368, 35371, 35441, 36557; 36688,
        37163, 37656, 37804, 39371, 39637

Saccharides and photorespiration 21619; 27284, 28477; 29220; 33153

Saccharides and resistances to CO_2 and water vapour transfer 38465

Saccharides and respiration 34409, 35932; 37576

Salinity of soil and algae productivity 26344; 30175; B33583; 39871-2, 40095

Salinity of soil and biliproteins 36617

Salinity of soil and carbon fixation pathways
        21736, 22484, 23274, 23786, B24070, 24986, 25035; 26446, 26550, B26551,
        27255, 28404; 29061, 29495, 30174, 30664, 31319, 32139, 32378, 32425, 32477;
        33304, 33554-5, B33583, 33777, 33858, 34764, 35084, 35249, 35688, 35701;
        36916, 37331, 37933-4, 38087, 39240, 39275, 39963, 40119, 40159, 40238

Salinity of soil and carotenoids 30067; B33583; 36617

Salinity of soil and chlorophyll
        22482, 23523, 24249, 24986; 25236, 25510, 26550, B26551, 26939, 28268;
        29410, 30067, 30175, 30633, 31855, 32259; 33292, 33562, 33583, 33900, 33935,
        35084, 35249; 36617, 37333, 37520, 38087, 38166, 38576, 39197, 40006-7

Salinity of soil and chloroplast (chromatophore)   24755; 28266; B33583, 33900,
        34986, 35793; 38576, 39248

Salinity of soil and ecosystem and plant productivity
        21508, 22485, 22937, 23113, B24070, 24204, 24249, 24470, 24628, 24986;
        25299, 25510, 25526, 26437, 26452, 26685, 26939; 29331, 29410, 30633, 31041,
        31086, 31217, 32200; 32995, 33306, B33583, 33935, 34444, 34648, 35033,
        B35165, 35781, 35943, 35980, 36143, 36252; 36617, 37027, 37796, 38385,
        38519, 39240, 39248, B39903, 40006-7, 40126, 40238

Salinity of soil and electron transport chain
        23274, B24070; 27255; 34764, 35688; 37333, 38166

Salinity of soil and gas exchange
        21508, 21699, 22344, 22484-5, 22791, 22965, 23112-3, 23522-3, B24070; 26344,
        26669, 26685, 26939, 27198, 27321, 28078, 28128, 28200; 28507, 29093, 29252-
        -3, 29331, 29495, 30139, 30633, 30968, 31086, 31217, 31681, 32139, 32378;
        32562, 32995, 33096, 33292, 33304, 33306, 33378, 33858, 34124, 34170, 34200,
        34444, 34562, 34648, 34764, 34988, 35069, 35084, 35249, 35625, 35688, 35793,
        35980, 36510; 36617, 36762, 36830, 36940, 37142, 37799, 37933-4, 38074,
        38087, 38166, 38324, 38519, 40126, 40159

Salinity of soil and photorespiration 29495, 30272, 31086; 33125, 35249

Salinity of soil and resistances to CO_2 and water vapour transfer
        22484, B24070; 26282; 30139, 32139; 33304, 33306, B33583, 34170, 34278;
        37108, 37799, 38166, 38519, 39224, 40126

Salinity of soil and respiration
        22484, 24001; 26669; 29495, 30139, 30968, 31086, 31217, 31681; 36830, 36940

Seasonal changes in carotenoids (continued)
        37232-3, 37257, 37263, 37417, B37528, 37647, 38221, 38290, 38581, 39083,
        39114, 39409-10, 39568, 40072, 40086, 40088

Seasonal changes in chlorophyll
        21543, 21563-4, 21626, 21732, 21760, 21793, 21877, 21949, 21965, 21970,
        22121, 22287, 22295, 22323, 22353, 22392, 22493-4, 22517, 22551, 22565,
        22763, 22804-5, 22841, 22911, 23053, 23107, 23149, 23174-7, 23259,
        B23262, 23269-70, 23275, 23277, 23425, 23536-7, 23560, 23663, 23706, 23739,
        23753, 23780, 23809, 23824, 23861, 23961, 23963, 24019, 24071, 24181, 24194,
        24202, 24282, 24285, 24308, 24330, 24404, 24512, 24528, 24530, 24541, 24562,
        24616, 24643, 24687, 24699, 24714, 24764, 24781, 24953, 24992, 25011, 25053,
        25102
        25210, 25218, 25270, 25390, 25422-3, 25507, 25644, 25679, 25685, 25743,
        25822, 25871-2, 25877, 25896, 25898, 25948, 25956, 26131, 26469-70, 26529,
        26552, 26617, 26707, 26800, 26849, 26857, 26918, 26998, 27021, 27068, 27124,
        27162, 27177-8, 27216, 27227, 27265, 27388, 27405, 27484, 27488, 27642,
        27666, 27733, 27766, 27885, 27926, 27971, 27980, 28017, 28073, 28117, 28156,
        28174, 28185, 28318, 28322, 28464
        28790, 28880, 28939-40, 29250, 29321, 29414, 29419, 29422, 29456-7, 29802-3,
        29809, 30017, 30057, 30186, 30295, 30357, 30400, 30464-5, 30572, 30581,
        30605-6, 30724, 30750-1, 30760, 30800, 30802, 30863, 31047, 31069, 31083,
        31143, 31213, 31385, 31403, 31499, 31533, 31559, 31579, 31727, 31855, 32009,
        32055, 32058, 32081, 32185, 32284, 32300, 32307, 32526
        32581, 32605, 32617, 32645-7, 32708, 32749, 33013, 33230, 33268-9, 33331,
        33379, 33562, 33641, 34318, 34438, 34457, 34463-4, 34483, 34493, 34581,
        34595-6, 34607, 34767, 34839, 34932, 35060, 35181, 35408, 35520, 35619,
        35643, 35799, B36081, 36154, 36528, 36534, 36571
        36642, 37138, 37232-3, 37257, 37263, 37321, 37338, 37346, 37417, 37490,
        37517, B37528, 37646, 37694, 37889, 38215, 38221, 38290, 38359, 38402,
        38463, 38725, 39076-7, 39114, 39207, 39409-10, 39539, 39568, 39573, 39908,
        39928, 40072, 40210

Seasonal changes in chloroplast (chromatophore)
        22565, 22573, 23265-6, 23706, 24404, 24770; 25821; 29238, 29569, 30295,
        31143, 31728; 32618, 33480, 33737, 35026, B36191; 37694, 38221, 38624

Seasonal changes in ecosystem and plant productivity
        21524, 21554, 21625, 21657-8, 21674, 21684, 21802, 21928, 21949, 22007,
        22108, 22118, 22144, 22173, 22196, 22246, 22271, 22302, 22352, 22379-80,
        22421, 22430, 22448, 22570, B22572, 22593, 22639, 22668, 22775, 22806,
        22878-9, 22908-9, 22917, 23002, 23037, 23039, 23063-4, 23081, 23091, 23145,
        23237, 23245, 23252, 23271, 23305, 23308, 23646, 23653, 23658, 23695, 23732-
        -3, 23776, 23795, 23805, 23895, 23910, 23961, 23983, 23991, 24006, 24019-20,
        24181, 24185-6, 24210, 24267, 24291, 24295, 24303, 24314, 24424-5, 24428,
        24430, 24558, 24563, 24567, 24585, 24641, 24666, 24705, 24773, 24841, 24880,
        24940, 24971-2, 25022, 25060, 25089, 25103, 25149-50
        25203, 25230, 25253-4, 25264, 25274-5, 25358, 25412, 25481, 25518, 25618,
        25737, 25771, 25792, 25826, 25886, 25909, 25946, 25948, 25967, 25997, 26017,
        26020, 26068, 26085, 26096, 26150, 26189, 26197, 26216, 26256, 26348,
        26372, 26501-2, 26506, 26652, 26724, 26747, 26783, 26787, 26803-4, 26841,
        26857, 26874, 26904, 26906, 26912, 27049, 27087, 27098, 27108, 27139,
        27150, 27162, 27216, 27220, 27228, 27336, 27384, 27404, 27604, 27625,
        27630, 27656, 27710, 27720, 27736, 27740, 27748, B27915, 28087, 28100, 28158,
        28175, 28197, 28265, 28283, 28300, 28322, 28350, 28413
        28516, 28533, 28624, 28711, 28800, 28813, 28864, 28913, 29053, 29057, 29116,
        29164, 29254, 29334, 29373, 29377, 29411, 29469, 29551, 29577, 29656, 29685,
        29703, 29783, 29791, 29794, 29967, 29975, 30011, 30060, 30083, 30095, 30195,
        30317, 30371, 30400, B30477, 30483, 30533, 30573, 30581, 30691, 30711, 30745,
        30802, 30824, 30869, 30874, 30891, 30917, 30967, 31019, 31070, B31140, 31199,
        31220, 31259, 31383, 31508, 31539, 31558, 31586, 31633, 31647, 31706, 31748,
        31817, 31822, 31826, 31843, 31875, 31921-2, 31952, 31954, 32027, 32151, 32197,
        32243, 32265, 32306, 32363, 32372, 32403, 32415, 32436, 32543
                                                                        (continued)

Seasonal changes in photorespiration   21803, 21951; 27425; 30751; 32573, 34834;
        37451

Seasonal changes in resistances to CO_2 and water vapour transfer
        21887, 21929, 22314, 22505, 22589, 23007, 23167, 23679, 23791, 24376, 24449;
        25971, 25999, 26015, 27049, 27269, 27432, 27575, 27609, 28102, 28350; 29053,
        29162, 29373, 29394-5, 29551, 30162, 30583, 30919, 31069, 31866, 32051,
        32194, 32369; 32573, 32693, 32777, 33451, 33694, 33857, 33985, 34008, 34110,
        34954, 35278, 35354-5, 35905, 35963, 36017; 36637, 36683, 37250, 37347,
        37441, 37638, 37753, 38075, 38389, 38401, 38997, 39328, 39511

Seasonal changes in respiration
        21564, 21626, 21803, 22165, 22302, B22572, 22778, 22832, 23145, 23257,
        23334, 23534, 23557, 23605, 23922, 23962, 24055, 24528, 24821; 25303, 25480,
        26046, 26370, 26572, 27774, 28086-7, 28102; 28272, 29016, 29443, 29569,
        30011, 30317, B30477, 30608, 30642, 30691, 30751, 30891, 31019, 31896,
        32182; 32573, 32882, 33013, 33565, 34100, 34373, 34471, 35040, 35092, 35233,
        36132; 36683, 36967, 37360, 37753, 37989, 38523, 39665, B39951, 40233

Simulation   see Model ...

Sink and source of photosynthates, CO_2, *etc.*
        21553, 21557, 21634, 21740, 21786, 21878, 22096, 22229, 22428-30, 22678,
        22777, 22782, 23649, 23779, 23912, 24000, 24275, 24325, 24359, 24363, 24448,
        24598, 24625, 24660, 24967, 24976, 25024
        25193, 25207, 25229, 25318, 25576, 25723, 25765, 25774, B25775, 25965, 25967,
        26028, 26032, 26121-2, 26243, 26320, 26433, 26489, 26537, 26688, 26775,
        27199, 27361, 27384, 27412, 27514-5, 27573, 27594, 27722, 27725, 27991,
        27995, 27999, 28053, 28098, 28296, 28336, 28339, B28340
        28578, 28624, 28872, 28913, 29202, 29381, 29482, 29506, 29645, 29789-90,
        29822, 29921, 29952, 29954, 29992, 30010, 30021-2, 30030, 30902, 30912, 30920,
        31086, 31214, 31277, 31281, 31306, 31399, 31790, 31792, 31823, 31879, 31978,
        32122, 32218, 32427, 32503, 32516
        32686, 32751, 32868, 33002, 33021, 33181, 33226, 33260-1, 33534, 33671,
        33846, 33860, 33871, 33928, 34721, 34923, 34932, 34936-7, 35033, 35283,
        35330, 35373, 35416, 35504, 35508, 35891, 35979, 35998, 36044, 36422
        36788, 36950, 36986, 37071, 37164, 37169, 37207, 37474, 37490-1, 37510,
        37617-8, 37848, 37863, 37888, 37897, 37941, 37953, 38168, 38465, 38488,
        38539, B38544, 38627, B38739, B38892, 39130, 39360, 39480-1, 39541, 39546,
        39755, 39764, 39793, 39852, 39858, 39864, 39905, 39993, 40143

Soil moisture and carbon fixation pathways
        23930; 27360; 28501, 32139; 33513, 33858, 35049; 36916, 36918, 37331

Soil moisture and carotenoids
        23375, 23753, 24148, 24527, 24540; 25930, 27398, 27855, 27925; 31014; 36669

Soil moisture and chlorophyll
        21880, 21974, 22519, 22763, 23753, 23961, 24019, 24148, 24181, 24527, 24540;
        25930, 26338, 26392, 27398, 27855, 27925, 28312; 30198, 30620, 31014, 31581;
        32749, 33014, 33680, 34306, 35049, 35661, 36016, 36394; 36669, 38863, 39601,
        40006-7

Soil moisture and chloroplast (chromatophore)
        21861, 21974, 24018, 24801; 31593; 33513, 35605, B35862

Soil moisture and ecosystem and plant productivity
        21511, 21555, 21621, 21625, 21787, 21845, 21862, 21880, 21919, 22074, 22112,
        22118, 22348, 22524, 22593, 22701, 22930, 22999, 23221, 23223, 23650, 23653,
        23669, 23678, 23784, 23790, 23813, 23900, 23961, 23991, B24132, 24181,
        24207, 24331, 24569, 24798, 25140
        25170, 25238, 25253, 25319, 25481, 25572, 25607, 25613, 25737, B25775, 25930,
        26124, 26216, 26337, 26501-2, 26509, 26537, 26569, 26605, 26723, 26729,
        26782, 26798, 26841, 26920, 27094, 27108, 27133, 27200, 27404, 27413, 27433,
        27457, 27656, 27686, 27710, 27984, 28098, 28196, 28241, B28482, 28495
                                                                      (continued)

Soil moisture and ecosystem ... (continued)
28566, 28788, 28873, 28899, 29009, 29022, 29120, 29221, 29496, 29550, 29790,
29843, 29881, 29886, 29968, 29975, 29985, 30168, 30195, 30203, 30249, 30371,
30451, B30477, 30733, 30821, 30824, 30829, 30902, 30917, 30995, 31014-5,
31041, 31086, 31148, 31279, 31408, 31463, 31496, 31608, 31753, 31823, 31825,
31849, 31922, 32033, 32047, 32147, 32265, 32363, 32436, 32504, 32530
32640, 33092, 33196, 33222, 33293, 33488, 33623, 33680, 33728, 34012, 34129,
34181, 34206, 34301, 34306, 34498, 34541, 34677, 34874, 34963, 35007, 35096,
35194, 35243, 35272, 35331, 35455, 35513-4, 35517, 35534, 35608, 35661,
35884, 35904-5, 36016, 36107, 36112, 36178, 36269, 36361, 36388, 36394,
36436
36637, 37082, 37168, 37190, 37323, 37334, 37343, 37368, 37439, 37462, 37473,
37544, 37637, 37838, 37899, 38082-3, 38134, 38293, 38399, 38423, 38779,
38990, 39035, 39068, 39223, 39419, 39561, 39659, 39889, 40006-7, 40240-2,
40299, 40326, B40331

Soil moisture and electron transport chain
22566, 23077, 23999, 24147, 24198-9; 31581, 32310; 33513, 35661, 36224

Soil moisture and gas exchange
21529, 21633, 21670, 21787, 21974, 22031, 22139, 22234, 22314, 22950, 23175,
23355, 23545, 23745, 23784, 23786, 23908, 23961, B24024, 24148, 24290,
24537, 25045
25419, 25613, 25642, 25930, 25951, 26096, 26124, 26392, 26464, 26729, 27009,
27139, 27289, 27398, 27656, 28066, 28196, 28462
28501, 28629, 28970-1, 29009, 29092, 29409, 29676, 29975, 29985, 30168,
30575, 30703, 30898, 31014, 31086, 31536, 31823, 31994, 32139, 32147
32986, 33335, 33513, 33664, 33858, 33920, 33928, 34129, 34181, 34368, 34689,
34886, 35049, 35098, 35243, 35476, 35484, 35632, 35661, 35789, 35886, 36013,
36207, 36388
36618, 37243, 37289, 37308, 37473, 37476-7, 37614, 37699, 37742, 37899,
37915, 38134, 38161, 38293, 38401, 38791, 38997, 39058, 39211, 39223, 39461,
39784, B39951, 39980

Soil moisture and photorespiration 26241

Soil moisture and resistances to CO_2 and water vapour transfer
21529, 21887, 22589, 22761-2, 23679, 23784, 23813, 25045; 25537, 25572,
25742, 25999, 26015, 26392, 26532, 27133, 27271, 27433, 28089, 28125, 28462;
28501, 28732, 28821, 29394, 30250, 31068, 31823, 31956, 31994, 32051, 32111,
32139; 33106, 33527, 33858, 34181, 34689, 35096, 35098, 35905, 36075;
37108, 37368, 37413, 37439, 37473, 37544, 37915, 38401, 38779, 38997, 39661,
39737, 39813

Soil moisture and respiration
23961, 24001; 26241, 27398; 28970, 29409, 29676, B30477, 31086; 33335, 33928,
34677, 34689, 35243, 35789; 37289, 37308, 39211

Solar radiation and canopy see Canopy, radiation distribution ...; Canopy, radiation profile ...

Specific leaf area see Growth analysis, specific leaf area ...

Spectral methods in photosynthesis research
21856, 21901, 22057, 22632, 22642, 22664, 22969, 22972, 23160, 23574, 24486,
24875; 25491, 25907, 26301, 26381, 26556, 28028, 28030, 28328-9; 28515,
28636, 28825, 29026, 29556, 29701, 30239, 30503, 30539, 30600, 30780, 31342,
31688, 32155; 32587, 32641, 32700, 32747-8, 32907, 33636, 33707, 33793-4,
34348, 34449, 34568, 34759-60, 34867, 34909, 34957, 35113, 35311, 35599,
35745, 35913; 36772, 36932, 36952, 36991, 37855, 37914, 37938, 38135, 38322,
38470, 38634, B38706, 38714, 39028, 39303, 39406, 39878, 39950, 40011

Stabilization of photosystems see Photosystem stabilization ...

Stand   see   Canopy ...; Ecosystem ...

Steady state and non-steady state   see   Oscillations. ...

Stem, petiole, morphology, structure and physiological activity in
          22932, 24323, 24336, 24526, 24759, 24974; 26931; 30675, 31606; 33570, 33573,
          33700, 35791; 39137

Stomata morphology and anatomy (number, dimensions, types, development, structure,
          etc.) (*cf.* also Leaf epidermis, stomata)
          21827, 22016, 22039, 22043, 22082, 22084, 22087, 22157, 22208, 22295, 22516,
          B22718, 22719, 22854, 22885, 23071, 23134, 23330, 23464, 23513, 23547,
          23757, 24449, 24645, 24704, 24742, 24795, 25031, 25045-6, 25111
          25194, 25251, 25364, 25404, 25406, 25445, 25604, 25653, 25663, 25810, 25881,
          25971, 26074-5, 26336, 26421, 26477, B26551, 26853, 26931, 27012, 27036,
          B27046, 27073, 27109, 27158, 27433, 27467, 27555, 27585, 28179, 28182, 28367,
          28399, 28456, 28463
          28532, 28613, 28622, 28703, 28866, 28925-7, 29008, 29069, 29247, 30131,
          30152, 30162-3, B30432, 30660, 30733, 30946, 31068-9, 31087, 31193, 31230,
          31251, 31270, 31440, 31550, 31604, 31769, 32194, 32447
          B32723, 32895, 33002, 33105-6, 33158, 33280, 33760-1, 33937, 34132, 34416,
          34636, 34743, 34835, B34865, 35016-7, 35049, 35197-8, 35222, 35234, 35350,
          35594, 35708-9, 35775, 35884, 35906, 35920, 36080, 36521-4
          36692, 36726, 36879, 36907, 37088, 37303, 37347, 37544, B37831, 38016-7,
          38134, 38372, 38640, 38747, 38791, 39010, 39059, 39259, 39442, B39537,
          39773, 39828, 39991, 40177, 40201, 40313

Stomata physiology (mechanism of action, reactivity, *etc.*)(*cf.* also Leaf epidermis,
          stomata)
          21849, 21916, 22084, 22116, 22186, 22190, 22380, 22652, 22960, 22964, B23006,
          23296, 23330, 23513-4, 23612, 23628, 23801, 23873, 24155-6, 24261, 24376-8,
          24426, 24688, 24739, 24742, 25035, 25111
          25194, 25251, 25271, 25604, 25928, 26074-5, 26238, 26265, 26436, 26612,
          26908, B26982, B27046, 27047, 27101, 27109, 27280, 27356, 27410, 27553,
          27555-6, 27573, 27578, 27846, 27936, 28182, 28209, 28244, 28275, 28463
          28504, 28703, 28768, 29020, 29179, 29312-3, 29395, 29793, 30076, 30163,
          30249, 30281, B30325, 30523, 30659, 30662, 30861, 31255, 31440-1, 31983,
          32015, 32121, 32156, 32369, 32435, 32521
          32843, 33106, 33285, 33288, 33396, 33505, 34682, 34743, 34786, 35139-40,
          35197-8, 35234, 35303, 35340, 35355, 35369, 35461, 36011, 36121, 36432,
          36472, 36521
          36701, B36781, 36917, 36951, 37224, 37303, 37319, B37831, 38134, 38401,
          38421, 38529, 38534, 38541, B38544, 38609, 39008-9, 39259, 39424, B39537,
          39825, 39891, 39906, 39911, 39954, 39991, 40079, 40339

Stomata role in photosynthesis
          22652, 23134, 23692, 23757, 23791, 23978, 24063, 24155, 24376, 24507, 25035;
          25194, 25271, 25404, 25446, 25604, 26075, 26121, 26238, 26265, 26436, 26802,
          27171, 27278, 28182, 28205, 28209, 28244, 28303, 28463; 28546, 29020, 29179,
          29615, 30861, 30905, 31057, 31250, 31255, 31442, 31925; 32772, 32777, 32871,
          33002, 33285, 33288, 33742, 33832, 33916, 34120-1, 34494, 34497, 34743,
          35140, 35234, 35303, 35963, 36135, 36418, 36472, 36521-3; 36630, 36951,
          B37831, 38016-9, 38134, 38389, 38401, 38663, 38791, B39537, 39599, 39906,
          39991, 40265, 40313

Stomatal diffusive resistance   see   Resistance, stomatal ...

Stroma of chloroplast
          21551, 21613, 21871, 21984, 22148, 22176, 22281, 22322, 22388, 22405, 22619,
          22967, 22994, 23009, 23129, 23141, 23154, 23319, 23607, 23670, 23680, 23706,
          23905, 23995, 24049, 24162, 24591, 24655, 24754, 25007
          25212, 25745, 25770, 25781, 25814, 25979, 26101, 27061, 27160, 28306
                                                                        (continued)

Stroma of chloroplast (continued)
        28699, 28733, 28870, 29578, 29808, 29871, 29892, 30428, 30659, 30856,
        30871, 30889, 32109, 32192
        32768, 32895, 33089, 33146, 33553, B33628, 33943, 33945, 34155, 34249,
        34525, 35644, B35862, 35936, 35956, 36240, 36486
        36611, B36768, 36980, 37055, 37154, 37362, 37929, 37960, 37976, 38301,
        38812, 38845, 38991, 39707, 39773, 39817, 40123, 40353

Sulphur oxides (and other sulphur compounds) see Pollution of air ...

Sunflecks in canopy see Canopy, radiation distribution

T

Taxons, algae productivity   27227; 30585; 33804; 40089

Taxons, biliproteins in   26158; 30552, 32102, 32258; 32617, 34660; 39804

Taxons, carbon fixation pathways in
        22049, 22958, 23256, 23930, 24310, 24555, 24980; 25431, 25437, 25685, 25692,
        25767, 25772, 26314, 26505, 27506, 28059, 28102, 28399, 28493; 29322, 30098,
        30273, 30474, 30538, 30980, 31910, 32025, 32373; 32555, 33642-5, 33858,
        34428, 34452, 34836, 35249, 35426, 35429, 36092-3, 36412, 36445; 36712,
        36814, 37187, 37713-4, 37905, 38062, 38088, 38288, 38323, 38340, 38358,
        39237, 39383, 39416, 39526, 39809, 39830, 40000

Taxons, carotenoids in
        22465, 23753; 25442, 25794, 25895, 26639, 27237, 27396, 27526, 28377; 28528,
        28715, 29208, 29256, 29342, 29780, 30017, 30592, 31406, 32142, 32258, 32289,
        32501; 32617, 33264, 33351, 33572, 33780, 34483, 34594, 34604, 35048, 35250,
        35962; 37232-3, 37346, 37684, 38221, 39736

Taxons, chlorophyll in
        22836, 23567, 23753, 24148, 24435, 24613; 25406, 25442, 25598, 25685, 25887,
        25931, 26521, 26617, 26674, 27227, 27237, 27396, 28377; 28528, 28715, 29342,
        29557, 29780, 30017, 30354, 30552, 30596, 30750, 30822, 31406, 32058, 32142,
        32258, 32289; 32617, 33014, 33349, 33351, 33572, 33602, 33967, 34452, 34483,
        35048, 35249-50, 35259, 35809, 35962, 36226; 36642, 36660, 36712, 36814,
        37232-3, 37271, 37346, 37569, 37693-4, 37747, 37889, 38221, 38493, 38628,
        38836, 38900, 39029, 39069, 39237, 39736, 40089, 40208

Taxons, chloroplast (chromatophore) in
        21610, 21723; 26260; 29033, 29780, 32102, 32221-2; 32836, 33256, 33602,
        34142, 34171, 35776; 36590, 36712, B36768, 36816, 37100, 37569, 37693-4,
        38221, 40078, 40231, 40356

Taxons, ecosystem and plant productivity of
        22603, 22668, 22829, 22878-9, 23466, 23500, 24196, 24438, 24509, 24558,
        24633, 24798, 24839, 24940, 25126; 25531, 25660, 25807, 25811, 26788, 26947,
        28013, 28228, 28470; 29057, 29254, 30483, 30750, 31759, 31761-2, 32151, 32325;
        33412, 34181, 34439, 34442, 34621, 35039, 35338, 35340, 35463, 35608, 36226;
        36656, 36846, 37063, 37240, B37727, 37795, 37867, 38493, 38860, 40220

Taxons, electron transport chain in
        22693, 23701, 24091, 24286; 26476; 28528, 29021, 29857, 30855, 31915, 32362;
        32948, 33869, 34488, 34983, 35547, 35752, 36556; 37632, 37643, 37870, 37901,
        38160, 38493, 38716-7, 39237, 39589

                                                                          (continued)

Temperature, high, and electron transport chain (continued)
        34527, 34529, 34998, 35076, 35125, 35804, 36506-7, 36519; 36632, 36807,
        36891, 37526, 37728, 38109, 38263, 39432, 39574, 39590, 39676, 39728

Temperature, high, and gas exchange
        21647, 21705, 21730, 21811, 22309, 22311, 22992, 23151, 23297, 23447, 23872,
        B24024, 24025, B24176, 24883, 25107, 25122, 25131; 25414, 25438, 25919, 25950,
        26004, 26370, 26485, 26854, 27278, 27792, 27825; 28818, 29527, 30088, 30299,
        30523, 30919, 31241, 31274, 31282, 31611, 31690-1, 31701, 32287, B32318,
        32502, 32523; 32639, 32772-3, 32866, 33150-1, 33431, 33687, 34125, 34664,
        34786, 34832-3, 35337, 35494, 35804, 36343, 36506, 36519; 36683, 36807,
        36901, 37093, 37141, 37998, 38222, 38312, 38761, 39115, 39409, 39524,
        39728, 39981, 40079, 40285

Temperature, high, and photorespiration   29747; 32773, 33150

Temperature, high, and resistances to CO_2 and water vapour transfer   30919, 32772,
        32866

Temperature, high, and respiration
        21705; 32772-3, 33150, 33687; 37463, 40285

Temperature, leaf   see Leaf temperature ...

Temperature, low, and algae productivity   28561, 29842; 39877

Temperature, low, and biliproteins   23728; 27259; 28917; 38945, 39499

Temperature, low, and carbon fixation pathways
        22019, 22250; 25278, 25738, 27236, 28351; 30236, 31530, 31908, 32297; 33156,
        33914, 34071-3, 35024, 36412; 37713-4, 37827, 37946-8, 39593, 39723, 39747,
        39809

Temperature, low, and carotenoids
        23080, 24203, 24317, 24583; 26131, 28239-40; 28917, 30974, 31139, 31559,
        31623; 33352, 34493, 34670, 36145, 36580; 36628, 37428, 38443, 38492,
        39747

Temperature, low, and chlorophyll
        21889, 22044, 22221-2, 22955, 23022, 23029, 23080, 23287, 23470, 23617,
        23728, 24307, 24389, 24583, 24902-3; 25311, 25474, 25480, 25931, 25956,
        26131, 26161, 26710, 27259-60, 27276, 27291, 27622, 27718, 28141, 28239-40,
        28379; 28526, 28917, 28963, 29177, 29291, 29337, 29354-5, 29596, 29621,
        30360, 30833-4, 30850, 30971, 30973, 31139, 31143, 31438, 31559, 31623,
        31876; 32645, 32704, 33317, 33352, 33880, 34105, 34493, 34609, 34670, 34766,
        34823, 35259, 35921, 36032, B36081, 36145, 36210; 36628, 36753, 37321,
        37428, 37705, 38443, 38492, 38576, 38680, 38695, 39054, 39133, 39319, 39702,
        39747, 39877, 40308

Temperature, low, and chloroplast (chromatophore)
        24808, 24925; 25277, 25532, 26103-5, 26487, 27922; 28840, 29299, 29596,
        31120, 31139, 31143, 31243, 31530, 31616, 31727, 31876, 31955; 33911, 34249,
        34597, 35223, 35921, 35999, 36145, 36210; 36898, 37113, 37843, 38129, 38624,
        38680, 38919, 39054, 39133, 39241, 39432, 39489, 39679, 39771, 40078, 40353-4

Temperature, low, and ecosystem and plant productivity
        22617, 23305; 25480-1, 25523, 25527, 27428; 29201, 29496, 31530, 31711;
        32986, 35330, 36304; 37041, 37207, 38314, 38680, 38875, 39074, 40018,
        40022

Temperature, low, and electron transport chain
        21567, 21712, 22051, 22142, 22409, 22856, 23080, 24133, 24312, 24468, 24840;
        25202, 25494, 25738, 26104, 26335, 27031, 27276, 27345, 27629, 28447-8;
                                                                        (continued)

Temperature, low, and electron transport ... (continued)
       28547, 29107, 29237, 30182, 30297, 31075, 31143, 31727-8, 31876, 31955,
       32486-7; 33248, 34107, 34232-4, 34587, 34609, 34766, 35547, 35648, 35921,
       36210; 37843, 37919, 38129, 38970-1, 39432, 39677, 39679, 39702-3, 39728,
       40325, 40343

Temperature, low, and gas exchange
       21705, 21999, 22019, 22140, 22221-2, 22231, 22993, 23013, 23015, 23080,
       23382, 23447, 23545, 23556, 23571, 23872, 24290, 24389, 24427, 24758, 24837;
       25224, 25252, 25414, 25480, 25523, 25527, 26313, 26854, 27185, 27311, 27320,
       27337, 27750, 27988, 28092, 28141, 28379; 29337, 29726, 30172, 30529, 30631,
       30684, 30743, 31258, 31711, 31727-8, 32297, 32353; 32868, 32986, 33912,
       33915, 34375, 34450, 34518, 35260, 35296, 35337, 35494, 35791; 36805, 37041,
       37113, 37142, 37150, 37347, 37566, 38374, 38411, 38416, 38492, 38508,
       38875, 39074, 39133, 39409, 39703, 39705, 39728, 39809, 39843, 39888, B39951,
       39981, 40142, 40158, 40209, 40330

Temperature, low, and photorespiration   34073; 37113, 37946, 38508

Temperature, low, and resistances to CO_2 and water vapour transfer
       30392, 30743; 34375, 35260; 37113, 37347, 38508, 39809

Temperature, low, and respiration
       21705, 23013, 23080, 23545, 24837; 25252, 25480; 29537, 29957, 31711; 34450,
       34518, 36435; 36805, 37113, 37150, B39951, 40142, 40193, 40209, 40285

Temperature, physiological, and algae productivity
       21953, 22457, 22464, 22867, 24349, 24440, 24579; 25694, 26002, 26058, 26163,
       26273, 26407, 26540, 26741, 26849, 28118, 28317, 28471; B28531, 28740, 29655,
       30397, 30724, 32273; 33340, 33656, 33722, 33896, 33977, 34733, 34768, 35597,
       35880, 36216; 37058, 37278, 37673, 38417, 39823

Temperature, physiological, and biliproteins   26168, 27259; 38945, 39499

Temperature, physiological, and carbon fixation pathways
       21604, 21975, 23289, 23343, 24404, 24553, 24795, 24806, 24984; 25216, 25431,
       25696, 25721, 25787, 25920, 26003, 26692, 27073, 27422, 27435, 27709, 27928,
       28342, 28386, 28405; 28645, 28652, 28931, 29641, 30097, 30684, 30686, 30907,
       30959, 31166, 31240, 31251, B32003, B32515; 32664, 33130, 33857, 34631,
       34950, 35024, 35141-2, 35192, 35294, 35701, 36115, 36414; 37331, 37360, 37628,
       37947, 38429, 39009, 39275, 39603, 39641-4, 39963, 40159, 40213

Temperature, physiological, and carotenoids
       22510, 22841, 23086, 23878, 24317, 24751; 25479, 25969, 26003, 27066, 28240;
       33351-2, 36498; 36628, 39408, 40352

Temperature, physiological, and chlorophyll
       21706, 21709, 21953, 22214, 22449, 22463, 22475, 22510, 22779, 22841, 22911,
       22953, 23086, 23560, 23729-30, 23874, 23878, 24051, 24194, 24285, 24699,
       24895; 25473-4, 25479, 25841, 25898, 25900, 25994, 26003, 26099, 26168,
       26246, 26258, 26658, 26706, 26976, 26990, 27124, 27259, 27270, 27276, 27522,
       27547, 27718, 27930, 28148, 28240, 28317; 29067, 29139, 29177, 29461, 29727,
       30007, 30354, 30866, 31110, 31141, 31251, 31878, 32065, 32287; 32581, 32667,
       32876, 33351-2, 33625, 34839, 34940, 34945, 34998, 35483, 35922, 36518;
       36628, 36753, 37007, 37017, 37173, 37400, 37444, 37893, 38091, 38680, 38865,
       39172, B39191, 39408, 39524, 40352

Temperature, physiological, and chloroplast (chromatophore)
       22449; 26003, 27922; 28924, 30668, 30866, 31878; 33615, 34129, 35921-2,
       36291; 37113, 38680, 38682, 39040

Temperature, physiological, and gas exchange (continued)
30822, 30898, 30919, 30927, 30943-4, 30946, 31019, 31067-8, 31090, 31240-1, 31245, 31256, 31258-9, 31460, 31470, 31514, 31585-6, 31613-4, 31711-2, 31732, 31861-6, 31887, 31916, 31946, B32003-4, 32063, 32065, 32093, 32181, 32287, 32312, B32318, 32327, 32364, 32412, 32428, B32515
32562, 32664, 32714, 32787, 32804, 32813-4, 32866-7, 33062, 33084, 33107, 33112, 33150, 33231, 33262, 33278, B33281, 33305, 33324, 33335, 33346, 33363, 33433, 33435-6, 33441, 33465, B33510, 33614, 33621, 33663-5, 33694-5, 33703, 33740, 33759, 33969, 33984, 34017, 34123, 34193, 34218-9, 34260, 34288-90, 34302, 34370, 34374-6, 34427, 34429, 34451, 34453, 34485, 34497, 34512, 34518, 34547, 34562, 34644, 34646-7, 34649-50, 34689, 34701, 34755, 34787, 34792, 34800, 34853, 34886, 34888, 34912, 34945, 34961-2, 34971, 35069, 35074, 35095-8, 35129, 35171-3, 35175, 35243, 35260, 35262, 35273, 35339, 35344, 35406, 35484, 35494, 35559, 35616, 35622, 35624, 35643, 35664, 35674, 35765, 35791, 35807, 35895, 35909-10, 36037, 36059, 36067-9, 36090, 36115, 36132-3, 36199, 36388, 36424, 36427, 36480, 36508, 36550
36602, 36604, 36629, 36655, 36659, 36672, 36686-8, 36753, 36758, B36781, 36799, B36859, 37054, 37067, 37113, 37123, 37141-2, 37171, 37173, 37189, 37242, 37268, 37287, 37289, 37298, 37325, 37338, 37347-8, 37352, 37360, 37397, 37400, 37434, 37444, 37506, 37515, 37580, 37649, 37654, 37657, 37699, 37709, 37737, 37744, 37772, 37774, 37799, 37857, 37874, 37893, 37915, 37981, 37988, 38141, 38181, 38209, 38222, 38259, 38287, 38360, 38401, 38410, 38416- -7, 38436, 38444, 38448, 38464, B38477, 38493, 38529, 38589, 38603, 38632, 38697, 38703, 38710, B38739, 38761, 38794, 38869, 38934, 38974, 38996, 39094, 39211, 39235, 39309, 39381, B39394, 39461, 39500, 39524, 39552, 39639-40, 39728, 39751, 39759, 39794, 39812, 39823, 39843, 39867-9, 39873, 39875, 39932, B39951, 40077, 40079, 40103, 40144, 40159, 40166, 40193

Temperature, physiological, and photorespiration
22071, 23714, 24795; 25721, 25858, 26037, 26785, 26941, 26988, 27460, 27754, 27929, 28342; 29169, 29514, 29693, 29848, 30097, 30271, 30288-9, 30457, 30684, 30686, 31251, 31256; 33150, 33324, 3363, 35339, 36115; 36687, 37084, 37113, 37515, 37744, 38020, 38287, 38436, 39869

Temperature, physiological, and resistances to CO_2 and water vapour transfer
22024, 22314-5, 22380, 22650-1, 22761-2, 22920, 23609, 23653, 23679, 23992, 24377, 24688, 24795, 24826
25327, 25604, 25653, 25873, 25999, 26240, 26246, 26265, 26282, 26514, 26524, 26865, 26909, 26941, 27049, 27271, 27360, 27574, 27912, 28067, 28341-2, 28367, 28391
28732, 29015, 29020, 29319, 29394, 29766, 30078, 30271, 30289, 30455, 30457, 30547, 31240-1, 31251, 31256, 31268-9, 31613, 31861-2, 31864-5, 32065, 32364, 32369, 32485
33000, B33281, 33305, 33396, 33614, 33694, 33857, 33985, 34333, 34375, 34633, 34647, 34689, 34747, 34814, 34945, 35095-6, 35098, 35260, 35295, 35537, 35910, 35939, 36424
36659, 36683, 36732, 37113, 37171, 37268, 37347, 37360, 37413-5, 37649, 37915, 38209, 38217, 38401, 38436, 38529, 38534, 38737, 38915, 38934, 38996, 39094, 39552, 39812, 39875, 39991

Temperature, physiological, and respiration
21705, 22165, 22210, 22419, 22832, 22950, 22992, 23069, 23105, 23298, 23545, 23587, 23695, 23764, 23865, 24001, B24271, 24291, 24528, 24535, 24795, 24837, 24919, 24934, 25047, 25052
25283, 25291, 25858, 25873, 26037, 26046, 26073, 26085, 26112, 26299-300, 26370, 26386, 26461, 26506, 26578, 26988, 27301, 27478, 27521, 27638, 27687, 27935, 28092, 28106
28527, 28703, 28970, 29065, 29068, 29154, 29223, 29319, 29425, 29443, 29605, 29848, 30011, 30218, 30271, B30477, 30515, 30804, 31240-1, 31470, 31711, B32003, 32312, 32412, 32491
32562, 32969, 33150, 33162, 33192, 33262, 33385, 33740, 34219, 34269, 34370, 34485, 34689, 34754, 34792, 34945, 34966, 35074, 35129, 35169, 35243, 35559, 35674, 36199, 36303, 36427, 36515-6

(continued)

Thylakoid, granum (continued)
          37918, 37921, 37960-1, 37976, 38015, 38024, 38054-5, 38101, 38150, 38208,
          38301, 38342, 38382, B38415, 38421, 38425, 38449, B38475, 38513, 38540,
          38556, 38561, 38614, 38730-1, 38832-3, 38837, 38840-1, 38846, 38896, 38898,
          38910, 38922, 38963, 38971, 38999-9001, 39017, 39023, 39082, 39197, 39221,
          39227, 39353, 39432-3, 39494, 39514, 39531, 39539, B39551, 39564, 39604,
          39623, 39630, 39632, 39646, 39745, 39762, 39805, 39817, 39835, 39929, 39987,
          40009, 40052, 40093, 40177, 40246, 40260, 40290, 40294, 40337, 40353-4,
          40366-7

Tissue cultures, carbon fixation pathways in
          24137; 26628; 30233-4, 30274, 30778, 31008, 31235, 31913; 33425, 33663, 34710
          35851, 36241, 36495; 36701, 37208-9, 37513, 37962, 37975, 38033, 38696,
          38770-1, 38965, 39079, 39153, 39334, 39441, 40002

Tissue cultures, carotenoids in
          23404, 23675, 23883, 24068; 25798, 27165; 29898, 30234, 30594, 30948, 32153,
          32365; 32819, 33629, 33952, 34608; 38364, 39585

Tissue cultures, chlorophyll in
          21701, 21771, 22205, 22304, 22735, 23404, 23629, 23883, 24068, 24414, 25108;
          25265, 25498-9, 25798, 26629, 28477; 28686, 28736, 29219, 29283, 29450,
          29898-9, 30056, 30234, 30241, 30363, 30467, 30594, 30948, 31012, 31038,
          31112, 31116, 31640, 31655, 31744, 32326; 32819, 33219, 33602, 33790, 33952,
          33968, 34087, 34346, 34579, 34608, 34948, 35687, 36084, 36494-5; 36701,
          36958, 37189, 37733, 37962, 37975, 38033, 38149, 39441, 39699, 39805, 40002

Tissue cultures, chloroplast in
          21791, 24414, 24888; 25486, 27002; 28736, 30068, 30467, 31012, 31743-4,
          32365, 32400; 32776, 33602, 33679, 34579-80, 36351, 36544; 37548,
          38397, 38513, 39805

Tissue cultures, electron transport chain in
          22304; 27471; 30594, 31012, 31038, 31116; 32819, 33663, 33790, 34608, 35036;
          37724, 37856, 39441, 40301

Tissue cultures, gas exchange in
          21771, 21994, 23629, 24068, 24568; 26629, 27471, 28477; 28772, 29219, 30056,
          30234, 30778, 30948, 31012, 31038, 31116, 31235, 31453, 31913; 32823, 33219,
          33663, 33665, 33790, 34710, 35057, 36051; 36597, 36701, 36811, 37189, 37208-
          -9, 37962, 37975, 38770-1, 38967, 39079, 39307, 39337, 39441, 39699

Tissue cultures, growth of 21771, 21994; 30056, 31112, 31116; 35932, 36494-5; 38149

Tissue cultures, photorespiration in   23629; 26628-9; 28772, 31913, 32529; 32822;
          39474, 40344

Tissue cultures, respiration in
          34800, 35163, 35265, 35267, 35471, 35932, 35950; 37576, 39699

Transient phenomena in carbon fixation pathways 28303; 28989; 37712

Transient phenomena in chlorophyll biosynthesis B38477

Transient phenomena in electron transport chain 28784, 30799, 32524; 40340

Transient phenomena in gas exchange
          22152, 22794, 23183-4, 23286, 24134, 25131; 25597, 25787, 25916, B26220,
          26835, 26863, 28029, 28193, 28244; 28745, 28989, 30000, 31280, 31944, 32207,
          32523-4, 32532; 32799, 33068, 33430, 34570, 35482, 36540, 36555, 36568,
          36584; 36937, 37084, 37452, 37929, 38339, 39774, 40330, 40349

Transient phenomena in photorespiration 29169, 32524; 37881, 40344

Transient phenomena in resistances to CO_2 and water vapour transfer 39892

Transpiration and photosynthesis
    21670, 21713, 21765, 21773, 21995, 22042, 22075, 22112, 22171, 22210, 22284,
    22314, 22343-4, B22358, 22507, 22651, 22686, 22831, 22951, 23112-3, 23444,
    23628, 23666, 23707, 23775, 23784, 23900, B24132, 24175, 24376, 24588,
    24675-6, 24689, 24718, 24776, 24820
    25309, 25419, 25484, 25540, 25613, 26065, 26227, 26265, 26282, 26329, 26373,
    26446, 26535, 26541, 26546-7, 26684, 26690, 26708, 26785, 26947, 27271,
    27289, 27409-11, 27578, 27595, 28428
    28546, 28732, 28768, 28843, 29103, 29179, 29212, 29409-10, 29816, 29944,
    30273, 30457, 30628, 30695, 30703, 30705, 30733, 31068, 31228, 31252, 31255,
    31303, 31464, 31887, 32023, 32220, 32226, 32252, 32541
    32573, 32633, 32871, 32973, 32996-7, 33050, 33280, 33304-6, 33388, 33441,
    33547, 33694-5, 33857, 33920, 34295, B34354, 35096, 35512, 35561, 35773,
    35886, 35920, 36480
    36618, 36659, 36910, 36916, 36918, 36985, 37021, 37243, 37397-9, 37649,
    37915, 38075, 38077, 38252, 38399, 38640, 38678, B38740, 38996, 39036,
    39147, 39259, 39270, 39567, 39599, 39667, 39831, 39874, 40264

Tritium oxide see Deuterium oxide, tritium oxide ...

U

Ubiquinones see Quinones ...

Ultraviolet radiation see Irradiance, spectral composition ...

Uncouplers of electron transport chain (*cf.* also Antibiotics and electron transport
    chain)
    21531, 21559, 21588, 21632, 21745, 21766, 21797, 21859, 21959, 21997, 22103,
    22106, 22136, 22154, 22392, 22425, 22452, 22491, 22548, 22625, 22689, 22705-
    -7, 22766, 22801, 22903, 22913, 22933, 22936, 23066, 23184, 23192, 23480,
    23484, 23541, 23617, 23641, 23722, B23933, 23989, 24011, B24024, 24034-5,
    B24176, 24187, 24192, 24224-5, 24298, 24367, 24418, 24471, 24497, 24525,
    24720, 24790, 24860, 24869, 24911, 24954, 24999, 25029, 25096
    25177, 25222, 25320, 25322, 25385-7, 25675, 25914, 25980, 26210, 26231,
    26376, 26411, 26473, 26523, 26544, 26556, 26718, 26790, 26792, 26902, 27093,
    27184, 27264, 27276-7, 27335, 27354-5, 27453, 27465, 27486, 27641, 27655,
    27872, 27907, 27916-7, 27996, 28203, 28235, 28452, 28483
    28562, 28582-3, 28636, 28640, 28739, 28741, 28747, 28867, 28884, 29037, 29081,
    29211, 29279, 29285-6, 29540-1, 29599, 29634-5, 29665, 29696, 29730, 29885,
    29942, 30008, 30062, 30093, 30115, 30126, 30135, 30210-1, 30251, 30259, 30291,
    30309, 30402, 30429-30, 30470, 30558, 30561, 30590, 30661, 30720, 30726,
    30765, 30771-2, 30806, 30831, 30910, 30922, 30984-5, 31075, 31124, 31152-3,
    31179, 31469, 31523, 31541, 31549, 31564, 31617, 31664, 31683, 31688, 31717,
    31746-7, 31818, 31828, 31873, 31959, 32042, 32068, 32135, 32204, 32407-8,
    32460, 32496, 32498, 32535-6
    32636, 32725, 32734, 32755, 32807, 32908, 32936, 33086, 33114, 33402, 33687,
    33744, 33748, 33753, 33886, 33929, 33986, 33990-1, 34026-7, 34102, 34138,
    34250, 34405, 34422, 34424, 34431, 34575-6, 34601, 34769, 34897, 35068,
    35076, 35305, 35307, 35573, 35681, 35915, 35955, 36094, 36397, 36496
    36764, 36833, 36862, 36892, 36993, 37047, 37137, 37210, 37245, 37280, 37372,
    37764, 37875, 37920, 37923, 38172, 38191, 38202, 38250, 38394, B38477, 38482,
    38531, 38623, 38651, 38733, 38771, 38773, 38808, 38926, 39075, 39102, 39104,
    39252, 39296, 39373, 39590, 39666, 39674-5, 39720, 39730, 39762, 39842,
    39870, 39915, 39955, 39987, 40038-9, 40120, 40124, 40164

V

Virus diseases see Phytopathological effects ...

Vitamin K_3 see Quinones ...

Volume changes in chloroplast (chromatophore) see Chloroplast and chromatophore
        volume changes

Volume changes in leaf and other organs
        21787, 22187, 22351, 22855, 22920, 22999, 24269, 24801; 25550, 25785, 25999,
        26276, 26612, 26909, 27109, 27432, 27636, 28489: 30665, 30675, 31516; 33111,
        33524, 33985, 35604; 39058

Volume of plant organs, measurement see Leaf volume ...

W

Warburg effect see O_2 and gas exchange

Water, heavy see Deuterium oxide, tritium oxide ...

Water saturation deficit
        21550, 21556, 21713, 21765, 21823, 21861, 22074, 22097, 22118, 22283-4,
        22314-6, 22327, 22348, 22380, 22419, B22572, 22584, 22676, 22752, 22920,
        22937, 23090, 23263, B23264, 23295-6, 23299, B23302, 23307-8, 23330, 23468,
        23587, 23608-9, 23665, 23696, 23813, 23887, 24063, B24271, 24302, 24377,
        24389, 24444, 24537, 24624, 24645, 24775, 24801, 25045, 25140
        25170, 25236, 25251, 25407, 25604, 25828, 25885, 25902, 25943, 26096, 26123,
        26204, 26265, 26296, 26392, 26437, 26494, 26828, 26947, 26993, 27139, 27280,
        27289, 27302, 27339, 27496, 27579, 27625, 27749, 27847, 28300, 28341
        28509, B28531, 28731-2, 28768, 29179, 29203-4, 29394, 29725, 29769, 30069,
        30111, 30272, 30310, 30481, 30861, 31086, 31875, 31899, 31983, 32023, 32187,
        32197
        32713, 33022, 33084, 33261, 33287, 33396, 33398, 33527, 33680, 33840, 33928,
        34120, 34129, 34170, 34181, 34239, 34314, 34747, 35354-5, 35512, 35537,
        35607, 35645, 35815, 35906, 35952, 36013, 36056-7, 36205, 36207, 36431
        36601, 36732, 36753, B36781, 36794, 37416, 37496, 37544, 37681, 37838,
        37897, 38161, 38246, 38399, 38534, B38544, B38722, B38740, 39248, 39501,
        39662, 39935, 40326

Water splitting mechanism see O_2 evolution mechanism and kinetics

Wind (air-flow rate) and CO_2-exchange in water algae 32984

Wind (air-flow rate) and ecosystem and plant productivity
        21633, 21928, 23304, 23346, 24744; 25659, 26199, 28197; 29064, 29550, B29724,
        30979, 31496, 31804, 31827, 31841, 32033; 33546, 33559, 34714, 35594-5,
        36343, 36388; 37439, 37462, 37696, 37995, 38550, B39523, 40288

Wind (air-flow rate) and gas exchange
        22487, 23444, 23727; 26199, 26708, 26745, 27139, 28087; B29724, 31793, 32296;
        36069, 36343; B36781, 38411, B39951

Wind (air-flow rate) and resistances to CO_2 and water vapour transfer
        22584, 22762, 23846; 25451, 25661, 26114, 26199-200, 27749, 28257; 29551,
        29722, B29724, 29855, 31793; 33287, 33559, 34038, 34886, 35303, 35529, 35594-
        -5; 37260

Wind (air-flow rate) and respiration   31793

Wind measurement   see   Aerodynamic methods ...

X

Xanthophylls
        21509, 21564, 21642, 21663, 21678, 21719, 21743, 21747, 21891, 21940, 22086,
        22323, 22465, 22580, 22618-9, 22694, 23115, 23169, 23200, 23259, 23399,
        23471, 23568, 23781, 24092, 24148, 24163, 24317-9, 24413, 24479, 24489,
        24541, 24777, 24892, 24909
        25376, 25424, 25838, 25897, 26225-6, 26370, 26400, 26666-7, 26707, 26740,
        26830, 26975, 27060, 27066, 27110, 27118-9, 27165, 27237, 27288, 27405,
        27527, 27816, 27863-4, 27874, 28060, 28356-7
        28569, 28607, 28651, 28708, 28713-6, 28770, 29172, 29240-1, 29324,
        29396, 29422, 29480, 29614, 30017, 30102, 30207, 30237, 30302, 30356-7, 30452,
        30594, 30619, 30730, 30760, 30802, 30851, 30911, 30923, 31144, 31361, 31403,
        31406, 31414, 31503-4, 31559, 31574, 31583, 31673, 31694, 31742, 31811, 31831,
        31961, 32016, 32038, 32055, 32133, 32142, 32185, 32190, 32284, 32379, 32424,
        32441, 32481
        32595, 32770, 32966, 33030, 33057, 33263-6, 33284, 33336, 33351, 33472,
        33589, 33684, 33795, 33862, 34030, 34163, 34175, 34214, 34399-400, 34457,
        34533, 34581, 34583, 34595-6, 34707, 34752, 34779, 34845, 34869, 34948,
        35048, 35358, 35379, 35421, 35427, 35432, 35449, 35519, 35653, 35696, 35763,
        35844, 35854, 36004, 36154, 36180, 36338, 36447, 36456
        36906, 37000, 37039, 37051, 37203, 37309, 37336-7, 37449, 37480, 37613,
        37620-1, 37647, 37686, 37732, 37734, 37739-41, 37864, 38102, 38110, 38228,
        38230, 38333, 38439, 38443, 38458, 38476, 38479, 38490, 38511, 38561, 38778,
        38819, 38842, 38849, 38904, 38913, 39039, 39246, 39316, 39436, 39477, 39540,
        39588, 39623, 39625, 39901, 39968, 39986, 40164, 40174, 40319

Xanthophylls of algae   .
        21595, 21678, 22166, 22241, 22450-1, 22465, 22538, 22599, 22616, 22767,
        22865, 22926, 22930, 23213, 23393. 23489, 23953, 24233, 24350, 24451, 24709,
        24829, 24978, 25016-7, 25055-6
        25247, 25442, 25586, 25648, 25794-5, 25876, 25895, 25975, 26025, 26127,
        26186, 26219, 26270-1, 26326, 26518, 26639, 26678, 26915, 27511-3, 27743,
        28094, 28191, 28377
        28569, 28698, 28760, 28767, 28770, 29209, 29256, 29268, 29525, 29677, 29802-3,
        29924, 30137, 30193-4, 30238, 30356-7, 30366, 30475, 30582, 30592, 30629,
        31188-90, 31203, 31405, 31430, 31583, 31650, 31810, 31812, 32016, 32037, 32142,
        32190, 32258, 32441-2, 32500-1
        32643, 32966, 33006, 33263, 33266, 33336, 33535, 33780, 33795-6, 33811,
        33834, 33862, 34215, 34380, 34595-6, 34604, 34607, 35380, 35763, 35768,
        36154-5, 36447-8, 36587
        36660, 36906, 37051, 37369, 37522, 37684, 37784, 37789, 37864, 38086, 38102,
        38333, 38476, 38521, 38913, 39477, 39901, 39949, 40054, 40174, 40305, 40363

Xanthophylls of photosynthetic bacteria
        21904, 22101, 22167, 22626, 22679, 22954, 23421, 24056, 24092, 24550; 25380,
        25555-6, 25746, 26328, 26363, 26681, 26819-20, 26915, 26942, 27173, 28094,
        28455; 28551, 28876, 28892, 28947-9, 29025, 29149, 29666, 29828, 29913,
        29947, 29995, 30079, 30134, 30269, 30672, 30788-9, 31349, 31490, 31534, 31741,
        32017, 32149, 32169; 32685, 32924, 33203, 33326, 33707, 33862, 34363-4,
        34603-4, 34687, 34744, 34791, 35066, 35236, 35323, 35386, 35899, 35901,
        36119, 36189, 36484; 36709, 36797, 37181, 37231, 37447, 38239, 38305, 38332-
        -3, 38368-70, 38476, 38821, 39440, 39901

This cumulative plant index contains references to Volumes 6 to 10. The index presents a selection of plant genera and types interesting as experimental material for physiological, ecological and agricultural studies. Latin scientific names of plant genera and English names of plant groups and types are the main items which present the reference numbers. References from the individual volumes are distinguished by a semicolon (;) or a paragraph.

A

Abies   21629, 21633, 21705, 21902, 22009, 22086, 22778, 23044, 23082, 23091, 23430-
-1, 23500, 23818, 23830, 24057, 24336, 24716; B26099, 26348, 26515, 27130,
27432, 27683, 28114, 28353; 29069, 29071, 29415, 29561, 29569, 30262, 30264,
31053, 31090, 31236; 32698, 33422, 34284, 34747, 34806, 35808, B35862, 36049,
B36191, 36201, 36551; 37407, 38285, B38477, 38493, 39077, B39951, 39991,
40360

Acacia   25521, 25592, 27769, 28139; 28987, 29489, 31697, 31769, 31839, 32324, 32394;
35008, 36050, 36445; 36724, 36913; 37049, 37814, 38285, 39917

Acer   21705, 21808-9, 22139, 22165, 22171, 22190-1, 22647, 22734, 22854, 22893,
22973, 23422, 23530, 23551, 23605, 23679, 24336, 24528, 24601, 24716, 24817-8,
25019, 25078; 25186, 25206-7, 25431, 25511, 25540, 25710, 25742, 25896, 25989,
25998-9, B26099, 26534, 26952, 27021, 27035, 27242, 27362, 27730, 28308,
28352, 28470; 28987, 28990, 29206-7, 29212, 29243, 29254, 29320, 29394-5,
29467, 29480, 29614, 29812, 30263, 30460, 30721-2, 30764, 31081, 31090, 31269,
31411, 32111, 32175, 32185, 32188, 32194; 33451, 33617-8, 33936, 33984, 34480,
34504-5, 34924, 35269, 35525, 35691, 35808, 35949, 36057, 36190, B36191,
36241, 36403; 36593, 36809, 36818, 37220-1, 37407, 37413-5, 37603, 37693,
37695, 37874, 38140, 38285, B38477, 38679, 38920, 39002-4, 39821, 39971

Acetabularia
21598-601, 21785, 21890, 22089, 22280, 22417, 23018, 23230, 23626-7, 23697,
B23767, 23969, 24353, 24865-8; 25228, 25598, 26340, 26583-5, 27023, 27155,
27233, 27646, 28453; 28574-7, 28579, 28609-10, 29649, 30020, 30667, 30701,
30923-4, 31847, 32240-2; 32920, 33762, 33774, 34396, 34578, 34917, 34948,
35211, 35356, 35400, 36045; 36730, 37006, 37162, 37281, 37656, 37717, 37925,
37973, 38467, 38545-6, 38573, 39049-50, 39197, 39970, 40023-4, 40314

Aesculus
24817-8; B26099, 28362; 29249, 29614, 29698, 31238, 32175; 33595, 34246,
35006, 35269, 36139, 36190, B36191; 39971, 39991

Agave   22316, 22508, 24872; 25445-6, 25926, 26303, 27271; 29616, 31066, 31069; 33440-
-1, 34649, 35097, 35442; 37177, 37399, 38083, 38185, 38285, 38915, 39830,
B39903, 40358

Alder   see *Alnus*

Alfalfa   see *Medicago*

Algae (*cf.* also *Acetabularia*, A. blue-green, A. brown, A. green, A. red, *Anabaena*,
Anacystis, *Ankistrodesmus*, *Chlamydomonas*, *Chlorella*, *Chrysophyta*, Diatoms,
Dinoflagellatae, *Dunaliella*, *Euglena*, *Nostoc*, *Porphyridium*, *Scenedesmus*, *Ulva*)
21546, 21575, 21595, 21681, 21707, 21726, 21772, 21776, 21805, 21829, 21883,
21910, 21925, 21952, 22048, 22121, 22238, 22257, 22268-9, 22292-4, 22357,
22387, 22389, 22450, 22454, 22458, 22493-4, 22538, 22545, 22592, 22599, 22628,
22634, 22644, 22654, 22743, 22745-6, 22756, 22763, 22791, 22807, 22846-7,
22867, 22898, 22921, 22929, 22943, 22963, 22976, 23016-7, 23088, 23091, 23096,

(continued)

Algae (continued)

(continued)

Algae (continued)
        38145, 38169, 38171, 38202, 38248, 38270, 38277, 38306-8, 38318, 38344, 38378,
        38402, 38409, 38417, 38437, 38454, 38473, 38476, 38514, 38520, 38524, 38554,
        38559, 38585, 38667, 38686, 38692, 38694, 38752, 38755, 38782, 38794, 38802,
        38807, 38826, 38864, 38932, 38973, 39018, 39048, 39055, 39060, 39083, 39096,
        39108, 39122, 39155, 39177, 39208-11, 39255-7, 39290-1, 39308, 39314, 39317,
        39327, 39342-3, 39346, 39370, 39378, 39412, 39414, 39418, 39455-6, 39509,
        39576, 39633-4, 39658, 39668, 39696, 39698, 39714-5, 39722, 39744, B39759,
        39781, 39787, 39789, 39794, 39803, 39823, 39825-6, 39836, 39844, 39854, 39859,
        39871-2, 39896, 39901, 39908, 39919, 39937, 39939, 39942, 39959, 39969, 39996,
        40019, 40065, 40084, 40095, 40116, 40168, 40172, 40174-5, 40195, 40269, 40305,
        40344

Algae, blue-green (*cf.* also *Anabaena, Anacystis, Nostoc*)
        21547, 21562, 21572, 21575, 21617, 21620, 21628, 21682, 21774, 21805, 21830,
        21836, 21846, 21908, 21911-2, 21920, 21925, 21981, 22054, 22104-5, 22121,
        22155, 22225, 22294, 22349-50, 22363, 22442, 22453-4, 22461, 22464, 22490,
        22497, 22515, 22517, 22521, 22538, 22546, 22592, 22637, 22642, 22644, 22658-
        -60, 22664-5, 22693, 22745-6, 22763, 22767, 22803, 22853, 22867, 22876, 22961,
        23010, 23083, 23097-8, 23107, 23110-1, 23128, 23153, 23165, 23178, 23268,
        23275, 23322, 23383-4, B23407, 23434, 23453, 23456, 23474, 23505, 23546,
        23606, 23728-30, 23750, 23788, 23824, 23843, 23868, 23870, 23914, 23931-2,
        23954, 23969, 24061, 24072, 24118-9, 24134, 24154, 24169, 24200, 24212,
        B24250, 24254, 24258, 24273-4, 24308, 24345, 24421, 24434, 24450, 24472,
        24498, 24518-20, 24550, 24572, 24579, 24586-7, 24589, 24611-2, 24618, 24644,
        24648, 24695, 24707, 24722-4, 24756, 24767, 24809-10, 24816, 24843, 24909,
        24929, 24944-5, 24960, 24964-5, 24982, 24998, 25005, 25033, 25112
        25163, 25165, 25248, 25270, 25280, 25286, 25345, 25361, 25391, 25394, 25402,
        25425, 25439, 25447, 25495, 25509, 25548, 25585, 25587, 25601, 25624, 25635,
        25651, 25665, 25699, 25743, 25759-60, 25770, 25847, 25849, 25875, 25883,
        25958, 25963, 25996, 26006, 26016, 26048, 26058, 26072, 26130, 26148, 26158-
        -60, 26171, 26179, 26186, 26203, 26207, 26219, B26220, 26242, 26250, 26254-5,
        26268, 26271, 26288, 26305-6, 26340, 26356, 26380-1, 26407, 26457, 26478,
        26520, 26528, 26570-1, 26576, 26697, 26718-9, 26725, 26805, 26823, 26916,
        26978, 26994, 27033, 27086, 27100, 27104, 27122, 27151-2, 27206, 27223, 27226,
        27252, 27281, 27297, 27299-300, 27312, 27393, 27418, 27426-7, 27440, 27453,
        27490, 27534, 27548, 27566, 27572, 27616, 27642, 27708, 27763, 27766-7,
        27823, 27825, 27827, 27858, 27898, 27911, 27940, 27943-4, 27958-60, 27964,
        27986, 28004, 28017, 28026, 28072, 28079, 28094, 28111, 28126, 28142, 28151,
        28159, 28173-4, 28184, 28190, 28198, 28234, 28267, 28331, 28354, 28412
        28513, 28535, 28568-9, 28592, 28631, 28720, 28737, 28750, 28760, 28795, 28853,
        28888, 28936, 28951, 28960, 28966-7, 28979, 28982, 29014, 29021, 29127, 29143-
        6, 29151, 29152, 29160, 29347, 29371, 29374, 29378, 29438, 29448, 29498,
        29525-6, 29528-9, 29552, 29556, 29589, 29601, 29610, 29670-2, 29683, 29691,
        29729, 29782, 29802-3, 29812, 29818, 29836, 29853, 29866, 29873, 29918, 29937-
        8, 29950, 29956, 30032, 30051, 30059, 30075, 30169, 30185, 30196, 30231,
        30248, 30311, 30365, 30381, 30401, 30407, 30421, 30442-3, 30470, 30472, 30500,
        30502, 30539, 30552, 30592, 30606, 30622, 30624, 30696, 30718, 30746, 30768,
        30772-3, 30799, 30813, 30863-5, 30872, 30971, 30974, 30983, 31011, 31031,
        31092, 31100, 31136, 31141, 31145-6, 31177, 31184, 31203, 31212, 31215, 31278,
        31327-8, 31360, 31391, 31402, 31412, 31415, 31437, 31481, 31521, 31603, 31638,
        31648, 31650, 31663, 31758, 31772, 31909, 31912, 31937, 31960, 31963-5, 31988,
        31992, 32007, 32021, 32053-4, 32069, 32071, 32102, 32113, 32157, 32205, 32305,
        32356, 32368, 32401, 32409-10, 32441, 32448, 32454, 32464, 32470
        32564-5, 32599, 32608, 32614, 32621, 32669, 32679-81, 32767, 32775, 32789,
        32794, 32803, 32911, 32937-8, 32979, 32993, 33070, 33083, 33133-4, 33148,
        33163, 33169, 33183, 33218, 33266, 33294, 33333, 33399, 33420, 33422, 33442,
        33446, 33464, 33482-3, 33521, 33535, 33568-9, 33584, 33650, 33656-7, 33749,
        33757-8, 33780, 33811, 33904, 33923, 33927, 34004, 34021, 34030-1, 34081,
        34197, 34199, 34216, 34222, B34248, 34329, 34382, 34394-5, 34418, 34430,
        34434, 34466, 34555, 34604, 34626, 34629, 34634, 34660, 34669, 34696, 34698,
        34717, 34726, 34768, 34780, 34798, 34832-3, 34881, 34885, 34893, 34902,
        34904, 34971, 34976, 34991-4, 35015, 35043, 35058, 35075, 35078, 35138,
        35147, 35167, 35182, 35195, 35206, 35209, 35244, 35306, 35323, 35328-9,
                                                                (continued)

Algae, blue-green (continued)
          35378, 35452, 35456, 35502, 35584, 35587, 35654, 35663, 35684, 35688, 35696,
          35699, 35706, 35763, 35811, 35935, 35965, 35974, 36002, 36028, 36054, 36071,
          36110, 36143, 36179-82, 36248, 36251, 36279, 36294, 36336-7, 36369, 36375,
          36446-7, 36455, 36466, 36490, 36505-7
          36610, 36624, 36633-5, 36651, 36731, 36771, 36836, 36843, 36848, 36858, 36891,
          36897, 36902, 36928, 36965, 37030, 37040, 37081, 37097, 37179, 37195, 37310,
          37329, 37453-4, 37466, 37478, 37483, 37487, 37524, 37562, 37581, 37601,
          37619, 37622, 37651, 37673, 37764, 37769, 37784, 37788-9, 37849, 37868,
          37883, 37885, 37927, 38070, 38086, 38102, 38138, 38160, 38212-3, 38248,
          38263, 38271, 38304, 38325, 38336, 38347-8, 38352, 38359, 38447, 38458-9,
          B38475, 38476, 38499, 38507, 38514, 38548, 38554, 38593, 38628, 38630, 38659,
          38693, 38761-3, 38782, 38945, 38947-8, 38972, 38982, 38985, 39017, 39024,
          39046, 39101, 39183, 39198, 39210, 39308, 39326, 39332, 39344, 39346-7,
          39359, 39361, 39385, 39476-7, 39555, 39573, 39608, 39610, 39685, 39744,
          39757, 39766, 39769, 39774, 39776, 39834, 39854, 39876, 39901, 39910, 39942,
          40034-6, 40065, 40153, 40155, 40157, 40163, 40174, 40178, 40283-4, 40305,
          40327, 40369

Algae, brown
          21562, 21903, 21945, 22166, 22256, 22258, 22261, 22451, 22462, 22538,
          22607, 22644, 22660, 22687, 22767, 22803, 22927, 23040, 23074, 23211, 23213,
          23335, B23407, 23475, 23534, 23606, 23728, 23827, 23941, 24073, 24154, 24169,
          24409, 24728-9, 24816, 24957, 24984, 25034, 25127
          25179, 25317, 25447-8, 25452, 25471, 25549, 25602, 25676, 25778, 25794,
          25883, 25975-6, 26025, 26038, 26063, 26186, 26203, 26313, 26521, 26549-50,
          26642, 26644-5, 26661-2, 26950, 27114, 27279, 27321, 27418, 27476, 27752,
          27763, 27921, 28017, 28129, 28142, 28147, 28151, 28325
          28601, 28698-9, 28771, 28853, 28940, 28981, 29178, 29336-7, 29460,
          29525, 29655, 29775, 29825, 29864, 29893, 29938, 30151, 30173, 30314, 30435,
          30592, 30616, 30627, 30968, 31047, 31062, 31084, 31132-3, 31289, 31430, 31675,
          31810, 32173, 32348, 32356, 32472, 32517, 32544
          32679-81, 32859, 32911, 33011, 33072, 33121, 33128, 33147, 33243, 33266,
          33312, 33384-5, 33813, 33915, 33922, 34016, 34058, 34143, 34251, 34258,
          34468, 34473, 34594, 34596, 34604, 34626, 34629, 34637, 34638, 34669, 34741,
          34824, 34903, 34912, 34914, 34955, 34979, 35069, 35109, 35119, 35585, 35713,
          35729, 35735, 35756, 35804, B35862, 36098, 36149, 36152, 36334, 36585
          36663, 36889, 36992, 37043-4, 37142, 37336, 37351, 37659, 37784, 37797-8,
          37943, 38112, 38178, 38327, 38476, 38499, 39181, 39211, 39233, 39480-1,
          39525, 39880, 39901, 40073, 40174, 40210-1

Algae, green (cf. also Acetabularia, Ankistrodesmus, Chlamydomonas, Chlorella, Duna-
          liella, Scenedesmus, Ulva)
          21584, 21610, 21620, 21737, 21755-6, 21767, 21769, 21805-6, 21871, 21896,
          21905-7, 21925, 21945, 21978, 22121, 22256, 22277, 22294, 22305, 22337, 22361,
          22364, 22442, 22450-1, 22453-4, 22462, 22515, 22538, 22548, 22578, 22591-2,
          22642, 22644, 22659-60, 22702-3, 22717, 22745-6, 22757-8, 22764, 22767, 22790,
          22794, 22803, 22828, 22846, 22866, 22928, 22965, 22991, 23040, 23042, 23097-8,
          23107, 23165, 23212, 23229, 23366, 23456, 23474, 23485, 23534, 23537, 23554,
          23567, 23596, 23606, 23685, 23728, 23824, 23847, 23868, 23876, 23931,
          23987, 24029, 24053, 24072, 24169-72, B24250, 24273, 24304, 24350, 24365-6,
          24409, 24420, 24456, 24472, 24498, 24500, 24572, 24579, 24589, 24608, 24611,
          24613, 24617-8, 24683, 24713, 24723, 24735, 24743, 24749, 24770, 24808, 24909,
          24929, 24944, 24957, 24978, 25003, 25005-6, 25069, 25080, 25112, 25127
          25179, 25286, 25306, 25317, 25333, 25345-6, 25391, 25447-8, 25459, 25471,
          25486, 25490, 25500-3, 25553, 25598, 25602, 25624-5, 25633, 25676, 25743,
          25770, 25794, 25847, 25849, 25876, 25882-3, 25933, 25936, 25960, 26035,
          26038, 26051, 26063, 26111, 26119, 26130, 26162, 26179, 26186, 26203, 26206,
          B26220, 26271, 26313, 26321-4, 26344, 26346, 26356, 26407, 26429, 26435,
          26437, 26504, 26518, 26549-50, B26551, 26570, 26643, 26645, 26662, 26683,
          26752, 26771, 26797, 26807, 26823, 26826, 26838, 26873, 26889, 26896, 26962,
          26989, 26994, 27051, 27057-8, 27086, 27206, 27222-3, 27226-7, 27281, 27310,
          27312, 27316, 27321, 27349-50, 27357, 27367, 27377, 27399, 27418, 27438-9,
          27476, 27534, 27567-9, 27588-92, 27642, 27652, 27700, 27727, 27743, 27756-7,
                                                                    (continued)

Allium 22484, 22508, 22719, 22747, 24527, 24716, 25011; 25511, 25930, B26099, 26151,
26476, 28075, 28102, 28139, 28267; 28743, 30267, 30995, 31568-9, 31854, 32029,
32521; 33730, 35006, 35426, 35840, 35912, 36054, 36308; 36994, 37241, 38285,
38421, 38540, 38685, 39077, 39259, 39492, 39678, 39761, 39773, B39903, 39991,
40339

Almond   see *Amygdalus*

Alnus   21705, 22086, 22333, 22685, 24336, 24765, 25009; 25569, 25592, 25742; 28987-
-8, 29467, 30263; 33422, 35006; 36637, 37031, 37049, 37259, 37688, 38285,
B39951

Aloe   29217, 30174; 34480, 34482, 35442; 37177, 38088, 38237, 38784, 39830, 40358

Alpine plants
23167, 23303-4, 23545, 25015; 25659-61, 26854, 27107, 28358; 28714-6, B29063,
29064-5, 29685, 29721-2, 30392-4, 30523-4, 30943, 31392, 31670, 32415; 33547,
34747, 35129, 35403, 36333, 36535; 36656, 36839, 37049, 37204, 37714, 38414,
38875, 39006, 39678, 40318

Amaranthus
22033, 22183, 22249, 22328, 22489, 22620, 22698, 22768, 22863, 22958, 23032-
-3, 23694, 23714, B23767, 23977, 24546, 24716; 25408, 25420, 25431, 26315,
26360, 26398, 26403, 26469-70, 26475, 26539, 27414, 27506, 27530, 27585,
28005, 28368; 28875, 29085, 29231-2, 29489, 29681-2, 29862, 29987, 30078,
30498, 30951, 31348, 31393, 31417-21, 32324, 32547; 32766, 32987, 32996, 33080,.
33388, 33914, 34254, 34256, 34513, 35186, 35263, 35403, 35427-8, 35430-3;
36711, 36871, 36956, 37022, 37084, 37331, 37344, 37392, 37506, 37827, 37863,
37904, 38082, 38121, 38185, 38260-1, 38385, B38477, 38990, 39107, 39123,
39142, 39236, 39239, 39259, 39809, 40360

Amygdalus 38285, 38607

Anabaena 21519, 21589, 22305-6, 22478, 22561, 22665, 23043, 23058, 23110-1, 23724,
23884, 24013, 24251, 24293, 24405, 24456, 24703
25286, 25508, 25579, 25629, 26084, 26179, 26472-3, 26527, 26544, 26575, 26608,
26866, 27258, 27440, 27572, 28072, 28105, 28354, 28446
28528, 28541, 28631,   28889, 28960, 29014, 29145, 29211, 29239, 29371,
29579, 29672, 29683-4, 29729, 29873, 30135, 30321, 30401, 30444, 30877-8,
30970, 31092, 31108, 31136, 31176, 31188-90, 31328, 31423, 31590, 32035,
32054, 32070, 32072, 32082, 32283, 32448, 32470
32599-600, 32656,   32681, 32715-6, 32774, 32859, 32948, 33027, 33062,
33070, 33083, 33133, 33169, 33200-1, 33218, 33422, 33442, 33535, 33610,
33656, 33718, 33749, 33833, 34051, 34161-2, 34197, 34277, 34405, 34418,
34555, 34629, 34798, 34967, 34992-4, 35015, 35043, 35078, 35148, 35216,
35315, 35477-8, 35587, 35761, 35780, B35862, 36033, 36096, 36099, 36182,
36209, 36455
36624, 36634-5, 36657, 36808, 36836, 36858, 36897, 36982, 37022, 37080, 37252,
37310, 37329, 37402, 37466, 37601, 37633, 37651, 37764, 37868, 37885, 37923,
37927, 38070, 38150, 38159, 38296, 38325, 38393, 38434, 38451, 38548, 38628,
38762-3, 38782, 38899, 38929, 38955, 39025, 39101, 39110, 39276-7, 39292,
39308, 39345-7, 39371, 39400, 39476, 39610, 39685, 39757, 39776, 39790-2,
39834, 39885, 39931, 39994

Anacardium 31004

Anacystis 21859, 21865, 21889, 22274, 22407, 22558, 22609, 22859, 23187, 23390,
23730, 23969, 24011, 24073, 24447, 24495-6, 24930, 24966
25197, 25493-4, 25532, 25586, 25866-8, 25980-1, 26168, 26575, 26682, 26948,
26957, 27048, 27156, 27257, 27259, 28008, 26072, 28075, 28213
28535, 28631,   28879, 28917, 29026, 29194-6, 29529, 29672, 29680, 29683,
30299, 30322-4, 30531, 30539, 30564, 30845, 30926, 30970, 31051, 31092,
31205-6, 31275, 31437-8, 31679, 31828, 31904, 32357-9, 32362, 32464-5

(continued)

Anacystis (continued)
        32668, 32714,   32792, 32859, 32988, 33218, 33233-4, 33257-8, 33360, 33442,
        33483, 33679, 33880, 33883, 34344, 34559-61, 34628-9, 34745, 34998-9, 35015,
        35177, 35242, 35307-8, 35755, 35890, 35944, 36065, 36273, 36279, 36365,
        36429, 36455, 36504
        36612, 36621, 36653-4, 36663, 36706, 36897, 36975-7, 37057, 37210, 37248,
        37320, 37448, 37452, 37524-7, 37566, 37601, 37652, 37676, 37705, 37786-7,
        38102, 38154, 38160, 38220, 38325, 38386, 38419, 38447, 38453, 38502, 38527,
        38554, 38667-9, 38688, 38757, 38862, 38978, 38982, 39103-5, 39457, 39497,
        39499, 39553, 39574, 39645, 39648, 39744, 39829, 39880, 39931

Ananas 22098, 22152, 23786; 25327, 25446, 25797, 27182, 27506, 28308; 28703, 29215-
        -7, 29489, 30927-8, 30994, 31077; 33547, 34649, 34950; 37177, 37317, 37564,
        38185, 38285, 38629, 39678, 39725, 39889

Ankistrodesmus
        21865, 22324, 22550, B23407, 24273, 24579; 25286, 26147, 27591; 28764, 28982,
        29185, 29945, 30864, 31031, 31184, 31218, 31909, 32206, 32215, 32448; 32814,
        33070, 33721, 34555, 34994, 35043, 35216, 35587, 35654, 35819, 36182, 36455;
        36635, 37329, 37868, 38050, 38070, 38708, 38782, 39101, 39346, 40005

Antirrhinum
        22148-9, 23129, 23624, 24135-6, 24351; 25891, 26200, 26700, 27114, 28139;
        29186, 30361-2, 31413, 31662, 31664-5, 32211; 32948, 33967, 34372, 34850,
        34882, 35420-1, 35698-700; 36963, 37770-1, 38459, B36475

Apium 27673; 31895; 36255; 39678

Apple see *Malus*

Apricot see *Armeniaca*

Aquatic macrophytes (*cf.* also *Elodea, Phragmites, Typha*)
        21847, 21973, 22000, 22297, 22305, 22731, 22625, 23188-9, 23476, 23596, 23632,
        23646, 23683, 23814, 23819, 23859, B24024, 24099, 24303, 24440, 24477, 24700,
        24773, 24969, 24979, 25003, 25005, 25103, 25130, 25149-50
        25210, 25253, 25282, 25374, 25436, 25518, 25523, 25826, 25878, 26038, B26099,
        26417, 26549, 26724, 26767, 26824, 26888, 27013, 27025, 27060-1, 27310,
        27371, 27439, 27458, 27470, 27611, 27671, 27710, 27741,   27773-4, 27989,
        28229-30, 28303, 28359
        28567, 28687, 28711, 28748, 28837, 28901, 28975, 28982, 29018-9, 29049, 29175,
        29193, 29277, 29362, 29469, 29605, 29675, 29764, 29880, 29989, 30339, 30490,
        30623, 30645, 30746, 30843, 31039, B31140, 31239, 31259-60, 31301-2, 31307,
        31328, 31609, 31766, 32383, 32403, 32518
        32570, 32643, 32645, 32752, 32803-4, 32830-1, 32953, 33364, 33379, 33412,
        33422, 33428, 33570, 33624, 33675, 33694, 33703, 33759, 33804, 33852, 33893,
        34017, 34030-1, 34035, 34052-3, 34213, 34216, 34304, 34351, 34387, 34421,
        34472, 34486-7, 34621, 34672-3, 34715, 34915-6, 34924, 35048, 35099, 35113,
        35160, 35168-9, 35171-4, 35244, 35298, 35334, 35401, 35425, 35597, 35607,
        35643, 35727, 35797, 36103, 36113, 36386-7, 36441, 36553
        36657, 36708, 36723, 36775, 36876, 36938, 36967, 37023-5, 37054, 37106,
        37348, 37543, 37619, 37713, 37780, 37889, 37924, 38040, 38068, 38079, 38085,
        38414, 38445, 38579, 38715, 38818, 38842, 39073, 39188-9, 39262, 39416,
        39527-9, B39537, 39716-8, B39759, 39768, 39886, 39900, 39932-4, 39976, 40027,
        40170, 40172, 40176, 40198, 40258, 40344, 40358

Arabidopsis
        21870, 22204-5, 22432, B23767, 23845, 24847-8; 25265, 27268, 28139; 29283,
        32221-2, 32481; 36590, 39558, 39712-3, 40009, 40300, B40331

Arachis 21550, 22352, 22685, 23039, 24424, 24639, 24787, 24820, 24861, 24971,
        25039; 25184, 25404-7, 25445, 25965, 26093, 26476, 26750, 28038, 28096,
        28228, 28347, 28392; 28872, 29334, 29603, 31038, 31193, 31262, 32034, 32236,
        32253; 32686, 32765, 32967, 33394, 33871-3, 34847, 35103, 35222, 35286, 35744;
        37298, 37449, 37807, 38285, 39037, 39069, 39259, B39537, B39903, 40304

Arbor vitae see *Thuja*

Armeniaca
       22747, 23297; 28796, 29443, 32526, 32539; 33547, 34274, 34649, 35481; 38401,
       38870, 39867

Armoracia 28267; 32948

Artichoke see *Cynara*

Ash see *Fraxinus*

Asparagus 22252, 22747; 27110; 31833; 35426, 36445; 37189

Aspen see *Populus*

Atriplex
       21652, 21748, 21773-4, 21810-4, 21825, 22033, 22156, 22249, 22318, 22328,
       22484, 22544, 22698, 22983, 23033, 23113, 23692, 23714, B23767, 24657-8,
       24804, 24806, 25018
       25360, 25431, 25437-8, 25457, 25521, 25538, 25732, 25807, 25835, 25923, 25944-
       -5, 26155, 26402, 26437, 26539, 26719, 26872, 27179, 27359, 27929, 28288,
       28489
       28645, 28774, 28818, 29015-6, 29085, 29330, 29383, 29787-8, 29862, 29865,
       30951, 31070, 31240-1, 31464, 31690-1, 31769-70, 31993, 32462, 32477
       32766, 32828, 32867, 32889, 32996, 33062, 33304-6, 33440, 33547, 33551, 33696,
       33843, 33914, 34013, 34219, 34256, 34659, 34717, 34926, 35186, 35273, 35286,
       35425, 36445
       36678, 36853, 37022, 37084, 37161, 37268, 37331, 37701, 37767, 37827, 37858,
       37863, 38125, 38285, 38385, B38477, 38519, 38785, 39093, 39240, 39242, 39275,
       39424, 39524, 40018, 40300, 40360

Avena 21916, 21948, 22028, 22068, 22107, 22130, 22156, 22348, 22566, 22611, 22615,
       22672, 22700, 22719, 23063, 23157, 23355, 23395-6, 23482-3, 23486, 23524,
       23586, 23588, 23716-7, 23788, 23859, 24001, 24199, 24284, 24302, 24424, 24508,
       24539, 24564, 24716, 24725, 24731, 24750, 24792-3, 24861, 24870, 24907, 24940,
       24993-7, 25136
       25340, 25431, 25516, 25545, 25624, 25697, 25729, 25745, 25874, 25877, 25961,
       26003, 26073, 26096, 26277, 26279-80, 26476, 26536, 26539, 26605, 26688,
       26919, 26947, 26991, 27239, 27287, 27439, 27530, 27885, 27931, 27997, 28032,
       28086, 28116, 28139, 28157, 28228, 28235, 28267, 28360-1, 28406
       28709, 28743, 28836, 28941, 28973, 29107, 29113, 29120, 29123, 29289, 29363,
       29461, 29471, 29478, 29813, 29849-50, 30077, 30140, 30267, 30668, 30690,
       31277, 31323, 31393, 31607-8, 31669, 31682, 31768-9, 31952, 31977, 32060-1,
       32092, 32387-90, 32402, 32440
       32765, 32769, 32847, 32956, 32973, 33007, 33062, 33158-9, B33343, 33375,
       33388, 33849-50, 33871-2, 33951, 33981, 34129, 34140, 34225, 34295, 34685,
       34815, 34924, 35142, 35410, 35555, 35590, 35614, 35711, 35925, 36019, 36067-
       -8, 36224, 36299, 36370, 36440, 36534
       36896, 37060, 37236, 37298, 37336, 37342, 37344, 37512, 37585, 37701, 37733,
       37758, 37793, 38026, 38120-1, 38183, 38234, 38421, 38450, 38540, 38547,
       38561, 38666, 38785, 39259, 39376, B39537, 39562, 39678, 39701, 39764, 39890-
       -3, B39903, 39991, 40125, 40143, 40164, 40250, 40360

Avocado see *Persea*

B

Bacteria, photosynthetic (*cf.* also *Chlorobium, Chromatium, Halobacterium, Rhodopseu-
       domonas, Rhodospirillum*)

(continued)

Bacteria, photosynthetic (continued)
        21690, 21703, 21815, 21829, 21843, 21912, 21938, 22020-1, 22055, 22090-1,
        22154, 22167, 22188, 22267-8, 22390, 22458, 22499, 22515, 22580, 22644, 22658-
        -60, 22679, 22695, 22767, 22903, 22913, 22931, 23020, 23022, 23120, 23124,
        23142-3, 23161, 23203, 23312, 23347, 23391, B23407, 23434, 23437, 23474, 23484
        23501, 23525, 23615, 23667, 23870, 23914, 23965-6, 24031, 24098, 24189, 24227,
        24306, 24327, 24367, 24409, 24454, 24459, 24671, 24682-3, 24723, 24784, 24814,
        24955, 25051, 25057, 25067, 25119
        25263, 25380, 25478, 25488, 25490, 25495, 25554, 25582, 25585, 25587, 25606,
        25614, 25628, 25631, 25669, 25733-4, 25770, 25773, 25816, 25835, 25855,
        25888, 25957, B26220, 26247, 26394, 26404, 26410, 26504, 26539, 26609, 26687,
        26727, 26805, 26915, 26926, 26942-3, 26974, 26978, 26985, 26994, 27068,
        27151-2, 27173, 27297, 27340-2, 27374, 27419, 27441, 27449, 27509, 27519,
        27531, 27535, 27537, 27615, 27627, 27631, 27822, 27841-2, 27867, 27945,
        28071, 28163, 28178, 28335, 28409, 28451, 28466, 28484
        28506, 28642-3, 28700, 28719, 28729, 28750, 28760, 28824, 28876, 28888, 28948,
        28951-2, 29014, 29025-6, 29035, 29073, 29098, 29124, 29126, 29134, 29149,
        29174, 29188, 29199, 29230, 29265-6, 29358, 29371, 29439, 29452, 29483, 29525,
        29618, 29683, 29748-9, 29779, 29796, 29798, 29818, 29896, 29973, 30066, 30079,
        30116, 30208, 30227, 30248, 30382, 30399, 30535, 30625, 30635, 30658, 30696-8,
        30702, 30718, 30762, 30780, 30934, 31072-4, 31138, 31191, 31219, 31253,
        31274, 31290, 31308, 31360, 31369, 31386-7, 31398, 31415, 31437, 31455,
        31458, 31624, 31730, 31741, 31789, 31810, 31929, 32114, 32251, 32282, 32409
        32602, 32614, 32619, 32623, 32679-81, 32760, 32764, 32779, 32803, 32859,
        33063, 33114, 33185, 33190, B33191, 33203, 33253, 33294, 33300, 33355,
        33380, 33403, 33406, 33473, 33494, 33509, 33514, 33552, 33612, 33688, 33702,
        33710-1, 33807, 33833, 33863, 34009, 34020, 34023, 34131, 34133, 34237,
        34270, 34281, 34296-7, 34362-4, 34393, 34435, 34460, 34469, 34583, 34603,
        34659, 34669, 34686-7, 34726, 34762, 34772, 34795, 34817, 34841, 34863,
        34902, 34957, 34968, 35078, 35166-7, 35238-9, 35274, 35306, 35311, 35333,
        35378, 35386-7, 35456, 35475, 35501, 35531, 35663, 35686, 35702, 35729,
        35855, 35868, 35900-1, 35937, 36001-2, 36118-9, 36188-9, 36247, 36399-400,
        36471, 36513, 36536, 36546, 36578
        36624, 36649-50, 36771-2, 36797, 36860, 36944, 37004, 37033, 37069, 37079,
        37180, 37206, 37231, 37248, 37446, 37632, 37660-3, 37690, 37760, 37914,
        38095, 38189, 38321, 38332, 38336, 38374, B38415, B38475, 38476, 38571,
        38648-9, 38665, 38706, 38811, 38906, 38972, 39204-5, 39222, 39255, 39290,
        39322, 39443, 39468, 39476, 39581, 39657, 39822, 39901, 39909, 39961, 40012,
        40157, 40247, 40279

Bamboo see *Bambussa*

Bambussa B24024; 27506

Banana see *Musa*

Barley see *Hordeum*

Bean see *Phaseolus*

Beech see *Fagus*

Bermuda grass see *Cynodon*

Beta   21563,21676, 21765, 21779, 21782, 21841, 21847, 22007, 22028, 22066, 22108,
        22123, 22171, 22245, 22318, 22345, 22347, 22419, 22462, 22487, 22565, 22647,
        22651, 22666, 22711, 22719, 22721, 22775-6, 22806, 22829, 23002-3, 23170,
        23221, 23252, 23259, 23330, 23448, 23461, 23549, 23588, 23625, 23648-9, 23714,
        23726, 23737, B23767, 23773, 23814, 23887, 23925, 23990, 24033, 24196, 24204,
        24442, 24468, 24504, 24537, 24552, 24557, 24643, 24683, 24716, 24730, 24801,
        24812, 24839, 24894, 24840, 25137-8, 25158
        25176, 25226, 25331, 25360, 25431, 25457, 25521, 25568, 25571, 25585, 25656,
        25693, 25709-10, 25714, 25723, 25772, 25872, 25877, 25935, 25940-1, 25948,
                                                                              (continued)

Beta (continued)
26039, 26063, 26121-2, 26151, 26209, 26235, 26237, 26260, 26267, 26281, 26360,
26419, 26471, 26476, 26539, 26605, 26733, 26767, 26783, 26816-7, 26954, 26999,
27010, 27133, 27180, 27327, 27466, 27478-9, 27592, 27673, 27692, 27715, 27748,
27779, 27918, 27998-9, 28039, 28112-3, 28139, 28252, 28315, 28432, 28478,
28489
28497, 28503, 28521, 28527, 28743, 28751, 28780, 28809, 28838; 28910, 28957,
29002, 29009, 29018, 29053-4, 29090, 29321, 29555, 29559, 29644-5, 29746,
29766, 29932, 29964, 29975, 30076, 30132, 30150, 30203, 30412-3, 30663,
30692, 30801, 30842, 30861, 30896, 30917, 30951, 31086, 31187, 31249, 31319,
31357-8, 31411, 31494, 31547, 31554, 31633, 31782, 31800, 31825, 31849,
31859, 31894-5, 31984, 32008-10, 32069, 32079-80, 32122, 32322, 32538
32573, 32699, 32701, 32865, 32874, 32888, 32948, 32987, 33062, 33310, 33398,
33562, 33590-1, 33650, 33658, 33671, 33808, 34012, 34182, 34412, 34506,
34545, 34625, 34650, 34670, 34735, 34857, 34875, 34940-1, 35006, 35104,
35155, 35159, 35341, 35488, 35505, 35528, B35862, 35871, 35933, 36032, 36058,
36224, 36357, 36375, 36431
36616, 36624, 36811, 36956, 37022, 37192, 37333, 37344, 37520, 37618, 37701,
37710, 37729, 37752, 37754, 37863, 37866, 37895, 37953, 38026, B38139, 38161,
38185, 38262, 38285, 38294, 38317, 38363, 38366, 38419, B38475, 38520, 38701,
38725, 38953, 38962, 39029-31, 39183, 39190, 39202, 39218, 39377, 39461,
B39523, 39632, 39638, 39672, 39678, 39680, 39820, 39881, 39884, B39903, 40009,
40250, 40303, 40306, 40344-5, 40358

Betula 21705, 21760-1, 22686, 22734, 22893, 22948, 23257, 23500, 23521, 23666, 23679,
24057, 24240, 24601, 24739, 24817-8, 24855, 25018, 25025; 25710, 25998-9,
B26099, 27021, 27060, 27484, 28160, 28470; 29212, 29467, 30262, 30476, 30721-
-2, 30760, 31090, 32081, 32094, 32175, 32415; 32891, 33264, 33559, 34101,
34471, 34747, 35006, 35014, 35194, 35269, 35358, 36190, B36191, 36408; 36593,
36818, 37220-1, 37407, 38209, 38285, 38442, 39159, 39920, B39951, 39971

Birch see *Betula*

Blackberry see *Rubus*

Blueberry see *Vaccinium*

Bluegrass see *Poa*

Brassica
21841, 21921, 22123, 22647, 22660, 22741, 22747, 22797, 23101, 23223, 23356-
-7, 23457, 23526, 23577, 23580, 23691, 23798, 23876, 23920, 23952, 23972,
24092, 24285, 24291, 24468, 24564, 24666, 24716, 24842, 24925, 25158
25511, 25772, 25812, 25935, 25982, 26151, 26267, 26355, 26360, 26476, 26814,
27016, 27024, 27041, 27222, 27265, 27300, 27312, 27439, 27561-2, 27673, 28005,
28228, 28315, B28482, 28487
28527, 28926, 29555, 29603, 29743, 29979, 30080, 30152, 30206, 30276-7, 30451,
B30477, 30711, 30731, 30995, 31160, 31257, 31567, 31586, 31616, 31657, 31895,
31908, 31983-4, 31992, 32106, 32122, 32211, 32304
32765, 32948, 32956, 32965, 33059, 33184, 33348, B33510, 33565, 33604, 33643,
33790, 33808, 33958, 34159, 34312, 34448, 34712, 34835, 34924, 35006, 35029,
35268, 35512-4, 35545, 35659-60, 35849, B35862, 35948, 36033, 36058, 36137-8,
36527
36642, 36676, 36871, 36963, 37022, 37150, 37298, 37339, 37548, 37563, 37732,
37735, 37863, 37987-9, 38142, 38285, B38475, 38512, 38597, 38749, 38933,
39048, 39063, 39107, 39123, 39142, 39283, 39312, 39380, 39508, 39678, 39723,
39858, 39899, B39903, 39928, 40141, 40191, 40201, 40207, 40323

Brinjal see *Solanum*

Bristlegrass see *Setaria*

Broadbean see *Vicia*

Broccoli see *Brassica*

Bromegrass see *Bromus*

Bromus 21862, 22765, 23061, 24704; 26145, 28406; 29663, 30483, 30547, 32052; 32950,
       36076, 36203; 36692, 37475, 38776, 39123, 39127, 39664, 40133

Brussels sprouts see *Brassica*

Bryophyllum
       22203, 22318, 22962-3, 23480, 23539, 23926, 24319, 24656-7, 24659; 26299,
       26578, 27948, 28146, 28308, 28388, 28489; 28787, 28896, 29092, 29274, 29912,
       30218, 30952, 31057-8; 33890-1, 33960, 34196, 34585, 34952, 35087, 35403,
       35442, 36146, 36253-4; 36939, 37022, 37858, 38907-8, 38961, 39830, 40010,
       40358

Buckwheat see *Fagopyrum*

C

Cabbage see *Brassica*

Cacti 24660, 24675-6; 25360, 25445, 25787, 25862, 28049, 28066, 28143; 29638-9,
       29816-7, 30583, 31068; 33062, 33440, 33547, 33663-5,   33857-8,   35095,
       35930, 36093; 37073, B38477, 38641, 38996-7, 39324, 39687, 39830, 40180

Cajanus 21521-2; 26274, 27389; 29811, 31840; 33934; 37298, 37940, 39569

Calluna see Heath plants and communities

CAM plants (*cf.* also *Agave, Ananas, Bryophyllum,* Cacti, Succulents)
       21884, 21932, 22098-9, 22116, 22152, 22203, 22508, 23011, 23013, 23300, 23374,
       23480, 23714, 23786, 23926, 23930, 24656-60, 24676, 24797, 24806, 25035-6
       25359, 25787, 25862, 25926, 26690-2, 27073, 27198, 27271, 27359, 27542, 27769,
       27948, 28065-6, 28143, 28389-90, 28403, 28404-5, 28407
       28625, 28787, 29216, 29912, 30332-3, 30516, 30664-6, 30822-3, 30928, 31057-8,
       31066, 31993, 32025, 32425-6
       32840, 33547, 33777, 33809, 33857, 34298, 34352-3, B34354, 34368-9, 34585,
       34684, 34950, 35097, 35192, 35442, 35625, 35950, 36442-5
       36647, 36672, 36713, 36916-8, 36939, 36997, 37177, 37317-8, 37399, 37701,
       37858, 38088, 38185, 38236-7, B38238, 38341, 38400, 38640-3, 38788-9, 38996-7,
       39132, 39382, 39724-6, 39830, 40010, 40180, 40212-4, 40236, 40238, 40358

Camellia see *Thea*

Canarygrass see *Phalaris*

Cannabis 21709, 23829, 24508, 24527, 24540-1; 25511, B27915; 29189, 31905; 37298

Caper-bush see *Capparis*

Capparis 36445

Capsicum
       21676, 21700, 22647, 22734, 22747, 22817-8, 23101, 23694, 24645, 25128, 25146;
       25168, 25225, 25521, 25947, 26186, 26232, 26290, 26311-2, 26775, 26956, 27142,
       27673, 27799, 27883, 27995, 28139; 28732, 29481, 29794-5, 30834, 31831-2,
       31834, 32119, 32229; 33057, 33832, 34403, 34624, 34678, 35006, 35033, 35446,
       35776, 36124; 36963, 37039, 37150, 37572, 38047, 38285, B38477, 38644, 38727,
       38849, 39760, 40033, 40360

Caraway see *Carum*

Carex 21762, 22380, 22950, 24015, 24185, 24624; 25254, 25887, 26490, 26532, 26854,
27013-4, 27139, 27604, 27814; 28843, B29063, 29064-5, 29332, 29685, 29703,
30392-4, 31208, 31307, 31392, 32185, 32257, 32518; 33050, 33331, 33728, 33756,
33804, B33926, 34052, 34351, 34825, 34924, 36037, 36132, 36136, 36201, 36205,
36408; 36844, 37054, 37346, 37382, 37889, 38285, 38508, 38776, 39077, 39262,
39354, 39821

Carica 38133, 38865, 39407, 39653, 39678

Carpinus
21674, 21808-9, 22841, 22854, 23232, 23234, 23560, 23829, 24567, 24601;
25896, 25898, 26490, 26787-8, 27362, 28160; 29206-7, 29320, 29394, 30096,
30460, 31163; 33262, 33451; 37230, 37413-5, 39002-4, 39991

Carrot see *Daucus*

Carthamus
22283, 24322; 25990; 30995, 31729; 37298

Carum 36255

Carya 22196, 23818; 25821, 26534; 29254, 29320, 29467; 33632, 33984, 34639-40,
35282, 36124; 36818, 39540

Cashew see *Anacardium*

Cassava see *Manihot*

Castanea
23521; 28546, 29249, 31508; 33477, 33567; 36795, 38285

Castor bean see *Ricinus*

Cat's tail see *Typha*

Cattail flag see *Typha*

Cauliflower see *Brassica*

Cedar see *Cedrus*

Cedrus 21633, 21925; 25186, 26348; B31990; 32698; 38900

Celery see *Apium*

Cerasus 21809, 22747, 22843, 23026; 29846; 36613, 38607

Cereals see *Avena, Hordeum, Oryza, Panicum, Secale, Sorgum, Triticum, Zea*

Chenopodium
22028, 22106, 22963, 23580, 23750, 23892, 23914, 24442, 24716, 25103; 26360,
27196, 27230, 27404, 28297, 28315; 28591, 28875, 29267, 29548, 29681-2,
29788, 29891, 30056, 30173, 30681, 30746, 30947, 31092, 31099, 31129, 32008,
32214, 32329, 32547; 32766, 32948, 32996, 33080, 33310, 33391, 33399, 34506,
34924, 35102, 35456, 35947, 36165; 36607, 36810-1, 37022, 37331, 37344, 37754,
37863, 37872, 37962, 38262, B38477, 38813, 39069, 39123, 39252

Cherry see *Cerasus*

Chestnut see *Castanea*

Chick pea see *Cicer*

Chicory see *Cichorium*

Chinese cabbage see *Brassica*

Chlamydomonas
21575, 21613, 21685, 21737-8, 21749, 21751, 21835, 21865, 21871, 21888,
21931, 21994, 22078-9, 22117, 22145, 22164, 22239, 22317, 22442, 22465, 22501,
22660, 22683, 22730, 22790, 22842, 22899, 22956-7, 23000, 23041, 23202, 23272,
23278-80, B23407, 23543, 23548, 23588, 23637, 23685, 23717, 23750, B23767,
23768, 23820, 23843, 23868, 23879-80, 23885, 23942, 23954, 23969, 24041, 24053,
24169, 24184, 24387, 24397-8, 24401, 24403, 24499, 24529, 24572, 24580, 24603,
24613, 24621, 24683, 24743, 24756, 24909, 24929, 24956, 24962, 25012, 25122
25286, 25370, 25378-9, 25395, 25429, 25463, 25504, 25524, 25570, 25616,
25690, 25693, 25726, 25849, 25882, 25925, 25935, 26024, 26051, 26144, 26171,
26295, 26340, 26371, 26414-6, 26430-1, 26557, 26573, 26575, 26659, 26774,
26823, 26829-30, 26994, 27077, 27206, 27246-7, 27451, 27591, 27603, 27690,
27695-7, 27766, 27858, 27891, 27962, 28074-5, 28127, 28174, 28223, 28267, 28452
28513, 28554, 28646-7, 28690-1, 28723, 28758-9, 28775, 28795, 28808,
28849-50, 28910, 28912, 29012, 29118, 29177, 29185, 29208, 29275, 29384,
29437, 29459, 29466, 29560, 29602, 29624-6, 29661, 29673, 29683, 29736, 29759,
29824, 29839, 29918, 30024-5, 30042, 30085, 30308-9, 30487-9, 30495-7, 30539,
30599-601, 30622, 30624, 30718, 30752, 30803, 30836, 30944, 31031, 31063,
31085, 31104-5, 31170, 31298, 31336, 31405-6, 31643, 31668, 31674, 31721-3,
31726, 31844, 31907, 31914, 31917, 31930, 31975, 32029-30, 32113, 32190,
32285, 32360-1, 32406, 32432, 32448, 32464-5, 32473
32599, 32681, 32716, 32811, 32827, 32872, 32902, 32932-3, 32948,
33026, 33070, 33083, 33126-7, 33163, 33170-1, 33212-3, 33218, 33265, 33282,
33311, 33453, 33545, 33686, 33783, 33816, 33861, 34054-5, 34066-7, 34185,
34303, 34305, 34328, 34343, 34490-1, 34531, 34618, 34750, 34801, 34817,
34883-4, 34917, 34994, 35027, 35043, 35143, 35191, 35211, 35241, 35244,
35544, 35755, 35763, B35862, 35873, 35967, 36182, 36341-2, 36455, 36476,
36541-2
36598, 36624, 36633-5, 36793, 36897, 37153, 37155-6, 37194, 37275, 37329,
37420, 37452, 37478, 37607, 37630, 37650, 37673, 37684, 37718, 37737, 37757,
37764, 37858, 37937, 37942, 38070, 38291, 38379-82, B38475, 38481-2, 38484,
38598, 38619-20, 38806, 38808, 38836, 38963, 39240, 39308, 39358, 39370,
39430, 39478, 39535, 39571, 39733, 39835, 39854, 40068, 40136, 40151, 40269,
40283-4, 40298

Chlorella
21620, 21643,   21703, 21706, 21718, 21720, 21734-5, 21750, 21789, 21806,
21871, 21888, 21891-4, 21986, 22028, B22052, 22062, 22083, 22125, 22181, 22211,
22236, 22248, 22265, 22291, 22305-6, 22353, 22359-60, 22442, 22454, 22461,
22484, 22515, 22522, 22558, 22580, 22605, 22642, 22660, 22664, 22692, 22710,
22719, 22808, 22956-7, 22997, 23029, 23091, 23153, 23184-5, 23229, 23248-9,
23253, 23272, 23326-7, 23365, 23384, 23390, B23407, 23434, 23452-3, 23485,
23488-9, 23511, 23576, 23588, 23593, 23614, 23632, 23642, 23647, 23668, 23714,
23718, 23726, 23728, 23744, 23750, 23768, 23794, 23850, 23868, 23870, 23881,
23914, 23924, 23954, 24021, B24024, 24025, 24043, 24048, 24053-4, 24073, 24101,
24117, 24137, 24154, 24169, B24176, B24250, 24258, 24273, 24320, 24384, 24397,
24401, 24421, 24459, 24499-9, 24572, 24577, 24583, 24637, 24650, 24667, 24683,
24702, 24719, 24724, 24768, 24804, 24815-6, 24845, 24864, 24910-1, 24962-3
25163, 25174, 25179, 25208, 25243, 25331, 25335, 25345, 25366-7, 25381-3,
25511, 25527, 25533-6, 25630-2, 25675, 25730, 25752, 25759, 25782, 25836,
25840, 25861, 25868, 25875, 25882, 25915, 25935, 26018-9, 26021, 26024, 26063,
26094, 26097, 26111, 26171, 26211, B26220, 26288, 26293-4, 26298, 26302,
26340, 26437, 26454, 26504, 26517, 26538-9, 26549-50, 26639, 26683, 26693,
26718, 26720, 26755-6, 26780-1, 26823, 26859, 26861, 26873, 26943, 26985,
27026, 27084, 27086, 27122, 27206, 27222, 27227, 27297, 27314-5, 27399, 27439,
27566, 27591-2, 27612-3, 27704, 27718, 27766, 27790-1, 27796, 27843, 27858-9,
27950, 27963, 27994, 28056, 28072, 28075, 28115, 28120, 28124, 28128, 28159,
28171-2, 28204, 28223, 28234, 28278, 28319, 28328-9, 28334, 28354, 28395,
28418

(continued)

Chlorella (continued)
28550, 28726, 28742, 28762-3, 28795, 28808, 28814, 28867, 28888, 28893-5,
28921-2, 28934, 28961, 29026, 29096, 29159-60, 29167, 29185, 29188, 29208,
29278-9, 29308-9, 29347, 29382, 29418, 29436, 29466, 29500, 29610, 29637,
29662, 29683, 29733-6, 29873, 29912, 29918, 29986, 30136, 30156-8, 30223-4,
30347, 30365, 30401, 30416-7, 30452, 30539, 30613, 30622, 30624, 30635,
30669, 30708-10, 30718-9, 20757, 30782, 30799, 30827, 30886, 30892, 30926,
30949, 30962, 30983, 31011, 31104, 31141, 31170, 31212, 31280, 31282-3,
31285, 31329, 31342, 31412, 31472, 31532, 31544, 31582, 31631-2, 31720,
31723, 31726, 31738, 31778, 31782, 31835, 31957, 32050, 32075, 32101, 32125,
32162, 32166, 32238-9, 32293, 32357, 32361, 32396, 32410, 32416-7, 32419,
32432-3, 32453, B32515, 32524, 32534
32622, 32649,   32669, 32681, 32687, 32728, 32748, 32767-8, 32792,
32794, 32798, 32803, 32813-4, 32816, 32822, 32859, 32945-8, 32966, 33062-3,
33068, 33070, 33072, 33083, 33131, 33146, 33217-8, 33265, 33314-6, 33323-4,
33351-2, 33355, 33360, 33430, 33475, 33511, 33545, 33625, 33652, 33660,
33721, 33749, 33765, 33784, 33795-6, 33814, 33856, 33875, 33880, 33911,
33912, 33987, 34007, 34025, 34082, 34185, 34197, 34303, 34328, 34361, 34363-
-4, 34404, 34417-8, 34447, 34468, 34529-31, 34589, 34607, 34628-9, 34634,
34658-9, 34664, 34691, 34717- 9, 34759, 34773, 34775, 34797, 34817,
34837-8, 34846, 34893, 34905-6, 34914, 34967, 34992, 34994, 35024, 35043,
35126, 35144-5, 35191, 35195, 35206, 35250, 35252, 35304-5, 35319-20, 35336,
35343, 35366, 35378, 35475, 35485, 35494, 35524, 35635, 35729, 35755, 35761,
35763, 35836, 35878-9, 35881, 35916, 35940, 35945, 35957, 36043, 36081,
36097, 36105-6, 36145, 36176, 36197, 36239, 36276, 36342, 36365, 36409,
36411, 36462, 36540, 36569-70
36598, 36624, 36634-5, 36689, 36808, 36828, 36859, 36897, 36921, 36937, 36981,
37051, 37129-30, 37194, 37264, 37315, 37373, 37420, 37549, 37574, 37604,
37639, 37673, 37705, 37718, 37739, 37741, 37784, 37835, 37858, 37860-1, 37868,
37902-3, 37990, 38002, 38099, 38154, 38162, 38292, 38321, 38338, 38344, 38433,
38447, 38451, 38460, B38475, B38477, 38503, 38548, 38621, 38633-4, 38690,
38757-9,   38836, 38912,   38946, 39018, 39064, 39102, 39106, 39113,
39115, 39176, 39183, B39191, 39222, 39347, 39373, 39459, 39516, 39533-4,
39598, 39647-8, 39663, 39721, 39728, 39848-9, 39877, 39930-1, 39975, 39982,
40019, 40047-8, 40059, 40076, 40089, 40101, B40105, 40163, 40247, 40269,
40283-4, 40330, 40332, 40349, 40372

Chlorobium
21938, 22416, 22549, 23347, 23909, 24706; 25478, 25562-3, 25584, 26630, 26695,
26926, 28232; 28878, 28905, 28992, 29484, 29913, 30138, 30398, 31137, 31845,
31900, 32036; 32619, 32673, 32681, 32764, 32915, 32977-8, 33294, 33402-3,
33612, 33688, 33902, 34194, 34229, 34296-7, 34356, 34522, 34537, 34817, 34852,
35167, 35239, 35323, 35333, 35387, 35411, 35501, 35702, 35855, 35968, 36188-9,
36513; 36624, 36866, 36944, 37085, 37231, 37877, 37632, 37994, 38240, 38336,
38374, 38593, 38796, 38906, 38952, 38972, 39323, 39468, 40157, 40186

Chromatium
21583, 21618-9, 21938, 22219, 22520, 22667, 23420-1, 23563, 23708, 24681,
24690-1, 24924
25249, 26677, 26702, 26959, 27262, 27326, 27517, 27868-9, 28135-6, 28238
28506, 28582-3,   28592, 28600, 28832, 29074, 29230, 29431-2, 29454, 29479,
29630-1, 29851, 29913, 29997, 30386, 30626, 30774, 31033, 31059, 31092,
31094, 31157-9, 31437, 31534, 31576, 32036, 32127, 32250
32563, 32585,   32657, 32673, 32681, 32683-5, 32764, 32915, 32925, 33119,
33188, 33294, 33325, 33402, 33468-9, 33515, 33552, 33612, 33661, 33688,
33833, 34130, 34230, 34277, 34281, 34297, 34355-6, 34522, 34566, 34658,
34817, 34852, 34968, 35066, 35107, 35193, 35239, 35323, 35387, 35397, 35501,
35573, 35702, 35855, 35937, 36025, 36119, 36129, 36357, 36513
36624, 36650, 36699, 36919, 36944, 37117, 37180, 37181, 37231, 37370, 37446-7,
37632, 37663, 37822, 38239, 38241, 38336, 38374, 38506, 38649, 38682-4,
38796, 38906, 38921, 38952, 38972, 39252, 39385, 39451, 39468, 39542, 39769-
-70, 39837, 39909, 40028, 40157

Chrysanthemum
29086, 29858, 30683, 31208; 32569, 33082, 34332; 36602, 36604, 36956, 37123, 37217, 37239, 38259

Chrysophyta
21505, 21562, 22866-7, 23449, 23611, 24120, 24743; 25215, 25247, 25286, 25346, 26130, 26179, 26437, 26867, 27120-1, 27406, 27592, 28126, 28129; 28568, 28767, 29127, 29658, 29677, 29928, 29937, 30046, 30582, 30906, 31031, 31184, 31328, 31758, 31788, 31918, 31965, 31988, 32101, 32448; 32599, 32621, 32624, 32898-9, 32907-8, 32921, 32993, 33070, 33371, 33446, 33545, 33856, 34016, 34555, 34604, 34884-5, 34991-4, 35043, 35216, 35587, 35654, 35715, 35763, B35862, 36182, 36420, 36455, 36474; 36634-5, 36929, 36946-7, 37329, 37478, 37684, 37715, 37797, 37893, 38050, 38070, 38145, 38394, 38454, 38476, 38521, 38548, 38782, 39308, 39669, 39854, 39901, 39924, 39942, 39944, 39969, 40174

Cicer 23168; 25965, 27237, 27389; 31395, 31568; 32686, 34834, 35225-6, 35603, 35891; 37298, 39431, B39537

Cichorium 22747, 24716; 28743, 29829; 39304

Citrullus 21705, 21781, 22719, 22982; 27880, 28267; 31769, 32337-8; 36328; 37339, 38517, B39551, 40358

Citrus 21683, 21694, 21705, 22028, 22102, 22191, 22298-9, 22323, 22650, 22747, 22782, 22914, 23372, 23781, 23803, 23846, 24069 , 24332, 24336, 24448, 24537; 25233, 25250, 25330, 25376, 25551, 25634, 26476, 26511, 26646-7, 26909, 27049, 27085, 27191, 27472, 27506, 28318; 28546, 28930, 29020, 29048, 29370, 29414, 29944, 30080, 30187, 30205, 30250, 30571, 30737-9, 30996, 31181, 31496, 31692, 31769, 31781-2, 31961, 32491, 32506; 32797, 32877, 33463, 33589, 34940, 35073, 35140, 35810, 35813-4, 36198, 36231-2; 36644, 36812, 36851, 36920, 37150, 37260, 37428, 37496, 37746, 37775, 37880, 37939, 38043, 38285, 38461, 38464-5, 38468, B38477, 38675, 38719-20, 38791, 38871, 39246, 39492, B39523, B39537, 39586-7, 39760, B39786, B39903. 39925

Clover   see *Trifolium*

Cocksfoot see *Dactylis*

Cocoa see *Theobroma*

Coconut palm see *Cocos*

Cocos 22010, 23570; 27069, 27220, 27506, 28384; 30858, 30979, 30994; 33409, 35426

Coffea 21928-9, 22518, 24336; 27289, 27601, 27951; 29563, 30699; 34743; 36865, 38609, B39903, 40293

Coffee tree see *Coffea*

Colocynthis 23913

Coniferous plants (*cf.* also *Abies, Cedrus, Cupressus, Juniperus, Larix, Metasequoia, Picea, Pinus, Pseudotsuga, Taxus, Thuja, Tsuga*)
21705, 22073, 22086, 22224, 22255, 23082, 23091, 23135, 23257, 23318, 23666, 23943, 24001, 24372, 24604, 25007; B25637, 26515, 26883, 27260, 28062, 28139, 28341; 28719, 28838, 28863, 28987, 29071, 30787, 31019-20, 31053, 31154; 32681, 33296, 33646, 35115, 35156, 35808, 36346, 36515; 36735, B37528, 37772, 38164, 38285, 38493, 38754, 38900, B39191, 39597

Corchorus 26245; 29240; 34050, 34151

Corn see *Zea*

Cornelian cherry   see *Cornus*

Cornus 21809, 22028, 22190, 22719, 23605; 25822, 25897, 26534, 27362, 28283; 29243, 29254, 29467, 29480, 31269, 31411, 32185, 32191; 33559; 36984, 37413, 37416, 37874, 38285, 39077, 39328, B39537

Corylus 29207; 33062; 37220, 39501

Cotton see *Gossypium*

Cottonwood see *Populus*

Cowberry see *Vaccinium*

Cowpea see *Vigna*

Crabgrass see *Digitaria*

Crataegus
    21809; 27362, 28283; 30911; 32948, 33602; 36963, 37416, 37569, 37693, 39077

Cucumber see *Cucumis*

Cucumis 21703, 21732, 21957, 22051, 22082, 22313, 22401, 22797, 22872, 22897, 22978, 22999, 23222, 23442, 23527, 23581, 23690-1, 23716-7, 23853, 23972, 24130, 24177-80, 24389, 24468, 24557, 24918, 24987, 25075, 25158
25176, 25258, 25345, 25368, 25666, 25747, 25877, 26073, 26079, 26103, 26125, 26476, 26599, 26769, 26779, 27142, 27291, 27430, 27439, 27667, 27922, 27927, 28182, 28239-40, 28369
28571, 28597, 28653-4, 28732, 29058-9, 29150, 29258, 29329, 29392, 29450, 29555, 29596-7, 29711, 30235-6, 30244, 30298, 30374, 30396, 30529, 30790-2, 31001, 31099, 31160, 31586, 31602, 31876, 31880-1, 32041, 32073, 32103, 32154, 32382, 32547
32581, 32776, 32783, 32948, 33094-5, 33205, 33301-2, 33432, 33503, 33589-91, 33593, 33775, 33797-8, 33875, 34036, 34098, 34232-3, 34250, 34256, 34266, 34275-6, 34426, 34438-9, 34450, 34523, 34666, 34784, 34790, 34924, 34940, 35029, 35604, 35620, 35749-50, 35781, 35787, 35800, 35809, 35923, 35943, 36060, 36158, B36191, 36199, 36221, 36288-90, 36327, 36367
36619, 36628-9, 36785, 36805, 36827, 36837, 37102-3, 37150, 37183, 37352, 37396, 37410, 37863, 38121, 38222, 38285, 38360, 38416, 38443, 38597, 38660, 38840, 39062, 39411, B39523, 39654, 39678, 39686, 39760, 39998, 40032, 40049, 40092, 40179, 40360

Cucurbita
    21673, 21791, 21847, 22110, 22734, 23025, 23528, 23575, 23750, 24334, 24413, 24830-1; 25511, 25614, B26099, 26151, 26186, 26476, 26718, 26720, 26961, 27162, 27439, 27500, 27566, 27730, 28117, 28139, 28245, 28315; 29218, 29902, 30173, 30300, 30431, B30477, 30514, 30621, 31122, 31160, 31411, 31446, 32202; 32948, 33470-1, 34320, 34506, 34556, 34871-2, 34924, 35006, 35180, 35604, 35729, 36215, 36293, 36437; 36924, 37150, 37187, 37339, 37380, 37733, 39062, 39259, 39524, 39814, 40292

Cupressus 25186

Currant see *Ribes*

Cyanobacteria see Algae, blue-green

Cycas 32681, 32948

Cydonia 38607

Cynara 21695; 29538

Cynodon 21508, 22953, 23245, 24266; 25679, 27895, 28100; 31022, 31530; 32766,
        33130, 33696, 34100, 34249, 34907, 34926, 35344; 36692, 37022, 37063, 37560,
        38385, B38477, 38776, 39107, 39127, 39917

Cyperus 22660, 23033, 23657; 26539, 26923, 27325, 27506; 29047, 29865, 30951; 32752
        32766, 34193, 34279, 35101-3, 35425; 37981, 38158, B38477, 38702, 38786,
        39107

Cypress see *Cupressus*

D

Dactylis
        21621, 22030, 22919, 23061, 23375, 24436, 24438; 25817-8, 26065, 26569,
        27296, 27333, 27396, 27480, 27819, 28005, 28397; 28973, 30140, 30829, 31208,
        31495, 31510, 31760; 32686, 32766, 32868, 33488, 34458, B35862, 36165, 36183,
        36205, 36456; 36692, 37609, 38793, 38979-80, 39682-3, 40117, 40142, 40144,
        40360

Dallis grass see *Paspalum*

Date palm see *Phoenix*

Daucus 21841, 21948, 22636, 22713, 22719, 22747, 23026, 23425, 23448, 23632, 23675,
        23807, 23951, 24203, 24716, 25136; 25306, 25479, 25486, 25935, 26151, 26157,
        26267, 27165, 27478, 27730, 28117; 28666, 29280, 29555, 30276-7, 30467,
        31038, 31855, 31933-4, 32153; 32722, 32808, 33089, 33572, 33590-1, 34477,
        35687, 35914, 35920, 36255; 37344, 37733, 38364, B38477, 38553, 38675, 38732,
        39183, B39523, 39585

Deciduous trees and shrubs (*cf.* also *Acer, Aesculus, Alnus, Amygdalus, Armeniaca,
        Betula, Carpinus, Carya, Castanea, Cerasus, Citrus, Cornus, Corylus, Cratae-
        gus, Eucalyptus, Fagus, Fraxinus, Hevea, Hibiscus, Juglans, Malus, Mangifera,
        Morus, Olea, Persea, Persica, Pirus, Platanus, Populus, Prunus, Quercus,
        Ribes, Robinia, Rubus, Salix, Sambucus, Sorbus, Syringa, Tamarix, Tilia,
        Ulmus, Vitis*)
        22016, 22028, 22086, 22492, 22734, 23499, 23599, 23605, 23814, 24093, 24147,
        24153, 24239, 24336, 24423, 24670, 24808, 24826, 25126
        25186, 25283, 25423-4, 25511, 25521, 25592, 25619-20, B25637, 25693, 25822,
        26533-4, 26674, 26977, 27062, 27222, 27506, 27730, 27939, 28255, 28283,
        28294, 28433, 28470
        28801, 28838, 28987, 29121, 29202, 29209, 29212, 29250, 29254, 29301, 29320,
        29398, 29467, 29480, 29587, 29614, 29721-2, 30060-1, 30465, 30683, 30695,
        30764, 30787, 30994, 31090, 31470, 31517, 31659, 31839, 31854, 31887-8,
        32058, 32185, 32263, 32325
        32646, 32667, 32681, 32948, 33167, 33260, 33294, 33346, 33440, 33559, 33602,
        B33628, 33669, 33696, 34101, 34278, 34677, 34690, 34826-7, 34924, 34945,
        35006, 35092, 35167, 35269, 35525, 35719, 35808, 35885, 35930-1, 36037,
        B36191, 36201, 36319, 36445, 36509, 36515
        36734, 36818, 36853, 36951, 36963, 37049, 37331, 37413-6, 37569, 37587,
        37638, 37678, 37693, 37780, 37909, 38147, 38285, 38371-2, 38504-5, 38679,
        38900, 38934, 39043, 39309, 39328, 39424, 39524, 39748, B39951, 40115, 40277

Desert plants and ecosystems
        21670, 22379, 22544, 22951, 23297-300, 23466, 23658, 23692, 23752-3, 24378,
        24629, 24935; 25303, 25538, 25653, 25922-4, 26155, 26580, 26803, 27179-80,
        27271, 27558, 27573-5, 27587, 27768, 27939, 28014, 28059, 28067, 28073,
        28407; 28498, B28545, 28588-9, 29116, 29281, 29529, 30505-6, 30515, 30675,
        30695, 30919, 30974, 31066, 31068-70, 31148, 31535, 31887-8, 32020, 32026-7,
                                                                        (continued)

Desert plants and ecosystems (continued)
    B32318, 32347, B32515; 32665-7, 32824, 33434-6, 33547, 33696, 33699, 33839-40,
    34239, 34511-2, 34613, 34945, 35096, 35098, 35258, 35885, 35912, 35930-1,
    36050; 36846, 36853, 37397, 37475-7, 37780, 38083, 38473, 38784, 39424, 39524,
    39678, 39688, 39830-1, 40277, 40300

Dewberry see *Rubus*

Diatoms 21543, 21582, 21682, 21711, 21775, 22109, 22158, 22287, 22357, 22362, 22387,
    22433, 22442, 22451, 22453-4, 22488, 22515, 22531, 22538, 22550, 22552, 22559,
    22578, 22580, 22592, 22644, 22660, 22745-6, 22763, 22790, 22807, 22848, 22865-
    -7, 22927, 22991, 23107, 23128, 23268, 23275, 23377, 23441, 23456, 23505,
    23542, 23569, 23606, 23632, 23808-9, 23824, 23843, 23931-2, 23940, 23981-2,
    24111, 24118, 24218, 24233, 24274, 24308, 24381-2, 24440, 24451, 24472, 24475,
    24572, 24579, 24581, 24589, 24616-8, 24727, 24732, 24816, 24929, 24956, 25005,
    25112
    25165, 25272, 25286, 25307, 25346, 25426, 25515, 25633, 25648, 25739, 25743,
    25763, 25847, 25849-50, 26025, 26127, 26179, B26220, 26261, 26270-1, 26371,
    26387, 26407, 26454, 26516, 26518, 26520-1, 26529, 26570-1, 26698, 26823,
    26849, 26939, 27067, 27206, 27226-7, 27330, 27367, 27377, 27406, 27416, 27418,
    27476, 27534, 27592, 27642, 27700, 27766, 27858, 27960, 27964, 27994, 28017,
    28026, 28076-7, 28088, 28128-9, 28174, 28190, 28204, 28226, 28354, 28376,
    28394, 28412, 28429
    28507, 28511, 28536, 28760, 28977, 28982, 29151, 29176-7, 29185, 29276,
    29354-5, 29428, 29434, 29466, 29479, 29588, 29655, 29677, 29691, 29802-3,
    29842, 29937, 29990-1, 30032, 30072, 30075, 30137, 30172, 30196, 30231, 30305,
    30381, 30407, 30450, 30592, 30606, 30616, 30624, 30865. 30906, 30987-8, 31031,
    31126, 31184, 31215, 31317, 31327-8, 31382, 31650, 31681, 31714, 31745, 31758,
    31810, 31889-90, 31909, 31965, 31988, 32077, 32107, 32138, 32274, 32288,
    32396, 32423, 32472
    32564-5, 32599, 32669, 32679, 32681, 32787, 32887, 32910, 32993, 33062,
    33070, 33083, 33118, 33127, 33163, 33169, 33183, 33224, 33266, 33333, 33366,
    33422, 33428. 33480. 33496. 33531, 33544-5, 33594, 33721, 33867, 33923,
    33927, 34002, 34016, 34021, 34030-1, 34058, 34069-70, 34160, 34191-2, 34197,
    34216, 34273, 34359, 34429, 34484, 34515, 34538, 34555, 34557, 34604, 34628-
    -9, 34634, 34669, 34768, 34793, 34885, 34914, 34953, 34991-5, 35043, 35204,
    35216, 35244, 35343, 35493, 35522, 35533, 35562, 35584, 35587, 35597, 35654,
    35811-2, 35880, 35946, 36061, 36132, 36369, 36420, 36474
    36633-5, 36657, 36660, 36868, 36873, 36886, 36897, 37070, 37188, 37196,
    37246-7, 37258, 37327-8, 37437, 37478, 37581, 37655, 37659, 37673, 37684,
    37764, 37784, 37797, 37868, 37943, 37945, 38050, 38060, 38070, 38200, 38374,
    38451, 38454, 38456-7, 38476, 38521, 38586, 38782, 38838, 39061, 39101,
    39196, 39210, 39308, 39321, 39325, 39346-7, 39425, 39485, 39559, 39668,
    39781, 39825, 39844, 39854, 39901, 39924, 39942, 40065, 40174, 40184, 40234,
    40305, 40308-9, 40327, 40371

Digitaria
    21874, 21999, 22502, 22698, 22833-4, 22837-8; 25445, 25913, 26441, 26444,
    27091, 28099-100, 28228; 29376, 29489, 29771, 29865, 30035-7, 30083, 30140-1,
    30951, 31393, 31447-8, 31450-2, 32047, 32152; 32653, 32766, 32867, 33130,
    33167, 33425, 33914, 34063, 34541, 34618, 35101-3, 35186, 35466, 35468;
    36692, 37336, 37344, 37391, 37627, 37863, 37981, 38125, B38477, 38588-9,
    38786, 39107, 39275, 39434, 39809, 39917, 40292

Dinoflagellatae, Dinophyceae
    21966, 22241, 22744, 22926, 22930, 23087, 23439, 23441, 23611, 24816, 25004,
    25055-6; 25346, 25763, 26025, 26179, 26326, 26359, 27511-3, 27952; 29077,
    29428, 29588, 29990, 30366, 30484, 30629, 31326, 31362-3, 31676, 31714, 31758,
    31812, 31901, 31965, 31988, 32195, 32442; 32564-5, 32599, 32859, 32887, 33006,
    33070, 33118, 33927, 34021, 34216, 34272-3, 34380, 34519, 34604, 34991-4,
    35043, 35380, B35862, 35880, 36061, 36117, 36474; 36634-5, 36906, 37246,
    37673, 37784, 37818, 38050, 38476, 38782, 39061, 39185-6, 39530, 39697, 39781,
    39825, 39854, 39942, 40174, 40178, 40244-5, 40305

Dioscorea   35426

Dogwood   see   *Cornus*

Douglas fir   see   *Pseudotsuga*

Dunaliella
        21736, 21909, 23440, 23941, 24030, 24982; 27429, 27572, 28354; 29177, 29210,
        30018, 30207, 30219-20, 31171, 32093, 32184, 32379; 32669, 32795, 32988,
        33001, 33371, 33545, 33687, 33689, 33721, 34157-8, 34223, 34538, 34978,
        34988-9, 35390, 36291, 36364-5, 36474; 36762, 37013, 37659, 37684, 38002,
        38275, 38514, 39240, 39405, 40073, 40159, 40224, 40269, 40305

E

Egg plant   see   *Solanum*

Elaeis   22123, 23448; 25772, 25774, E25775, 26267, 26922; 28838, 29496; 32888, 33229,
        33868

Elder   see   *Sambucus*

Elm   see   *Ulmus*

Elodea   22721, 22892, 23188, 24546, E25064; 25523, 25831, 26067, 26824, 27285, 27470;
        29262, E29468, 29469, 29616, 29778, 30316, 30490, 30767, 31327-8, 31541,
        31766, 32518; 32831, 33103, 33675, 33804, 34052, 34216, 34421, 34629, 34715,
        34859, 35048, 35099, 35113, 35653, 35727, 36453-4, 36553; 37924, B38477,
        38842, 38924, 38926, 39188-9, 39356, 39436, 39529, 39716, 39718, 40156, 40314,
        40358, 40371

Endive   see   *Cichorium*

Equisetum
        22543, 22658, 22660, 22979, 23870; 25253, 25693, 27222, 27730; 29021, 29205,
        29852-3, 32069; 32681, 32948, 33804, 33904, 34715, 34917, 34924

Ericaceae   see   Heath plants and communities

Eucalyptus
        23634, 24336, 24585; 25331, 25755, 25785, 26828, 26848, 27062, 27433, 27912;
        28838, 29018, 29119, 29136, 29162, 30494, 31270, 31769, 31861-6, 32394; 32888,
        33368, 34101, 34124, 34946, 34954, 35488, 35909-10, 36472; 36726, 36734,
        37087, 37814, 38285, B39951, 39991, 40239, 40265

Euglena   21562, 21575, 21703, 21741, 21804-5, 21888, 21925, 21935, 21976, 21992,
        22107, 22110-1, 22121, 22159, 22192, 22214, 22264, 22303-5, 22307, 22312,
        22442, 22453, 22462, 22469, 22521, 22535, 22596, 22616, 22644, 22660, 22711,
        22729, 22745-6, 22767, 22827, 22864, 22933, 23095, 23224, 23354, 23384, B23407
        23446, 23451, 23453-4, 23456, 23490, 23569, 23577, 23597, 23606, 23623, 23638-
        -9, 23728, B23767, 23824, 23843, 23851, 23868, 23915, 23954, 23968-9, 24027,
        24067, 24097, 24128, 24130, 24154, 24243, 24273, 24292, 24304, 24308, 24339-
        -40, 24346-8, 24357, 24397, 24409, 24459, 24466-7, 24544, 24579, 24608, 24613,
        24723, 24819, 24823, 25030, B25064, 25065-6, 25071, 25112, 25120, 25132-3
        25266, 25286, 25427-8, 25550, 25570, 25574, 25577, 25587, 25622, 25688-9,
        25706, 25749, 25763, 25770, 25853, 25935, 25939-40, 25942, 25995, 26051,
        26063; 26069-70, B26099, 26167, 26178-9, 26186, 26205, 26261, 26340, 26504,
        26539, 26683, 26768, 26823, 26842, 26885, 26926, 26990, 26994, 27076, 27249,
        27286, 27297, 27377, 27383, 27407, 27486, 27591-2, 27698, 27703, 27709, 27766,
        27775, 28007, 28111, 28122, 28174, 28217-8, 28223, 28234, 28243, 28248-50,
        28267, 28386, 28423, 28476   (continued)

Euglena (continued)
28760, 28804-5, 28828, 28923-4, 28966, 28980, 28982, 29014, 29017, 29183,
29318, 29341, 29364, 29371, 29374, 29412, 29465-6, 29479, 29516-7, 29560,
29637, 29773, 29801, 29803, 29815, 29840, 29918, 30169, 30208, 30260-1,
30307, ·30309, 30439-40, 30536, 30592, 30630, 30673, 30848, 30926, 30935,
31011, 31031, 31044, 31064, 31104-5, 31170, 31211, 31222-3, 31538, 31612,
31638, 31644, 31650, 31656, 31674, 31705, 31767, 31778-9, 31810, 31930,
31965, 32069, 32077-8, 32101, 32260, 32276, 32286, 32333, 32349-50, 32421-2,
32448, 32452, 32464-5
32565, 32580, 32599, 32621, 32679-81, 32701-2, 32854, 32876, 33003, 33047,
33070, 33113, 33225, 33235, 33256, 33266, 33294, 33389-90, 33417-8, 33423,
33461-2, 33475, 33596-8, 33648, 33674, 33678, 33810, 33841, 33875, 33942,
34058, 34164, 34188, 34327, 34360, 34557, 34604, 34629, 34644, 34652, 34669,
34688, 34692, 34694, 34817, 34867, 34885, 34917, 35043, 35071-2, 35143,
35167, 35190, 35206, 35241, 35244, 35289-90, 35382-3, 35479, 35567, 35596,
35628-30, 35655, 35689, 35729, 35731, 35755, 35799, B35862, 35992, 36028,
36031, 36056, 36182, 36249, 36414, 36455, 36459, 36574
36624, 36634-5, 36639, 36660, 36841-2, 36894-5, 36979, 36987-8, 37075-7,
37127, 37228, 37312, 37322, 37329, 37336, 37420, 37426, 37457, 37478, 37484,
37522, 37554, 37557-8, 37684, 37764, 37835, 37846, 37916, 38426, 38449,
B38475, 38476, 38515, 38548, 38950, 38991-2, 39273, 39308, 39346, 39415,
39504-6, 39526, 39608, 39833, 39880, 39901, 40063-4, 40090, 40151, 40174,
40178, 40310, 40369

Euphorbia
22249, 22328, 23986, 24980; 27550, 27552, 27585, 27769; 28799, 29018-9, 29057,
29232, 30750, 31697, 32324; 33054-5, 33440, 33653, 34980, 35160, 35186, 35425,
35442, 35906, 36201; 36766, 37331, 37387, 37572, 38776, 38784, 39107, 39237,
39830, 40237

Evergreen plants see Sempervirent plants

F

Fagopyrum 22028; 25511, 25920, 26476, 27730, 28139; 32948, 33655, 34410, 35245,
35403; 37344, 38644, 38954, 40041

Fagus   21633, 21705, 21902, 22065, 22208, 22254, 22619, 23091, 23138, 23818, 24601;
25205, 25887, 25989, 25998-9, B26099, 28117, 28315; 28622, 29344, 29398,
29467, 29489, 29614, 30263, 30476, 30594, 30881, 31696, 32197, 32257; 33700,
34227, 34285, 34471, 34607, 34860, 35200, 35808, B36191, 36201, 36345; 36593,
36809, 36905, 38285, B38477, 38631-2, 39077, 39821

Fern-palm see *Cycas*

Ferns 22016, 22147, 22434, 22475, 22644, 22965, 23460, 23464, 23661, 24013, 24371,
25008, 25025; 26322, 26895, 27252, 27730, 28139, 28254, 28267, 28365; 29835,
32263, 32377, 32394, 32458; 32679-81, 32948, 33103, 33823, 33879, 34485,
34669, 35096, 35098, 35477-8, 35746, 36345; 36880, 36905, 36936, 37001, 37336,
37400, 37408, 37482, 37582, 37766, 39276-7, 39494, B39537, 39803, 39991,
39994, 40147, 40357, 40371

Fescue see *Festuca*

Festuca 21853, 21926, 22668, 22918, 23061, 23136, 23375, 23795-7, 24435-6, 24438,
24563, 24746, 24795, 24961, 25031-2
25445, 25573, 26216, 27296, 27337, 27480, 27686, 28031, 28121, 28148, 28158,
28186, 28191, 28397
28973, 29064, 29217, 29581, 29725, 30393, 30483, 30566, 30829, 31021-2,
31208, 31307, 31383, 31433, 31510, 31759, 31761-2, 31870, 32097, 32144
(continued)

Festuca (continued)
  32766, 32867, 32912-3, 32997, 33022, 33130, 33216, 33416, 33756, 34100, 34181,
  35014, 35040, 35052, 35142, 35280, 35301, 35595, 35744, 36156-7, 36165, 36417-
  -9, 36456
  36692, 36722, 37021, 37382, 37544, 37560, 37609, 37701, 37912, 38044, 38285,
  38313, B38477, 38704, 38776, 38785, 38793, 39785, 40117, 40142, 40208, 40344-5

Ficus 22619, 23846, 24557, 25117; 25764, 27526-7; 28546, 29155, 29163, 30833, 31839;
  33589, 35006; 36951, 37217, 38065, 38285, 39162

Fig see *Ficus*

Fir see *Abies*

Flax see *Linum*

Forage crops (*cf.* also *Brassica*, Grasses, Leguminous plants, *Lupinus*, *Medicago*,
    Trifolium, *Vicia*, *Vigna*, etc.)
  21667, 22123, 22279, B23767, 24446, 24584; 27931, 28381; 28516-7, 28973,
  29120, 30378, 30380, 31130-1, 31244, 31498, 31759-62; 33658, 34074, 34541,
  35118, 35158; 36678, 36788, 37073, 37382, 37751, 38534, 39243, 39667, 40088

Forest (including undergrowth) plants and ecosystems (*cf.* also Coniferous plants, De-
    ciduous trees and shrubs, Ferns, *Fragaria*, Grasses, Heath plants and communi-
    ties, Lichens, Liverworts, Medicinal plants, Mosses, *Sphagnum*, *Vaccinium*, etc.
  21760, 21825, 22271-3, 22302, 22327, 22447, 22505, 22573, 22645, 22747, 22832,
  22841, 23135, 23138, 23232, 23234, 23313, B23407, 23412, 23534, 23578, 23582,
  23605, 23741, 23911, 23943, 24150, 24278, 24313, 24473, 24530, 24567, 24835,
  24942
  25195, 25331, 25423-4, 25472, 25539, B25637, 25693, 25742, 25811, 25887,
  25896, 25972, 26026, 26034, 26133, 26348, 26490, 26504, 26534, 26719, 26870,
  26883, 26895, 27797, 27894, 28068, 28102, 28294, 28470
  28801, 28857, 28890, 29071, 29561, 29614, 29628, 29703-4, 29791, 30060-1,
  30096, 30120, 30459, 30523, 30691, 31018, 31695, 31879, 32051, 32175, 32325,
  32394
  32749, 32891, 33209, 33358-9, 33608, 33736-8, 33984-5, 34061, 34290-1, 34330,
  34613, 34677, 34987, 35014, 35165, 35445, 35488, 35808, 35885, 35919, 36201,
  36319, 36360, 36534
  36593-4, 36734, 36962, 37150, 37220-1, 37346, 37413-6, 37693-5, 37780, 37814,
  38000, 38207, 38215, 38221, 38284, 38583, 39077, 39226, 39328, 39640, 40115,
  40277, 40288

Fountain-grass see *Pennisetum*

Foxtail millet see *Setaria*

Fragaria 22719, 22783, 23371, 23716; 26145, 26183, 26476; 29068, 29614, 29663,
  31208, 32200; 33107, 33415, 35202, 35450, 35776, 36201, 36213; 36613, 37113-4,
  37150, 37413, 37736, 38098, 38498, 38607, 39077

Fraxinus 22189-91, 22492, 22647, 22893, 23243, 23521, 23551, 23829, 24240, 24336,
  24817-8, 25077; 25540, 25592, 25901, 26095, 26499, 26534, 26801, 27062, 27248;
  28987-8, 29243, 29249, 29320, 29467, 29614, 30764, 31081, 31090, 31238, 31269-
  -70, 31692, 32175; 33062, 33089, 33440, 33559, B33628, 34246, 34924, 35006,
  35269, 36139, 36190, B36191; 36593, 37049, 38285, 38310, 38582, 38679, 38886,
  39900, 39971

Fruit plants and trees (*cf.* also *Ananas*, *Armeniaca*, *Carica*, *Cerasus*, *Citrullus*,
    Citrus, *Cocos*, *Cucumis*, *Cucurbita*, *Cydonia*, *Ficus*, *Fragaria*, *Malus*, *Mangifera*,
    Musa, *Oxycoccus*, *Persea*, *Persica*, *Phoenix*, *Pirus*, *Prunus*, *Ribes*, *Rubus*, *Sor-
    bus*, *Vaccinium*, *Vitis*)
  21625, 21640, B23262, B23264; 25608; 28546, 30017, 32011; 38465, 38509, 39678

Fungi (parasitic)
>   21783, 21910, 21984, 22085, 22283, 22331, 22470, 22655, 22662, 22687, 22752,
>   22804-5, 22851, 22872, 22963, 23385, 23687, 23713, 23846, 24114, 24216-7,
>   24452, 24632, 25072, 25128
>   25271, 25514, 25705, 26107, 26234, 26369, 26777, 26993, 27042, 27059, 27675,
>   27751, 27877, 27941-2, 28296, 28374
>   28623, 28705, 28898, 29711, 29859, 29965-6, 30008, 30445, 31181, 31229, 31557,
>   31784, 32064, 32402
>   32850, 32869, 33372, 33477, 33599, 33909, 34146, 34413, 34443, 34501-2,
>   34847, 34851, 34894, 34898, 35210, 36047, 36053, 36184
>   36748, 37504, 37748, 37867, 37972, 38390, 38463, 38718, 38749, 38777, 39082,
>   39366-7, 39464, 39732, 39925, 40074, 40132, 40263

G

Garlic see *Allium*

Gherkin see *Cucumis*

Ginger see *Zingiber*

Glycine 21538, 21549, 21556, 21590, 21701, 21887, 21925, 22028, 22080, 22084, 22123,
>   22162-3, 22279, 22352, 22489, 22685, 22698, 22735, 22761-2, 22787, 22806,
>   22830-1, 22923, 22998, 23002-3, 23007, 23085, 23174-6, 23235, 23250-2, 23282-
>   -3, 23339, 23367-8, 23448, 23504, 23608-9, 23645, 23647, 23732, 23735, 23737,
>   23782, 23784, 23849, 23877, 23883, 23972, 23980, 23999, 24088, 24122-5, 24137,
>   24259-60, 24337, 24364, 24408, 24424, 24444, 24508, 24537, 24546, 24649, 24705,
>   24774, 24820, 24824, 24839, 24841, 24861, 24968, 24970, 24985, 25026, 25103,
>   25109, 25136-8
>   25173, 25186, 25223, 25308-9, 25331, 25336, 25407, 25521, 25576, 25646, 25693,
>   25721, 25727, 25877, 25881, 25917, 25965, 25990, 26093, 26106, 26124, 26134,
>   26190, 26318, 26337, 26360, 26373, 26440, 26450, 26463, 26474, 26476, 26618-9,
>   26684, 26729, 26749, 26784, 26898, 27003, 27163, 27215, 27237, 27272, 27287,
>   27336, 27397, 27415, 27421, 27501-2, 27530, 27540, 27707, 27719, 27730, 27800,
>   27821, 27849, 27887, 27897, 27929, 27935, 28041, 28130, 28144, 28182, 28228,
>   28267, 28309, 28311-3, 28342-4, 28347-8, 28426-8, 28478, B28482, 28495
>   28505, 28595, 28689, 28785, 28788, 28838, 28986, 28989, 29003-4, 29104,
>   29113, 29376, 29381, 29475, 29482, 29705, 29868-9, 29881, 29887, 29898-9,
>   29984, 30021-2, 30073, 30077, 30080, 30221, 30276-8, 30455, 30530, 30833,
>   30912, 30952, 30980, 30994-5, 31071, 31107, 31340, 31348, 31377, 31393,
>   31439, 31464, 31494, 31570, 31592, 31606, 31731-4, 31823, 31849, 31953-4,
>   32062, 32073, 32218-9, 32304, 32312-3, 32364, 32374, 32466, 32488
>   32686, 32765, 32871, 32879, 32888, 32923, 32956, 33002, 33019-21, 33060,
>   33068, 33076, 33092, 33103, 33181, 33222, B33281, 33369, 33394, 33453,
>   33472, 33525, 33548-9, 33703, 33846, 33871-3, 33875, 33916-8, 33952, 33999-
>   -4000, 34003, 34083, 34096, 34098, 34136, 34195, 34228, 34384, 34416, 34467,
>   34506, 34524, 34544, 34549, 34609-10, 34794, 34800, 34831, 34936-7, 34940,
>   35061, 35103, 35141-2, 35184, 35340-1, 35403, 35458, 35476, 35564, 35611,
>   35632, 35744, 35765, 35809, B35862, 35866-7, 35902-5, 35977, 35998, 36004,
>   36058, 36066, 36069, 36077, B36191, 36207, 36243, 36303, 36325, 36348, 36372,
>   36378, 36388, 36402, 36412, 36421, 36425, 36497
>   36597, 36600, 36603, 36784, 36791-2, 36827, 36840, 36879, ·36910, 36996,
>   37010-1, 37059, 37060, 37124-5, 37257, 37185, 37235, 37239, 37298, 37334,
>   37339-40, 37344, 37360, 37386, 37401, 37449, 37480, 37491-2, 37579, 37704,
>   37748, 37792, 37804, 37807, 37848, 37863, 37866, 37898, 37906, 37908, 37926,
>   37950, 37961, 37982-3, 38081-2, 38123, 38166, 38195, 38285, 38312, 38465,
>   38535, 38650, 38658, 38718, 38760, 38779, 38805, 38828, 38858-60, 38951,
>   38964-5, 38967, 38990, 39032, 39040, 39052, 39080, 39146, 39214-5, 39219,
>   39395, 39419, 39442, 39510-1, 39522, 39546, 39570, 39621, 39637, 39649, 39660-
>   -1, E39786, 39793, 39813, B39903, 39904, 39928, 39946, 40008, 40079, 40138,
>   40160, 40166, 40217, 40249-50, 40344-5

Gossypium 21705, 21740, 21787, 22024-7, 22043, B22115, 22123, 22266, 22356, 22403, 22484-5, 22509, 22521, 22660, 22734, 22762, 22826, 22870, 22964, 23587, 23602--4, 23717, 23756-7, B23767, 23768, 23979-80, 24001, 24036, 24088, 24143, 24149, 24331, 24424, 24508, 24527, 24716, 24804, 24806, 24839, 25037, 25088 25331, 25411-2, 25511, 25521, 25572, 25772, 25786, 25835, 25971, 26086, 26093, 26260, 26263, 26267, 26651, 26777, 27036, 27059, 27102, 27212-3, 27216, 27290, 27318, 27415, 27559, 27625, 27675, 27845-7, 27981, 28078, 28123, 28139, 28196, 28245, 28431
28500, 28505,   29652, 28749, 28882, 29007, 29164, 29203-4, 29334, 29536, 29662, 29984, 30033, 30067, 30095, 30287, 30301-2, 30415, 30445, 30513, 30734, 30818, 30995, 31008, 31230-1, 31233, 31252, 31293-4, 31340, 31348, 31439, 31512, 31537, 31557, 31593, 31769, 31843, 31849, 32024, 32103, 32287, 32459, 32462, 32482
32706, 32803,   33019-20, 33096, 33125, 33261, 33344, 33388, 33500, 33506, 33547, 33684, 33827, 33871-2, 33928, 34040, 34319, 34416, 34450, 34508, 34648, 34794, 34856, 34898, 34954, 35016, 35079-80, 35146, 35221, 35260-1, 35263, 35302, 35324-5, 35416, 35440, 35893, 36063, 36124, B36191, 36202, 36528, 36532
36590, 36641, 36777, 37017-8, 37060, 37071, 37123, 37133, 37150, 37260, 37298, 37334, 37344, 37499, 37701, 37838, 37898, 38062, 38100, 38285, 38316, 38518-9, 38650, 38673, 38790, 38868, 38949, 38995, 39037, 39058, 39074, 39138, 39154, 39224, 39232, 39243, B39523, 39532, B39786, 39815, B39903, 39991, 40009, 40017, 40264-5, 40267

Gourd   see *Cucurbita*

Gram chick pea see *Vigna*

Grape fruit see *Citrus*

Grape vine see *Vitis*

Grasses *(cf.*also *Avena, Bromus, Carex, Cynodon, Cyperus, Dactylis, Digitaria, Festuca. Hordeum, Lolium, Oryza, Panicum, Paspalum, Pennisetum, Phalaris, Phleum, Poa, Saccharum, Secale, Setaria, Sorgum, Triticum, Zea)*
21527, 21552-3, 21667, 21684, 21705, 21739, 21760, 21762, 21794-5, 21814, 21847, 21862, 21871, 21942, 21944, 21973, 22041, 22053, 22074-6, 22119-20, 22123, 22156, 22184, 22249, 22251, 22328, 22344, 22366, 22379, 22570, B22572, 22617, 22631, 22635, 22668, 22684, 22700, 22716, B22822, 22823, 22833-6, 22838-40, 22868, 22896, 22916, 22919, 22950, 22983, 23001, 23033, 23061, 23064, 23081, 23136, 23138, 23221, 23231, 23234, 23245, 23313, 23338, 23405, 23447-8, 23457, 23522, 23586, 23620, 23714, 23741, B23767, 23894, 23903-5, 23992, 24006, 24015, 24044, 24088, 24159-65, 24181, 24185, 24214-5, 24231, 24266, B24271, 24276, 24310, 24314, 24419, 24424, 24431, 24446, 24509, 24512, 24563, 24588, 24624, 24759, 24861, 24935, 24968, 25023, 25049, 25060, 25123 25331,   25422, 25431, 25473-4, 25531, 25538, 25596, 25613, 25660, 25674, 25711, 25753, 25772, 25873, 25877, 25910, 25912-3, 25946, 25978, 26065, 26074-5, 26145, 26164, 26181, 26189, 26216, 26267, 26308, 26316, 26320, 26390, 26442, 26451-2, 26501, 26532-3, 26539, 26625, 26803-4, 26854, 26919, 26947, 26973, 27002, 27098, 27139, 27242, 27322, 27421, 27428, 27439, 27480, 27493, 27530, 27580-5, 27597, 27630, 27656, 27686, 27769, 27814, 27819, 27830, 27893-4, 28013, 28078, 28099-100, 28158, 28160, 28187-8, 28228, 28280, 28323, 28343, 28397, 28406, 28454, 28489
28838, 28875, 28970-3, 29024, 29057, B29063, 29064-5, 29129, 29181, 29187, 29232, 29332, 29376, 29407, 29409, 29419, 29456-7, 29480, 29515, 29542, 29573, 29581, 29638, 29663, 29669, 29676, 29685, 29704, 29744, 29863, 29865, 29911, 29985, 30009, 30044, 30048, 30077, 30083, 30098, 30104-5, 30140-1, 30270, 30273, 30354, 30379, 30393, 30459, 30483, 30533, 30573, 30633, 30763, 30847, 30945, 30951, 31070, 31131, 31217, 31383, 31392, 31411, 31464, 31476, 31510, 31530, 31633, 31697, 31748, 31759-62, 31766, 31849, 31920, 31993, 32047, 32098, 32117, 32144, 32151-2, 32186-7, 32189, 32259, 32324, 32363, 32477, 32542-3, 32547

(continued)

Grasses (continued)
32576, 32625, 32653, 32867, 32882, 32888, 32895, 32950, 33014, 33050, 33130,
33158-9, 33167, 33176, 33178-9, 33209, 33216, 33232, 33331, 33335, 33378-9,
33395, 33412, 33433, 33440, 33537, 33547, 33607, 33620, 33730, 33742, 33756,
33843, B33926, 34098, 34100, 34134-5, 34137, 34269, 34308, 34351, 34365-6,
34387, 34442, 34444, 34452, 34458, 34499, 34536, 34541, 34613, 34646-7,
34676, 34825, 34907, 34924, 34926, 35031, 35037, 35040, 35056, 35118, 35168,
35174, 35176, 35186, 35243, 35272, 35301, 35403, 35426, 35441, 35468-9,
35484, 35488, 35529, 35550, 35607-8, 35674, 35678, 35711, 35744, 35885,
35891, 35924, 36014, 36027, 36037, 36070, 36092, 36103, 36115, 36132-3,
36135, 36165, 36252, 36333, 36340, 36386-7, 36424, 36445, 36457
36656, 36659, 36692, 36694, 36697, 36708, 36733, 36839, 36885, 36956, 37022,
37073, 37084, 37086, 37116, 37239, 37269, 37289-92, 37323, 37331, 37344,
37382, 37475, 37500-1, 37511, 37552, 37571, 37574, 37609, 37612, 37657-8,
37701, B37727, 37751, 37826-7, 37857, 37863, 37870, 37889, 37901, 37912,
37981-3, 38044, 38056, 38125, 38158, 38185, B38211, 38285, 38313, 38341,
38358, 38385, 38420, 38422-3, B38477, 38516, 38551, 38704, 38716-7, 38776,
38785-6, 38793, 38799, 38829, 39069, 39072, 39089, 39262, B39394, 39434,
39460, 39512, B39523, 39589, 39664, 39690-2, 39808-9, 39917-8, 40119, 40127,
40145, 40199, 40219-23, 40277, 40292, 40337, 40361

Groundnut see *Arachis*

H

Halobacterium
21714-6, 22022, 22045, 22059-60, 22179, 22218, 22753, 22760, 23432, 23458,
23495-6, 24056, 24195, 25110
25191, 25296, 25298, 25337, 25351-2, 25369, 25461, 25547, 25561, 25803-4,
25927, 26180, 26283, 26365, 26379, 26439, 26611, 26636-7, 26728, 26819-21,
26850-1, 26963, 27305-7, 27545, 27626, 27828-9, 27836, 27890, 27905, 28042-3,
28150, 28293, 28298-9, 28378
28722, 28826, 28831, 28848, 28943, 29028-9, 29034, 29084, 29095, 29099-
-100, 29225, 29227-9, 29387-9, 29420, 29507, 29606-8, 29632, 29659-60, 29692,
29742, 29847, 29854, 29900, 29926, 29929, 29935, 30064-5, 30192, 30258, 30269,
30284, 30384, 30391, 30519-21, 30614-5, 30632, 30649, 30679, 30785, 31029,
31036, 31110-1, 31150, 31167, 31173-5, 31286, 31309-10, 31458, 31491, 31497,
31548, 31773, 31838, 31997, 32167, 32330, 32507
32559-9, 32655, 32727, 32734, 32781, 32793, 32800, 32878-9, 32901,
32903, 33046, 33048, 33075, 33099, 33228, 33267, 33275-7, 33313, 33326,
33328, 33376, 33421, 33437, 33443-4, 33636-8, 33650, 33681-2, 33752, 33833,
33862, 33903, 33924, 33947-9, 33966, 33972-3, 33993, 34015, 34028, 34077-8,
34084-5, 34221, 34263, 34270, 34345, 34391-2, 34398, 34449, 34516, 34590,
34604, 34748, 34900, 34977, 35120, 35135, 35185, 35189, 35199, 35208, 35228-
-9, 35253, 35312, 35486, 35551, 35565, 35568, 35718, 35722-3, 35782, 35805,
35817, 35829-31, 35913, 35926, 35951, 36006-8, 36131, 36160-1, 36317-8,
36321-2, 36349, 36373-5, 36380
36743, 36915, 36954, 36999, 37005, 37214, 37231, 37248, 37283-4, 37345, 37353,
37403-5, 37424, 37508, 37538, 37583, 37608, 37625, 37720, 37764, 37816, 37853-
-4, 37865, 37910-1, 38032, 38042, 38184, 38197, 38278, 38282-3, 38330, 38368-
70, 38403-5, 38432, 38491, 38494, 38595, 38712, 38815, 38817, 38821, 38937-8,
38986-7, 39011, 39021-2, B39220, 39228, 39302, 39320, 39340, 39362, 39385,
39396, 39470, 39495-6, 39575, 39670-1, 39729, 39735, 39780, 39782-3, 39878-9,
39914-6, 39978, 40094, 40110, 40148, 40150, 40171, 40338, 40370

Halophilous plants (*cf.* also Salt marsh and strand plants)
22251, 22380, 22544, 23902, 24070, 24378, 24755; 25510, 26033, 28071, 28138,
28151, 28288-9; 28728, 29061, 29489, 29495, 29865, 31070, 32373, 32477; 32562,
32681, 32895, 32996, 33554-5, 33696, 33900, 35186, 35401, 36445; 36679, 36831,
36843, 37022, 37331, 37333, 37348, 37476, 37893, B38477, 38576, 38746, 39240,
39243, B39537, 39809, 40126, 40198, 40300

Hawthorn see *Crataegus*

Hazel see *Corylus*

Heath plants and communities
          22034-5; 26739; 33925, B33926, 34190, 34880, 35792, 35928, 36037; 37168,
          37174, 37709

Hedera 21705, 24557; 25685, 27060-1, 27109; 29398, 29489, 30173, 30276-7, 30730,
          31471, 31619, 31692, 31769; 32772-3, 34260, 35006, 36018, 36345; 36806-7,
          37027, 37159, 37416, 38285, 39283, 39437, 40335

Helianthus
          21708, 21729, 21785,   21847, 21897, 21943, 21970, 21973, 21995, 22031,
          22197, 22246-7, 22273, 22415, 22651, 22734, 22772, 22968, 23026, 23036, 23052,
          23117, 23250-1, 23329, 23339, 23408, 23448, 23547, 23625, 23673, 23703, 23714,
          B23767, 23814, 24000-1, 24110, 24143, 24196, 24216, 24364, 24441, 24537,
          24595, 24598, 24610, 24627, 24716, 24839, 24863, 24928, 25136, 25138
          25251, 25331, 25445, 25511, 25521, 25640, 25833,25877,25954,25982,26007,26037,
          26048, B26099, 26132, 26237, 26264, 26267, 26360,26449,26579,26747,26779,26844,
          26852, 26991, 27163, 27287, 27468-9, 27524, 27546, 27592, 27595, 27728, 27755,
          27820, 27897, 27902-3, 27980, 28303, 28425, 28478, 28489
          28505, 28774,   29054, 29328, 29372-3, 29485, 29514, 29523, 29793, 29858,
          29912, 29932, 29982, 30011, 30074, 30106, 30132, 30146, 30199, 30253, 30353,
          B30511, 30542-4, 30622, 30808, 30852, 30892, 31168, 31257, 31267, 31411,
          31464, 31494, 31537, 31604, 31620, 31633, 31790, 31801, 31859, 31926, 31932,
          32073, 32111, 32220, 32304, 32310, 32343, 32462, 32469
          32577, 32629,   32690, 32788, 32828, 32867-8, 32948, 32969, 33068-9,
          33087, 33089, 33291, 33310, 33424, 33426, 33524, 33561, 33742, 33808, 33818,
          33827, 34037, 34271, 34325-6, 34585, 34659, 34794, 34877, 34940, 34954,
          35006, 35029, 35105, 35142, 35321, 35341, 35371, 35544, 35656, 35707, 35824,
          35842, 35972-3, 35979, 36058, 36060, 36184, B36191, 36206, 36225, 36333
          36617, 36732, 36763, 36765, 36983, 37084, 37101, 37224, 37271, 37298, 37334,
          37336-7, 37339-40, 37434, 37515-6, 37523, 37527, 37577, 37701-2, 37742,
          37752, 37863, 37898, 38134, B38139, 38185, 38257, 38285, 38421, 38429, 38439,
          38614, 38616, 38644, 38650, 38701, 38951, 38962, 39044-5, 39173, 39206, 39207,
          39260, 39270, 39333, 39350, 39380, B39523, B39537, 39595, B39711, 39740,
          39750,. 39751-2, 39773, B39786, 39828, 39860, 39897, 39900, 39946, 39991,
          40060-1, 40201, 40216, 40344-5

Hemlock see *Tsuga*

Hemp see *Cannabis*

Hevea B21979; 25627, 27506; 29018-9, 30929, 31596; 33054-5, 35634; 38036, 39420-3

Hibiscus 21511-2, 21659, 23023-4, 23066, 24739; 25170; 28838; 33718, 35600, 36386;
          36905, 37217, 39069

Hickory see *Carya*

Holly see *Ilex*

Hop see *Humulus*

Hordeum 21520, 21529, 21532, 21558, 21575, 21627, 21630, 21710, 21801-3, 21825,
          21860, 21870, 21984, 22001, 22011, 22028, 22123, 22156, 22202, 22240, 22281;
          22348, 22441, 22462, 22470, 22479, 22589, 22596-7, 22600-2, 22618-9, 22627,
          22674-5, 22696, 22737, 22800, 22804-5, 22829, 22833, 22836, 22840, 22945,
          22952, 22987, 22990, 23009, 23014, 23072, 23115-6, 23221, 23231, 23357, 23398,
          23402-3, 23428-9, 23443, 23468, 23493-4, 23529, 23531, 23547, 23555, 23586,
          23642, 23681, 23704, B23767, 23773, 23775, 23777-80, 23821-3, 23833, 23859,
          23944, 23954, 23960, 23969, 23973-4, 24001, 24016, 24055, 24076, 24079, 24158,
(continued)

Hordeum (continued)
　　24198-9, 24247, 24275, 24280, 24284, 24328, 24330, 24364, 24397, 24429-30,
　　24459-65, 24468, 24545, 24590, 24625, 24633, 24716, 24734, 24827, 24889-90,
　　24912, 24916, 24923, 24933, 24940, 24962, 25015, 25037, 25052, 25072, 25090,
　　25111, 25134, 25136
　　25185, 25199-200, 25224, 25232, 25236, 25256, 25413, 25430-4, 25445, 25456,
　　25496-7, 25521, 25564, 25585, 25693, 25715, 25748, 25772, 25838, 25858,
　　25912-3, 25918, 25988, 26053, 26063, 26068, 26076, 26085, 26092, B26099,
　　26113, 26181, 26191, 26202, 26205, 26215, 26234, 26256, 26267, 26331, 26360,
　　26432, 26444, 26474, 26476, 26531, 26539, 26545, 26562, 26572, 26575, 26586,
　　26594, 26626, 26699, 26758, 26802, 26925, 26927-8, 26930-1, 26940, 26947,
　　26980, 26995, 27022, 27027, 27043, 27082, 27137, 27234, 27238, 27276-7,
　　27284, 27287, 27292, 27346, 27439, 27443, 27488-9, 27503, 27538, 27584,
　　27592, 27614, 27634, 27687-8, 27690,   27729-30,   27751, 27771, 27784,
　　27851-2, 27854, 27922-4, 27997, 28032, 28085-6, 28139, 28189, 28200, 28205-6,
　　28220, 28267, 28281, 28285, 28406, 28408, 28463, 28487
　　28496, 28521,   28578, 28595, 28618, 28623, 28628, 28639, 28667-8, 28689,
　　28745, 28755, 28777, 28789, 28809-10, 28817, 28836, 28973, 29008, 29040,
　　29113, 29123, 29221-3, 29270, 29289, 29338, 29350, 29405, 29410, 29456, 29478,
　　29489, 29546-7, 29555, 29593, 29610-1, 29699, 29708, 29760, 29765, 29823,
　　29859, 29875, 29903-5, 29932, 29939, 30027-8, 30034, 30077, 30081, 30097-8,
　　30132, 30163, 30201-2, 30215, 30312, 30334, 30345, 30404, 30478, 30557, 30568,
　　30594, 30597, 30602-3, 30610-2, 30669, 30687-8, 30693-4, 30718, 30740-1,
　　30793, 30801, 30826, 30844, 30847, 30857, 30891, 30917, 30920, 30995, 31007,
　　31013, 31054, 31075, 31160, 31169, 31192, 31209, 31226-7, 31279, 31381, 31431,
　　31444, 31463-4, 31494, 31504, 31533, 31547, 31572, 31587, 31715, 31742, 31752,
　　31795, 31829, 31830, 31842, 31849, 31876-8, 31930, 31952, 31978, 32046, 32084,
　　32086, 32118, 32176, 32232, 32266, 32298-9, 32312, 32361, 32387, 32390, 32397-
　　8, 32431, 32477, 32483, 32530
　　32579, 32597, 32626, 32650-2, 32707, 32779, 32799, 32820, 32862, 32883,
　　32889-90, 32917, 32950, 32956, 32958, 32970, 33004, 33025, 33030, 33062,
　　33064, 33177, 33293, 33309, B33343, 33373, 33375, 33493, 33530, 33547, 33578-
　　-9, 33589-91, 33599, 33639, 33670, 33781-2, 33805, 33808, 33827, 33871-2,
　　33875, 33895, 33909, 33953-5, 33967, 33953, 34062-5, 34076, 34080, 34094,
　　34096, 34123, 34142, 34220, 34235-6, 34331, 34467, 34513, 34534, 34598,
　　34606, 34608, 34610, 34635, B34706, 34707, 34726, 34732, 34739, 34746, 34758,
　　34817, 34823, 34828, 34882, 34894, 34986, 35006, 35026, 35029, 35039, 35042,
　　35064, 35067-8, 35112, 35121, 35247-8, 35257, 35372, 35378, 35396, 35490,
　　35505, 35519, 35550, 35602, 35614, 35617-8, 35627, 35671, 35754, 35766,
　　35776, 35778, 35809, 35835, 35874-6, 35882, 35887, 35891, 35893, 35921-2,
　　35988, 36015, 36029-30, 36057, 36067-8, 36101, 36205, 36211, 36235, 36259,
　　36299, 36303, 36315, 36342, 36431, 36521-3
　　36590, 36624, 36631-2, 36658, 36690-2, 36728, 36732, 36748, 36767, 36837,
　　36867, 36899, 36957, 36973, 36986, 37007, 37022, 37034, 37056, 37060, 37150,
　　37182, 37298, 37309, 37334, 37336, 37340, 37342, 37349, 37356-7, 37366-7,
　　37504, 37534, 37539, 37559, 37589-91, 37687, 37701, 37703, 37738, 37742, 37747,
　　37761, 37771, 37792, 37803, 37863, 37866, 37872, 37898, 37918, 37928, 37930-1,
　　37935, 38119, 38130, B38139, 38183, 38186,   38228-30,   38252, 38285,
　　38294, 38350, 38353, 38376, 38421, 38433, 38436, 38455, B38477, 38479, 38489,
　　38497, 38530, 38557, 38609, 38617, 38663, 38680, 38705, 38723, 38728, 38730,
　　38741, 38804, 38823, 38877, 38896, 38898, 38988, 39042, 39064-5, 39069, 39081-
　　-2, 39112, 39156-7, 39168, 39174-5, 39183, 39194, 39218, 39261, 39282, 39285,
　　39292, 39316, 39318, 39334, 39366-7, 39397, 39517, 39548, 39562, 39646, 39656,
　　39678, B39711, 39727, 39764, B39786, 39801, 39843, 39889, B39903, 39927-8,
　　39989-91, 40046, B40105, 40113, 40122, 40125, 40144, 40165, 40225, 40250,
　　40306, 40312-4, 40335, 40360

Hornbeam see *Carpinus*

Horse chestnut see *Aesculus*

Horseradish see *Armoracia*

Horsetail see *Equisetum*

Humulus   23557, 23829, 24267, 24532, 24771, 25103; 25312, 28139; 38533

I

Ilex   22302, 22924, 23556, 24557; 30787; 34854, 35537, 35691; 37220, 39328

Ipomoea   22028, 22250, 24468, 24820, 24883; 25810, 26433-4, 26476, 26607, 27404,
        28245, 28355; 29427, 29789-90, 30030; 33168, 33549, 34264, 34807, 35330-2,
        36066, 36338, 36478; 37362, 38148, 38588-9, 39130, 39160-1, 39429, 39900

Ivy   see   *Hedera*

J

Jerusalem artichoke   see   *Helianthus*

Jointgrass   see   *Paspalum*

Juglans   22190, 22719, 22761; 25989, 26534, 28145; 29243, 31515; 32939, 35006,
        35282; 37874, 36285, 839903

Juniper   see   *Juniperus*

Juniperus   25507, 25592, 25807, 26857, 27348; 28987, 29250; 33378, 33440; 37049,
        37874, 38285, 38493, 3982}

Jute   see   *Corchorus*

K

Kalanchoë   21652, 21748, 21885, 22098-9, 22508, 22963, 23011, 23121, 23374, 23379,
        23481, 23764, 23928, 24126, 24656-8, 24742, 25118; 25445, 25787, 25862-3,
        26477, 27359-60, 28308; 29092-3, 29142, 29217, 29273, 29376, 29401, 29772,
        29865, 30665-6, 30937, 30952, 31999, 32109, 32458; 34585, 34684, 34952,
        35326, 35466; 36663-4, 36685, 37028, 37274, 37317-8, 38235-6, 38285, 38640-1,
        38788-9, 38961, 39132, 39216-7, 39380, 39725-6, 39830, 40237

Kale   see   *Brassica*

Kenaf   see   *Hibiscus*

Kohlrabi   see   *Brassica*

L

Lactuca   21613, 21671, 21744-5, 21871-2, 22091, 22131, 22141-2, 22233, 22372, 22483,
        22615, 22647, 22747, 22797, 22996, 23477, 23568, 23577, 23717, 23728-30,
                                                                    (continued)

Lactuca (continued)
        24042, 24130, 24166, 24321, 24416-7, 24468, 24471, 24488-91, 24716, 24750,
        24870, 24910, 24952, 25093, 25152
        25176-7, 25375, 25781, 25877, 25935, 26123, 26353, 26388, 26476, 26559, 27140,
        27245, 27300, 27439, 27454, 27456, 27566, 27571, 27602, 27690, 27738, 27745,
        27758, 27839, 27874, 27992, 28130, 28139, 28235, 28315
        28509, 28682, 28689, 28754, 28844, 28934, 28936, 29105, 29271, 29529, 29858,
        29907, 30182-3, 30291, 30718, 30801, 30950, 30971-2, 30974, 31028-9, 31684,
        31726, 31746-7, 31799, 31809-10, 31849, 31858, 31942, 32103, 32122, 32252,
        32412
        32578, 32736, 32829, 32896, 32948, 32996, 33048, 33060, 33078, 33089,
        33347, 33399, 33650, 33670, 33748-51, 33956-7, 34530, 34629, 34924, 35054,
        35291, 35456, 35511, 35752, 35838, 35859, 35982-4, 36126, 36167, 36175,
        36375, 36499
        36914, 37017, 37022, 37146, 37178, 37272, 37469, 37534, 37806, 37894, 38285,
        B38475, 38595, 38816, 38882, 38888, 38927, 39126, 39154, 39252, 39608, 39622,
        39629, 39631, 39678, 39887, 40033, 40201

Larch   see *Larix*

Larix   21633, 21705, 23091, 23319, 23500, 23607, 23830, 24240, 24716, 24958; 25371,
        26078, B26099, 26150, 26515, 27105, 27124, 27260; 28793, 29569, 30040, 30263,
        30736, 30889, 31053; 33221, 33268, 33920, 34654, 35808, 35949, 36049; 37036,
        37677, 38493, 38900, B39951, 39991

Lathyrus   23154, 23156, 23168; 29094, 31208, 31569; 32739, 32843, 34924, 25605;
        B38475, 39641-4

Leguminous plants (*cf.* also *Arachis, Cajanus, Cicer, Glycine, Lathyrus, Lens, Lupinus,*
        Medicago, Phaseolus, Pisum, Trifolium, Vicia, Vigna)
        21705, 22631, 22823, 23154-6, 23714, 24012; 25613, 26466, 26973, 27389, 27812;
        28782, 29380, 29832, 29923, 30376, 31228, 31771; 32739, 32762, 33547, 34049,
        34825, 35464, 35828; 36799, 37701, 38534

Lemna   see *Lemnaceae*

Lemnaceae   21509, 22472-3, 22644, 22819, 23005, 23189, 23494, 24145-6, 24166, 24234,
        24669; 25338, 25349, 25418, 25658, 25854, 26071, 26088, 26291-2, 26322-3,
        26377, 26701, 27101, 27372-3, 27470, 27730, 27809, 28303; 29385-6, 29616,
        30350, 30533, 30549-51, 30817, 30914, 30957, 31344, 31545, 31810, 32278,
        32447; 32571, 32645, 32681, 32780, 33419, 33600-1, 33739, 34052, 34060, 34145,
        34726, 35110-1, 35426, 35497-500, 36083-4, 36219; 36822, 37001, 37107, 37502,
        37511, 37697-8, 37708, 37934, 38126, 38276, 38492, 38513, 38522, 38593,
        38842, 39106, 39145, 39169, 39921, 40158, 40161-2, 40251, 40371

Lemon   see *Citrus*

Lens   23771-2, 24508; 25965, 27237; 31842; 36438; 37298

Lentil   see *Lens*

Lettuce   see *Lactuca*

Lichens   21628, 21759,21761, 21910, 22053, 22404, 22412, 22665, 22993, 23015, 23027,
        23069-70, 23121, 23298, 23305-8, 23342, 23423, 23716, 24173, 24614, 24834-5,
        25015, 25025, 25054
        25795, 25991-3, 26005-6, 26186, 26296, 26754, 27763, 27824, 27914, 28263
        28607, 28702, 28961-2, 29421, 29435, 29806, 30204, 30285-6, 30392, 30523-4,
        31006, 31052, 31388, 31392, 31500, 31540, 31798, 31882, 32146
        32990, 33084, 33215, 33249, 33264, 33476-7, 33491, 33507, 33660, 33756,
        33974, 34239, 34302, 34518, 34547, 34966, 35281, 35691, 36361, 36408, 36427,
        36572
        36695, 36750, 36880, 37208-9, 37487, 38132, 38181, 38410-1, 38414, 38444,
        38622, 38866, 39199-200, 39250, 39313, 39381, 39503, 39605, 39628, 39821, 40072

Lilac see *Syringa*

Linden see *Tilia*

Linseed see *Linum*

Linum 21880, 21882, 22240, 24075, 24077-8, 24141; 25845, 27142, 27387, 28185;
    28527, 28743, 30068, 30400, 31194, 31237, 31361; 34087, 35221, 35379; 36617,
    36752, 37150, 37298, 37339-40, 37877, 37951, 39039, B39903, 40319-20

Liverworts 24246; 25187, 26244, 27622, 27727; 29446, 29769, 30596, 31112; 32681,
    32989, 34700, 35018, 35463, 36051, 36057, 36060; 37820, 38149, 39253, 39383,
    39895

Locust see *Robinia*

Lolium 21624, 21692, 22039, 22096, 22112, 22504, 22516, 22676, 22779, 23061, 23063,
    23331, 23713, 23897, 23983-4, 24033, 24100, 24201, 24375, 24435-8, 24744-5,
    25045-7
    25274-5, 25324, 25364, 25445, 25832, 25842, 26865, 26874, 26979, 27333, 27337,
    27396, 27473, 27688, 27714, 27819, 28099, 28191, 28337-8, 28344, 28413
    28521, 28527, 28602, 28689, 28704, 28743, 28757, 28896, 28899, 28925, 28973,
    29122, 29604, 29819-20, 29883, 30140, 30167-8, 30257, 30310, 30463-4, 30483,
    30891, 31220, 31261, 31527, 31759-62, 31910, 32170-1, 32449-50
    32686, 32867, 32980, 33130, 33193, 33242, 33488, 33633, 33859, 34100, 34176,
    34181, 34408, 34458-9, 34541, 34649, 35011, 35158, 35176, 35373, 35508,
    35542-3, 35594-5, 36060, 36111-2, 36165, B36191, 36205, 36348, 36433, 36457
    36692, 36845, 36980, 37121, 37344, 37609, 37801, 38210, 38285, 38313, 38666,
    38968, 39097, 39127, 39259, 39434, 39565, 39683-4, 40117, 40142, 40144, 40254,
    40280

Lucerne see *Medicago*

Lupine see *Lupinus*

Lupinus 22285-6, 22593, 22755, 23026, 23562, 23588, 24007; 25311, 25566, 27237,
    27812, 27834; 28614, 28727, 29115, 29323, 29758, 30507-8, 31379, 31657; 32665-
    -6, 32691, 32948, 33214, 33216, 33791-2, 33961-3, 34506, 35029, 35255-6,
    35338, 35464, 35554, 36057; 36829, 36963, 37339, 38421, 38431, 39066-7,
    B39537, 40100, 40201, 40240-3

Lycopersicon
    21570-1, 21676, 21686, 21860, 21948, 21977, 22028, 22037, 22130, 22188,
    22719, 22747, 22818, 22860, 22999, 23075, 23101, 23106, 23323-5, 23467,
    23502, 23528, 23565, 23690, 23694, 23716, 23728-9, 23738, 23853, 23878,
    24071, 24094, 24129, 24229-30, 24299, 24372, 24468, 24527, 24645, 24716,
    24806, 24808, 24883-4, 25117
    25167, 25172, 25259-61, 25310, 25431, 25511, 25521, 25623, 25641, 25682,
    25693, 25696, 25754, 25771, 25835, 25877, 26048, 26073, 26157, 26186, 26229,
    26260, 26336, 26360, 26383-5, 26476, 26495, 26591, 26648, 26699, 26748, 26779,
    B27001, 27041, 27047, 27148, 27159, 27205, 27209, 27283, 27508, 27510, 27538,
    27543, 27579, 27643, 27679, 27690, 27730, 27771, 27777, 27793, 27884, 28130,
    28139, 28228, 28245, 28267, 28270, 28302, 28315, 28346, 28399
    28705, 28732, 28928-9, 28931, 28978, 29036, 29050, 29424,   29462-3,
    29586, 29603, 29951, 29954, 30010, 30080, 30090, 30199, 30294, 30377, 30427,
    B30477, 30537, 30656, 30797, 30833-4, 30883, 30920, 30972, 30974, 30976,
    31122, 31187, 31229, 31303, 31399, 31558, 31657, 31692, 31792, 31855, 31983,
    32339-41, 32373
    32568, 32639, 32741, 32759, 32796, 32948, 33010, 33093, 33219, 33642,
    33644, 33761, 33827, 33967, 33976, 34005-6, 34098, 34163, 34212, 34232-4,
    34241-2, 34250, 34252-6, 34271, 34425, 34448, 34506, 34649, 34683, 34851,
    34939, 34972, 35004, 35006, 35029, 35060, 35128, 35142, 35259, 35335, 35601,
    35613, 35633, 35776, 35971, 36060, 36213, 36329, 36335, 36571

                                                            (continued

Lycopersicon (continued)
      36590-1, 36642, 36736, 36803, 36898, 36898, 36963, 36974, 37022, 37027, 37039,
      37108, 37123, 37172, 37213, 37262, 37313, 37336, 37340, 37510, 37620, 37681,
      37713, 37724, 37812, 37863, 37866, 37886-8, 38062, 38129, 38163, 38285, 38511,
      38625, 38635, 38644, 38655, 38675, 38690-1, 38695-6, 38720, 38774, 38878-9,
      38902, 38933, 38975-6, 39029, 39107, 39134, 39163, 39183, 39187, 39218, 39251,
      B39523, B39537, 39596, 39677, 39679, 39702-3, 39755-6, 39749, B39786, 39889,
      B39903, 39991, 40000, 40003, 40018, 40092, 40253, 40292, 40329

Lycopodium 32680-1, 33264, 34917, 35014

M

Macereed see *Typha*

Maize see *Zea*

Malus 21625, 21640, 22028, 22085, 22418, 22661, 22678, 22855, 23045, B23262, B23264,
      23295, 23341, 23382, 23406, 23846, 23887, 24270, 24336, 24363, 24531, 24602,
      24641
      25162, 25328, 25683, 26013-4, 26246, 26269, 26723, 26840, 26843, 26869, 26996,
      27235, 27514-5, 27520-1, 27693, 27733, 28095, 28101, 28156, 28367
      28629-30, 28661, 28685, 29005, 29066, 29251, 29324, 29368, 29702, 29822, 30017,
      30081, B30511, 30620, 30748, 30917, 31187, 31197-9, 31591, 31711-2, 31750-1,
      31782, 32011, 32120
      32708, 32729, 32751, 32846, 33089, 33104, 33285-6, 33322, 33504, 33522,
      33532, 33697-8, 33733, 33860, 34463-4, 34480, 34483, 34494-5, 34497, 34665,
      34742, 34924, 35201, 35395, 35577-8, 35740, 35784, B35826, B35862, 36078,
      36086, 36120, 36371, 36560
      36613, 36742, 36776, 36780, 36945, 37150, 37381, 37488-9, 37519, 37734, 37802,
      37847, 37991, 38034, 38127-8, 38156, 38242, 38285, 38389, 38465, B38477,
      38532, 38558, 38572, 38607, 38777, 38829, 38867, 38869, 39029, 39563, 39732,
      39967, 40154

Mangifera 24751; 31841; 33129, 35251; 37150, 38106

Mango see *Mangifera*

Mangold see *Beta*

Mangrove communities 29116, 29327; 33166, 33292, 34887; 40252

Manihot 22845, 23448; 26267, 27011, 27431, 27506; 28606, 28838, 30705-6; 32758,
      32867, 32888, 35064, 35297; 37176, 37184, 38196, 38629, 40250

Manioc see *Manihot*

Maple see *Acer*

Marrow see *Cucurbita*

Meadowgrass see *Poa*

Medicago 21748, 21960, 22030, 22123, 22130, 22279, 22660, 22792, 23194, 23338,
      B23407, 23448, 23534, 23588, 23895-6, 24080, 24558, 24574, 24716, 24723,
      25037, 25137, 25155
      25331, 25445, 25450-1, 25511, 25546, 25687, 25693, 25714, 25737, 25772, 25852,
      26055, 26145, 26267, 26305, 26409, 26546, 27108, 27284, 27398, 27461-2, 27464,
      28078, 28139

                                                        (continued)

Medicago (continued)
     28813, 28827,   28838, 29018, 29021, 29237, B29468, 29663, 29853, 30019,
     30371, 30455, 30458, 30619, 30707, 30824, 30847, 30932, 30995, 31264, 31313-5,
     31408, 31784, 31849, 31859, 32067, 32069, 32147, 32547
     32582, 32686,   32762, 32888, 33223, 33294, 33338, 33488, 33547, 33894,
     33904, 34052, 34100, 34142, 34301, 34541, 35100, 35158, 35295-6, 35341,
     35359, 35488, 35729, 35775, 35895, 36124, 36166, 36391, 36445, 36458
     36623, 36669, 36700, 37150, 37300, 37339, 37382, 37456, 37468, 37864, 38062,
     38438, 38747, 38776, 38842, 38979-80, 39099, 39150, 39218, 39223, 39244,
     B39523, B39537, 39566-7, 39636, 39667, B39903, 39991, 40066, 40087, 40275,
     40289, 40292, 40344-5

Medicinal plants (*cf.* also *Carica, Cynodon, Hibiscus, Papaver, Ricinus, etc.*)
     22028, 22209-10, 22380, 22963, 23166, 23386, 23471, 23680, 24559, 24568;
     26853, 28069; 29055-6, 29342, 30584, 31416, 32324; 32665-6, 32766, 32819,
     32948, 33054, 33089, 33440, 34174, 34924, 35006, 35687, 35930, 35932; 36956,
     36963, 37239, 37344, 37614, 37975, 37997, 38059, 38300, 38701, B39537, 39678,
     39699, 40318

Melon see *Colocynthis, Cucumis*

Metasequoia 25186; 32681; 38493

Millet see *Panicum*

Morus 23727, 24325-6, 24528, 24605, 24701, 25095, 25097; 26534, 26775, 27207, 27725,
     27730, 28315; 29320, 31627-8; 34940, 34997, 35149, 36060, 36225, 36509; 38679

Mosses (*cf.* also *Sphagnum*)
     21626, 21699, 21737, 21759, 22053, 22056, 22231, 22270, 22543, 22644, 22767,
     22992-4, 23189, 23317, 23342, 23460, 23616, 23632, 23789, 23872, 23969, 24017,
     24091, 24137, 24554, 24557, 24833, 24961, 24989-90, 25025
     25252, 25700-1, 25858-60, 26088, 26166, 26186, 26300, 26370, 26601, 26952,
     26994, 27160, 27241, 27301-2, 27782, 28158, 28221-2, 28267, 28435
     28521, 28596, 29101, 29727, 29770, 29931, 30120, 30237-8, 30242, 30448, 30523,
     30596, 31065, 31067, 31204, 31378, 31392, 31694, 32012, 32230-1, 32551-2
     32676, 32679-81, 32694, 32803, 32837-8, 33028-9, 33052, 33410, 33660, 33756,
     33804, 33875, 34171, 34288-90, 34421, 34440, 34445, 34669, 34715, 34765,
     34886, 34917, 34924, 35014, 35094, 35129, 35244, 35463, 35525, 35928, 36013,
     36172, 36201, 36233-4
     36712, 36875, 36880, 37140, 37308, 37336, 37431, 37654, B37727, 38143-4,
     38335, 38414, 38445, B38477, 39383, 39665, 39935, 40016, 40196, 40358, 40369

Mulberry see *Morus*

Mung bean see *Vigna*

Musa 22747, 24468; 26225-6; 30994, 32033; 35426; 37150, 38285, 39775, B39903

Musk-melon see *Cucumis*

Mustard see *Sinapis*

N

Napier-grass see *Pennisetum*

Nicotiana
     21549, 21613,   21665, 21705, 21724, 21771, 21791, 21925, 21961, 21994,
     22023, 22050, 22066, 22069-70, 22097, 22296, 22313, 22396, 22484, 22529,
     22647, 22719, 22734, 22739-41, 22796, 22835, 22851-2, 22980-1, 23051, 23180,
                                                                    (continued)

Nicotiana (continued)
       23231, 23254-6, 23343, 23369, 23373, 23404, 23427, 23512, 23547, 23588, 23618,
       23624, 23629, 23643, 23647, 23714, 23717, 23725, B23767, 23788, 23870, 23887,
       23920, 23954, 23969, B24024, 24064, 24113, 24136, 24190, 24351, 24364, 24414,
       24424, 24442, 24476, 24536-7, 24655, 24683, 24693-4, 24716, 24748, 24754,
       24779, 24804, 24853, 25044, 25079, 25108, 25136-8
       25181, 25186, 25204, 25429, 25445, 25457, 25521, 25692-3, 25710, 25713, 25891,
       25912, 25935, 25982, 26037, B26099, 26187, 26208-9, 26297, 26340, 26349, 26360,
       26398, 26402, 26444, 26505, 26507, 26539, 26575, 26602, 26716, 26719, 26767,
       26809-10, 26814, 26886, 26899, 26997-8, 27063, 27111, 27142, 27164, 27190,
       27242, 27560, 27571, 27584, 27592, 27595, 27840, 27935, 27937, 27967, 28003,
       28032-3, 28039, 28139, 28146, 28216, 28224, 28260, 28267, 28274, 28303, 28385,
       28387, 28398, 28416-7, 28477-8, 28480
       28652, 28659,   28686, 28736, 28766, 28772, 28897, 28910, 29033, 29050,
       29108, 29217, 29268, 29322, 29356, 29376, 29464, 29560, 29586, 29683, 39698,
       29762, 29891, 29895, 29956, 30042-3, 30054, 30057-8, 30241, 30276-7, 30361,
       30363-4, 30472-4, 30498, B30511, 30546, 30569-70, 30647, 30662, 30714, 30778,
       30948, 30980, 31012, 31037. 31117-9, 31134-5, 31160, 31274, 31411, 31437,
       31439, 31480-1, 31489, 31588, 31640, 31663-6, 31743-4, 31769, 31782, 31785,
       31802, 31850-2, 31856, 31991, 32002, 32103, 32113, 32326, 32365-6, 32397,
       32420, 32462, 32516, 32529
       32586, 32589,   32822-3, 32890, 32948, 33062, 33089, 33103, 33135-6, 33150,
       33153-7, 33271, 33273, 33303, 33445, 33489, 33629, 33642, 33672-3, 33686,
       33701, 33767-8, 33798, 33815, 33836, 33919, 33950, 33958, 33967, 34007,
       34026-7, 34066-7, 34072, 34125, 34205, 34267, 34286, 34346, 34372, 34379,
       34388, 34438-9, 34470, 34488, 34506, 34513, 34539, 34572, 34579, 34608,
       34618, 34670, 34676, 34683, 34716, 34758, 34776, 34781-3, 34817, 34882,
       34940, 35006, 35035-6, 35051, 35100, 35142, 35152, 35162, 35164, 35341,
       35359, 35378, 35404, 35456, 35459, 35502, 35507, 35509, 35687, 35697-700,
       35704, 35752, 35794, 35851, 35898, 35959-60, 35983, 35996, 36032, 36146,
       36280, 36288-9, 36344, 36351, 36381, 36461, 36494-5, 36529, 36544, 36561
       36590, 36614-6, 36624, 36905, 36958, 36963, 37015, 37084, 37100, 37132, 37143-
       5, 37150, 37192, 37219, 37257, 37267, 37469, 37513, 37529, 37576, 37584,
       37686, 37856, 37858,   37862-3,   37963, 38061-2, 38227, 38256, 38285, 38288,
       38340, 38356, 38383-4, 38397, 38430, 38447, 38458-9, B38475, 38540-2, 38548,
       38566, 38617, 38731, 38966, 38982, 39057, 39062, 39069, 39153, 39183, 39218,
       39252, 39310, 39318, 39389, 39398, 39441, 39458, 39472-4, B39537, 39568,
       39805, B39903, 39991, 39999, 40001-2, 40080, B40105, 40169, 40202, 40306,
       40314, 40344-5, 40360

Nostoc 21628, 22590, 22853, 24709, 25145; 26179, 26306, 28072, 28354; 28528, 29037,
       29421, 29557, 29590, 29752, 29853, 30336, 31151-3, 31176, 31389, 31529, 31783,
       31882, 31919, 32070, 32470, 32540; 32674, 32681, 32943, 33083, 33201, 33294,
       33399, 33442, 33679, 33757-8, 33776, 33903-5, 34081, 35696, 35699, 35955,
       36096, 36279, 36575; 36858, 36928, 36930, 37080, 37444, 37849, 37882, 37884,
       38179, 38213, 38297, 38628, 39476, 39730, 39804, 39965

O

Oak see *Quercus*

Oat see *Avena*

Oil palm see *Elaeis*

Olea 21705, 22002-3, 22187, 22278, 23243, 24232, 24988; 26095, 26801; 33247, 33573,
       B33628, 34107-8; 38285, B39903

Olive see *Olea*

Onion see *Allium*

Orange see *Citrus*

Orchardgrass see *Dactylis*

Orchids 22175-6, 23787, 23801, 24114; 25802, 26368, 27114, 27744; 33892, 35426;
    36701, 37177, 37237

Ornamental plants (*cf.* also *Agave, Antirrhinum, Asparagus, Chrysanthemum,* Coniferous
    plants, *Cyperus,* Deciduous trees and shrubs, *Eucalyptus, Euphorbia, Ficus,
    Hedera, Hibiscus, Ilex, Lathyrus, Lupinus,* Orchids, *Pelargonium, Perilla,
    Rosa, Tradescantia, Tulipa, etc.*)
    21748, 21798, 21814, 21840, 21961, 21975, 21998, 22028, 22098, 22130, 22148-9,
    22183, 22314-5, 22333-4, 22380, 22508, 22537, 22643, 22647, 22662, 22719,
    22725-7, 22770-1, 22925, 22980, 23026, 23036, 23091, 23129, 23132, 23244,
    23265-6, 23350, 23434, 23450, 23503, 23618, 23647, 23747, B23767, 23817,
    23829, 23859, 23873, 24206, 24217, 24285, 24317, 24369-70, 24372, 24400,
    24506, 24672, 24760-1, 24788, 24915, 24919-21, 24948, 24969, 25117
    25360, 25376, 25429, 25496, 25511, 25531, 25590, 25652, 25659 , 25693, 25814,
    25822, 25935, 25962, 25977, B26099, 26157, 26186, 26260, 26362, 26405, 26461,
    26476, 26534, 26683, 26719, 26853, 26918-9, 26993, 27118-9, 27404, 27457,
    27467, 27506, 27573, 27592, 27711, 28133, 28139, 28198, 28292, 28401
    28502, 28620, 28853, 28870, 29039, 29232, 29250, 29313, 29373, 29425, 29612,
    29614, 29616-7, 29780, 29807, 29817, 29858, 29953, 29963, 30002, 30042,
    30346, 30392-3, 30665, 30683, 30771, 30787, 31231, 31233, 31348, 31434-5,
    31471, 31550, 31563, 31854, 32005, 32101, 32185, 32229, 32319-20, 32324,
    32397
    32560, 32664, 32766, 32866, 32948, 32996, 33089, 33248, 33349, 33440, 33565,
    33590-1, 33642, 33669, 33756, 33804-5, 33831, 33842, 33858, 33968, 34018-9,
    34067-8, 34101, 34175, 34507, 34549, 34608, 34788, 34814, 34817, 34924,
    35006, 35140, 35260-1, 35263, 35314, 35330-2, 35426, 35438, 35657, 35691,
    35703, 35776-8, 35785, B35862, 36063, 36217, 36257-8, 36379, 36388, 36407,
    36437, 36445, 36478, 36535, 36549
    36590, 36655, 36705-6, 36818, 36901, 36963, 37017, 37022, 37027, 37118, 37123,
    37187, 37217, 37336-7, 37418, 37474, 37505-6, 37678, 37713, 37747, 37889,
    38033, 38101, 38371-2, B38475, B38477, 38504-5, 38540, 38635, 38829, 38933,
    39117, 39162, 39202, 39241-2, 39248, 39390, 39398, 39428, 39441, 39458, 39463,
    B39523, B39537, 39676, 39678, 39753, 39775, 39830, 39887, B39951, 39979,
    40018, 40026, 40051-2, B40105, 40146, 40201, 40231, 40371

Oryza 21506, 21596, 22012, 22123, 22305, 22700, 22806, 22869, 22880, 22895, 22965-
    -6, 22998, 23002-4, 23059, 23091, 23235, 23250-2, 23352, B23407, 23448, 23657,
    23669, 23706, 23714, 23732-4, 23736-7, 23758, 23773, 23814, 23922, 24002,
    24142, 24283-4, 24324, 24388, 24452, 24508, 24510, 24609, 24646-7, 24661-2,
    24692, 24705, 24769, 24806, 24808, 24820-2, 24841, 24851-2, 24972, 25037,
    25086, 25136-7
    25331, 25410, 25435, 25457, 25686, 25772, 25809 , 25967, 26267, 26284-6,
    26307, 26474, 26476, 26539, 26549-50, 26581, 26652, 26736-8, 26806, 26904-6,
    27157-8, 27210, 27221, 27311, 27325, 27384-6, 27413, 27506, 27592, 27709,
    27721-2, 27835, 27971, 28087, 28092-3, 28096, 28109 , 28130, 28139, 28195-7,
    28228, 28326, 28347, 28461-2
    28787,28809, 29110-2, 29233, 29865, 29874, 30047, 30077, 30080, 30084, 30097-
    -9, 30105, 30131, 30351-2, 30565, 30648, 30933, 31139, 31234, 31419, 31509,
    31578, 31601-2, 31623, 31777, 32006, 32047, 32312, 32504-5
    32574, 32603, 32681, 32686, 32737-8, 32863-4, 32874, 32948, 33109-10, 33130,
    33158, 33162, 33607, 33871-2, 33918, 33921, 33999-4000, 34038, 34095-8,
    34118-21, 34123, 34152-4, 34200, 34261, 34397, 34615, 34738, 34786-7, 34907,
    34940, 34942, 34956, 35022-3, 35044, 35064, 35091, 35103, 35183, 35210,
    35232, 35257, 35270, 35341, 35363-4, 35428, 35430-4, 35437, 35647, 35759,
    36056-8, 36060, 36064, 36066, 36069-70, 36077, 36185, 36212, 36227, 36267,
    36303, 36412, 36497, 36500-1, 36524

(continued)

Oryza (continued)
36630, 36698, 36708, 36734, 37239, 37298, 37334, 37377-8, 37574, 37696, 37898, 37915, 37981-3, 37995, 38016-20, 38071, 38074, 38103, 38131, 38158, 38246, 38489, 38646, 38856, 38872-3, 38984, 38993, 39005, 39033-4, 39054, 39062, 39068-9, 39183, 39195, 39236, 39238, 39274, 39281, 39439, 39541, 39741, 39802, 39855-6, 39889, B39903, 39981, 40083, 40152, 40231, 40250, 40287, 40292, 40324

P

Paddy see *Oryza*

Palms see *Cocos, Elaeis*

Panicum 21926-7, 22123, 22249, 22266, 22281, 22397, 22489, 22698, 22700, B22822, 22835-6, 22878-9, 22983, 23001, 23032, 23038, 23072, 23231, B23407, 23465, 23586, 23813, 24040, 24115, 24164-5, 24310, 24769, 24850, 24861, 25048-9 25164, 25431, 25445, 25573, 25913, 26052, 26181, 26249 , 26308, 26315-6, 26360, 26474, 26618-9, 26786, 26919 , 26971-3, 26979, 27008, 27175, 27539, 27582, 27584, 27596-8, 27929, 28005, 28100, 28130, 28199, 28228, 28257, 28400 28838, 29008, 29085, 29187, 29217, 29407, 29534-5, 29860, 29862-3, 29865, 30099, 30140-1, 30233, 30235, 30951, 31400, 31447-8, 31451-2, 31465-6, 31633, 31993, 32047, 32259, 32312, 32324, 32335-6, 32462 32653, 32867, 32888, 32913, 32950, 32997, 33053, 33130, 33167, 33537, 33705, 33843, 33914, 34095-7, 34134-5, 34252-4, 34256, 34365-6, 34452-3, 34541, 34549, 34618, 34676, 34924, 34943, 35186, 35286, 35425-6, 35466, 35468, 35471-2, 35704, 35766, 36056, 36146, 36205, 36252, 36303, 36386, 36436 36692, 36694, 36966, 37020-2, 37084, 37216, 37392-3, 37827, 37863, 37905, 37981, 38125, 38134, 38285, 38313, 38340, B38477, 38588, 38785, 38793, 39093, 39247, 39263-6, 39275, B39523, 39809, B39903, 39918, 40074, 40285, 40345

Papaver 25581; 31235; 32768, 33264, 35265-7; 37791, 38117, 38884-5, 39079

Papaya see *Carica*

Paprika see *Capsicum*

Para-rubber tree see *Hevea*

Parasitic plants
23652; 26009-10, 26339, 26493, 27147, 27450; 29487, 29614, 30704, 31739, 31927; 33230, B35862, 36194; 37074, 37271, 38106, 38922, 39754

Parsley see *Petroselinum*

Parsnip see *Pastinaca*

Paspalum 21705, 22366, 22449, 24806; 25274, 26308, 27106, 27506, 28099; 30140, 31613, 32047; 32653, 32766, 32976, 33914, 34301, 34365-6, 35186, 35403, 36348; 36692, 37063, 37084, 37981, 38313, 38385, 38793, 39127, 39434, 39808-9

Pastinaca 22110; 25511; 29280, 29456; 32948, 36255; 36963, 37331

Pasture plants see Forage plants

Pea see *Pisum*

Peach see *Persica*

Peanut see *Arachis*

Pear see *Pirus*

Peavine see *Lathyrus*

Pecan see *Carya*

Pelargonium
   21828, 21849, 22016, 22028, 22098, 23129, 23180, 24442; 25521, 26260, 26362,
   26649, 26699, 27694, 27771, 28139, 28308; 29343, 29487, 29894, 31471, 31769;
   33015, 33827, 33967, 36422; 36674, 37770-1, 37863, 37880, 38134, B38477,
   38784, 39922, 39991

Pennisetum
   21941-2, 22183, 22249, 22670, 22910, 23231, 24139; 25445, 25521, 28099;
   29217, 29231, 29862, 30140, 30213, 30538, 30967, 31243-5, 31419, 32259;
   32653, 33073, 33130, 33614, 33855, 33914, 34134-5, 34942, 35007, 35186,
   35257, 35427-8, 35430-3, 35455, 35560, 35888, 36205, 36252; 36692, 37190,
   37721, 37780, 37827, 37933, 37981, 38185, 38285, 38428, 38822, 39069, 39088,
   39127, 39429, 39737, 39809, B39903, 39917

Pepper see *Capsicum, Piper*

Perilla 21702, 25073; 25453; 30080; 32613; 38933, 40360

Persea 22250, 24468; 30801, 30835, 31956; 35691; 37430, 39183, 39508, 39678

Persica 22014-5, 22028, 22153, 22747, B23262, B23264, 23687, 23846; 25329, 26183,
   26421, 26476, 27445; 28546, 29622, 29944, 31885, 32011, 32539; 33111, 34098,
   34480, 34483, 34742; 38345, 38607-8

Petroselinum
   22660, 22747, 23577, 23580, 24304, 24648, 24723, 25070; 26360, 27673; 29182-3,
   30244, 32069; 32948, 32962, 33556, 34717, 34812, 34938, 35743, 36255; 36707,
   37674, 39154, 39678

Phalaris 27720, 28099; 28973, 29676, 30483; 32950, 33703, 33728, 34181, 34351,
   36386; 37331, 37511, 37609, 38793, 39127

Phaseolus
   21518, 21528,   21533, 21560, 21603, 21606, 21616, 21647, 21676, 21683,
   21705, 21730, 21733, 21786, 21842, 21918, 21925, 21961, 21968, 22006, 22028,
   22041, 22066, 22113, 22130, 22156-7, 22171, 22200-1, 22240, 22273, 22275-6,
   22305-7, 22484-5, 22506, 22594, 22647, 22660, 22663, 22691, 22719, 22734,
   22747, 22777, 22780-1, 22789, 22915, 22937-8, 22963, 22975, 23060, 23077,
   23108-9, 23170, 23173, 23205-6, 23231, 23242, 23263, 23287, 23291, B23407,
   23434, 23450, 23494, 23508, 23513-4, 23523, 23547, 23608, 23716-7, 23750,
   B23767, 23793, 23836, 23859, 23866, 23870, 23874-5, 23898, 23900-1, 23958-9,
   23961, 23969, 23972, 23997, 24001, 24018-9, 24100, 24116, 24127, 24138, 24143,
   B24250, 24256, 24284, 24291, 24341, 24348, 24364, 24383, 24411, 24468, 24508,
   24537, 24542, 24546, 24592-3, 24598, 24626, 24628, 24631, 24636, 24716, 24747,
   24750, 24754, 24760-1, 24808, 24885, 24912, 24922, 24941, 25019, 25028, 25136
   25189, 25221, 25237, 25293, 25295, 25326, 25431, 25476, 25489, 25496, 25513,
   25521, 25528-30, 25624, 25654, 25748, 25796, 25829-30, 25899, 25929, 25935,
   25965, 26011, 26073, 26091, B26099, 26109, 26121, 26205, 26221, 26264, 26282,
   26289, 26310, 26351, 26360-1, 26397, 26399, 26462, 26476, 26513, 26524, 26535,
   26560, 26587, 26620, 26675, 26683, 26744-6, 26753, 26779, 26784, 26818, 26842,
   26852, 26911, 26935-8, 26964-5, 26994, 27036, 27043, 27078, 27106, 27109,
   27168, 27170-1, 27201, 27287, 27300, 27329, 27363, 27402, 27407, 27412,
   27422, 27439, 27494, 27496, 27533, 27565-6, 27592, 27647, 27690, 27699, 27717,
   27730, 27736, 27738, 27763, 27765, 27804, 27816, 27861, 27877, 27929, 27935,
   28005, 28027-9, 28053, 28090-1, 28106-7, 28117, 28146, 28179, 28181, 28207,
   28244, 28264, 28269, 28277, 28315-6, 28330, 28354, 28374-5, 28380, 28382,
   28432

(continued)

Phaseolus (continued)
28518-20, 28585,   28604, 28634, 28657, 28662-3, 28697, 28782, 28856, 28874,
28883, 28887, 28909-12, 28956, 29030-1, 29046, 29054, 29138, 29214, 29247,
29259-61, 29269, 29296-8, 29301,   29347-9,   29351, 29391, 29555, 29571,
29579, 29603, 29610, 29648, 29841, 29922, 29932, 29964, 30042, 30088-9, 30139,
30182-4, 30229, 30244, 30295, 30348, 30422-3, 30425, 30441, 30455, 30548,
30643, 30797, 30814, 30837, 30849, 30862, 30897, 30901, 30903-5, 30974, 30994-
-5, 31002, 31014, 31034, 31087-8, 31164-5, 31211, 31250-1, 31262, 31300,
31324, 31352, 31366, 31411, 31427, 31543, 31547, 31568, 31574, 31620-1, 31692,
31737, 31874, 31899, 31940, 31979, 31982-3, 32066, 32099-101, 32126, 32266,
32275, 32290-1, 32297, 32314, 32352, 32385, 32399-400, 32414, 32502, 32547
32590-1, 32593-5,   32648, 32660, 32680, 32731, 32835, 32868-9, 32889-90,
32929-31, 32948, 32966, 33035, 33062-3, 33079, 33150, 33205, 33219, 33244,
33272, 633281, 33290, 33298-9, 33392, 33429, 33481, 33526, 33542, 33561,
33590-1, 33639, 33742, 33778, 33805, 33808, 33870-3, 33875, 34066-7, 34111,
34147, 34170, 34198, 34232-3, 34277, 34294, 34306, 34409, 34418, 34420,
34480, 34482, 34493, 34501, 34627-9, 34703, 34742, 34817, 34821, 34927,
34929, 34934-5, 34940, 35019, 35021, 35161, 35241, 35278, 35283, 35321,
35359, 35368-70, 35374, 35385, 35423-4, 35541, 35545, 35613, 35641, 35656,
35677, 35710, 35729, 35767, 35773-4, 35591, 35893, 35897, 35979-80, 36023-4,
36044, 36140, 36146, 36166, 36193, 36195-6, 36270, 36274, 36280, 36384-5,
36431, 36435, 36478-9, 36487, 36516, 36519, 36576, 36582
36616, 36702, 36727, 36738, 36825-6, 36877, 36887, 36905, 36919, 36938, 36950,
36963, 37093-4, 37105, 37150, 37169, 37170, 37183, 37265, 37298, 37304, 37336-
-7, 37339-40, 37344, 37358, 37534, 37555, 37605, 37640, 37742, 37747, 37799,
37807, 37819, 37828, 37863, 37872, 37897, 37998-9, 38053, 38091, 38122,
38134, 38192, 38285, 38321, 38346, 38433, B38475, B 38477, 38519, 38616,
38655, 38678, 38775, 38809, 38819, 38834, 38846, 38853-4, 38859-60, 38874,
38919, 39053, 39057, 39070, 39107, 39109, 39154, 39179, 39232, 39234, 39254,
39376, 39399, 39417, B39537, B39537, 39543, 39552, 39556, 39624-5, 39709,
39734, 39797-8, 39825, 39852, 39864, B39903, 39977, 39991, 39993, 40006-7,
40020, 40062, 40080-1, 40091, 40102, B40105, 40114, 40131, 40143, 40177,
40192-3, 40201, 40286, 40290, 40325, 40360, 40371

Phleum 22071, 22279, 23237, 23375, 24277, 24436, 24438; 26145, 27819, 28158; 29663,
30483, 30829, 31760; 34458, 34743, 34924, 36165, 36333; 37331, 37382, 37609,
B38475, 39244, 40117, 40274, 40300

Phoenix 27666; 35426; B39903

Photosynthetic bacteria see Bacteria, photosynthetic

Phragmites
23910, 25103; 25253, 25909, 26017, 26919, 26923, 27013, 27366, 27506, 27926;
29362, 29675, 31307, 31817, 31965, 32047, 32518; 33364, 33411-2, 33536, 33703,
33804, 34035, 34052, 34216, 34442, 34487, 34621, 34715, 34907, 35168-9, 35171-
-4, 35607, 35674, 36386, 36553; 36657, 37054, 37330, 37890, 37981, 38579,
39262

Picea 21540, 21629,21633, 21705, 22041, 22086-7, 22447, 22584, 22734, 23082, 23135,
23257, 23319, 23666, 23791, 23818, 23828, 23834, 23856, 23891, 24057, 24336,
24404, 24556-7, 24604, 24606, 24785, 24837, 24934, 24958, 25115
25186, 25255, 25956, B26099, 26227, 26338, 26480, 26482, 26515, 26621, 26635,
26707-8, 26883, 26910, 27087, 27094, 27105, 27332, 27609, 27750, 27930, 28193,
28350
28717-8, 28793, 28914, 28927, 29071-2, 29250, 29569, 29572, 29704, 30040,
30085, 30262-4, 30283, 30476, B30511, 30563, 30574, 30579-81, 30736, 30800,
30876, 31128, 31695-6, 31727-8, 31946-7
32698, 32777, 32891, 32960, 33199, 33278, 33422, 33677, 33783, 33878, 33920,
34280, 34283-4, 34318, 34323, 34402, 34532, 34588, 34689, 34747, 35081,
35115, 35156-7, 35491, 35536, 35764, 35808, 35936, 35949, 36049, B36191,
36201, 36358, 36551-2
36821, 37002, 37036, 37304, 38194, 38285, 38472, 38493, 38583, 38602-3,
(continued)

Picea (continued)
      38737, 38886, 38900, 38960, 39077, 39493, B39537, 39539, 39665, 39765, 39900,
      B39951, 39991, 40142, 40360

Pigeon pea see *Cajanus*

Pine see *Pinus*

Pineapple see *Ananas*

Pinus 21633, 21660,21705, 21831, 21925, 22028, 22189, 22400, 22505, 22523, 22598,
      22660, 22681, 22714, 22719, 22734, 22778, 22932, 23091, 23257, 23313, 23500,
      23534, 23605, 23666, 23818, 23828, 23830, 23861, 23916-7, 24219, 24232, 24240,
      24336, 24432, 24530, 24557, 24670, 24958, 25025, 25089, 25116
      25186, 25239, 25371, 25475, 25521, 25575, 25578, 25597, 25742, 25785, 25807,
      25921, 25928, 26015, 26047, B26099, 26161, 26184, 26197, 26267, 26338, 26348,
      26375, 26464, 26476, 26492, 26509, 26515, 26811, 26883-4, 26988, 27028, 27166,
      27345, 27432, 27465, 27484, 27636, 27654, 27730, 27795, 27886, 28024, 28052,
      28057, 28160, 28166, 28474
      28616-7, 28803, 28806, 28838, 28987-8, 28990, 29071-2, 29119, 29137, 29250,
      29299, 29325, 29641, 29693, 29717, 29732, 29792, 29806, 30029, 30040, 30120,
      30249-50, 30262-4, 30556, 30563, 30608, 30642, 30691, 30715, 30787, 30850,
      30869, 30898-9, 31019, 31032, 31045, 31053, 31143, 31258, 31268, 31477, 31516,
      31536, 31634, 31946-7, 31962, 32023, 32198-9, 32271-2, 32369, 32525
      32644, 32681, 32698, 32713, 32802, 32806, 32810, 32891, 33043, 33066, 33077,
      33103, 33209, 33221, 33269, 33378, 33440, 33467, 33508, 33528, 33587, 33595,
      33693, 33725, 33845, 33878, 33920, 34174, 34287, 34307, 34318, 34402, 34416,
      34441, 34471, 34651, 34654-5, 34752, 34766-7, 35014, 35115, 35156, 35181,
      35224, 35358, 35453, 35559, 35808, B35862, 35885, 35949, 36049, B36191, 36310,
      36392-3, 36465, 36478, 36530-1, 36551-2
      36708, 36735, 36809, 36830, 36850, 37002, 37036, 37049, 37052, 37138, 37232,
      37301, 37304, 37394, 37646-7, 37808, 37825, 37880, 38059, 38174, 38194,
      38207, 38285, 38448, 38493, 38523, 38567-8, 38583, 38624, 38664, 38863,
      38880, 38886, 38900, 39043, 39077, 39159, 39368, 39435, 39438, 39507, 39765,
      39821, 39898, 39900, B39951, 39991, 39997, 40112, 40230, 40336, 40360

Piper 27506; 28976, 30880, 30994; 39678

Pirus 22468, 22606, B23262, 23263, B23264, 24023, 24468, 24602; 25931, 26476, 26818;
      29737, 30481, 32485; 34480, 34483, 35203, 35282, 35800; 38285, 38607, 39014-5,
      39943

Pistacia 36445

Pisum 21530-1,21545, 21565, 21576, 21579, 21594, 21607-8, 21613, 21662, 21690,
      21705, 21727-8, 21747, 21825, 21857, 21943, 21947-8, 21996-7, 22028, 22058,
      22067, 22120, 22123, 22156, 22183, 22193, 22237, 22281, 22335, 22338, 22347,
      22371, 22405, 22423-4, 22462, 22482, 22512-3, 22540-1, 22547, 22560, 22566,
      22595, 22615, 22647, 22656, 22660, 22685, 22698, 22747, 22891, 22894, 22941,
      22977, 22980-1, 22988, 23032, 23068, 23118, 23120, 23141, 23150, 23154, 23156,
      23159, 23168, 23199, 23205, 23215, 23343, 23461, 23486, 23491, 23630, 23660,
      23709, 23716, 23719, 23726, 23742-3, 23748, 23750, B23767, 23847, 23851, 23859
      23870, 23914, 23918, 23934-5, 23949, 23964, 23969-70, 23975, 24001, 24012,
      24020, B24024, 24048, 24050, 24084-5, 24104, 24106-7, 24154, 24167-8, B24176,
      24188, 24226, 24241, 24284, 24364, 24385, 24415, 24457, 24468, 24478-9,
      24484, 24508, 24537, 24545-6, 24716, 24721, 24750, 24812, 24832, 24854,
      24861, 24870, 24886, 24910-2, 24973, 24986, 25009, 25070, 25099-101, 25124-5
      25183, 25192, 25196, 25209, 25211-4, 25240-1, 25246, 25262, 25271, 25315,
      25362-3, 25377, 25398, 25413, 25496, 25526, 25605, 25650, 25703, 25819, 25877,
      25892-3, 25904, 25935, 25949-51, 25965, 25969, 26008, 26022, 26063, 26073,
      26102, 26143, 26152, 26175-6, 26201, 26256, 26267, 26334, 26340, 26360, 26476,
      26497, 26500, 26539, 26575, 26589, 26597, 26617, 26653-4, 26710, 26721, 26740,
      26773, 26994, 27018, 27020, 27074-5, 27126, 27140, 27181, 27203, 27214, 27237,

(continued)

Pisum (continued)
27245, 27284, 27391, 27403, 27451, 27523, 27637, 27640-1, 27659-60, 27674,
27676, 27711, 27746, 27817, 27848, 27856-7, 27863-4, 27866, 27870, 27917,
27943, 27955, 27979, 27985, 28002, 28074-5, 28104, 28139, 28199, 28207, 28211,
28228, 28261, 28276, 28285-6, 28437, 28466-7, 28489
28508,28521, 28524-5, 28540, 28542, 28555, 28557-60, 28585-7, 28590, 28595,
29613-4, 28680, 28738-9, 28743, 28755-6, 28781-3, 28886, 28918, 28934, 28936,
28991, 29021, 29081, 29106, 29114, 29157, 29165, 29238, 29285, 29323, 29404-6,
29447, 29449, 29477-8, 29499, 29509, 29603, 29646, 29700, 29706, 29709, 29848,
29875, 29906, 29934, 29939, 29946, 29992-3, 30013-4, 30034, 30038, 30066,
30103, 30110-1, 30114-5, 30150, 30182-4, 30190, 30200, 30214, 30235, 30251,
30267, 30278, 30303-4, 30309, 30327, 30329, 30360, 30369, 30376, 30395, 30436-
-7, 30548, 30727, 30733, 30749, 30761, 30769, 30794, 30797, 30801, 30816,
30867, 30937, 30950, 30963-4, 30982-5, 30998-9, 31043, 31075, 31092, 31161,
31191, 31195, 31210, 31238, 31274, 31281, 31322, 31333-5, 31356, 31380-1,
31424-6, 31454, 31503, 31523-6, 31542, 31552-3, 31568, 31708, 31742, 31807,
31858, 31930, 31938-9, 31959, B32004, 32029, 32049, 32228, 32236, 32284,
32309, 32312, 32315-6, 32418, 32427, 32456, 32489-90, 32512-3, B32515, 32547,
32551, 32553
32561, 32594, 32601, 32610, 32616, 32627-8, 32660-1, 32675, 32679, 32687-8,
32700, 32739, 32746-7, 32801, 32832-3, 32836, 32848-9, 32855, 32859, 32889,
32892, 32934-6, 32948-9, 33024, 33063, 33090, 33137, 33149, 33160-1, 33171-2,
33206-7, 33219, 33310, 33317, 33323, 33370, 33399, 33450, 33455-4, 33461-2,
33550, 33579, 33684, 33727, 33731, 33750, 33770-2, 33806, 33809, 33827,
33871-3, 33898-9, 33914, 33959, 33979, 33990-1, 34026-7, 34042, 34141,
34207-8, 34214, 34250, 34253-6, 34277, 34312, 34317, 34321-2, 34349-50,
34371, 34400, 34418, 34431-2, 34475, 34527, 34586, 34610, 34629, 34717,
34723, 34753, 34779-80, 34817, 34855-6, 34869, 34892, 34924, 34982-3, 35009,
35012, 35026, 35065, 35076, 35104, 35142, 35264, 35356, 35361-2, 35378,
35456, 35464, 35515, 35530, 35566, 35569-70, 35572, 35576, 35581, 35591,
35685, 35755, 35796, 35800, 35809, 35844, B35862, 35976, 35988, 35997,
36005, 36054, 36094, 36192, 36200, 36223, 36233, 36250, 36268, 36302-3,
36306, 36331, 36375, 36423, 36469, 36486, 36558-9
36645, 36661-7, 36708, 36725, 36739, 36770, 36773-4, 36802, 36828, 36849,
36854-5, 36878, 36883, 36898, 36898, 36953, 36963, 36993, 37027, 37042, 37047-
-8, 37050, 37053, 37065, 37083, 37139, 37150, 37186, 37275-6, 37288, 37311,
37326, 37333, 37336, 37339-40, 37371, 37411, 37421, 37458, 37461, 37553,
37634, 37640-1, 37643-4, 37670, 37706, 37712, 37807, 37845, 37863, 37880, 37917-
-8, 37979-80, 38003, 38021-2, 38026-7, 38029-30, 38035, 38089, 38099, 38110,
38136, B38139, 38183, 38185, 38191, 38193, 38199, 38232, 38255, 38273, 38315,
38395, 38419, 38421, 38433, B38475, 38540, 38553, 38574-5, 38580-1, 38592,
38651, 38699, 38735-6, 38861, 38882, 38889, 38919, 38989, 39001, 39038, 39048,
39075, 39131, 39152, 39154, 39166-7, 39183, 39193-4, 39202, B39267, 39285,
39300, 39311, 39319, 39336, 39356-7, 39369, 39374-5, 39458, 39515, B39537,
39577-8, 39602, 39678, 39734, 39742, 39745-6, 39763, 39772-3, 39850, 39887,
B39903, 39973, 39987, 40014, 40016, 40039-40, 40047, 40068-70, B40105, 40107,
40109, 40111, 40128, 40201, 40203, 40218, 40259-60, 40302, 40358, 40360

Plane tree see *Platanus*

Platanus 26534, 27021, 28383; 29042; 36818, 38582

Plum tree see *Prunus*

Poa 21762, 22380, 23912, 24314, 24435, 24716, 25031-2; 25274, 26145, 26451, 28158,
28191, 28406; 28725, 29057, 29663, 29865, 29965-6, 30140, 31208, 31383, 32185,
32547; 32567, 32766, 33130, 33331, 33391, 33728, 34021, 34100, 35176, 35521,
35795, 36445, 36456; 36692, 37022, 37344, 37382, 37609, 38313, 39123, 40208,
40235, 40300

Poplar see *Populus*

Poppy see *Papaver*

Populus 21633, 22138, 22226-9, 22234, 22542, 22614, 22761-2, 23286, 23521, 23670,
   23709, 23829, 24045-6, 24105, 24183, 24240, 24336, 24528, 24601, 24739, 24817-
   -8, 24935, 25015, 25018
   25521,   25663, 25698, 25813, 25856, 25989, 25998, 26045-6, B26099, 26188,
   26200, 26489, 26515, 26534, 26835, 26987, 27125-6, 27331, 27705, 28114, 28160,
   28470
   28616, 29071, 29209, 29212, 29301-2, 29306, 29400, 29467, 29555, 29753,
   30087, 30130, 30563, 30787, 31090, 31238, 31270, 31403, 31599-600, 31769,
   31793, 32087, 32175, 32196, 32297, 32541
   32618, 32692, 33013, 33089, 33105-6, 33220-1, 33230, B33281, 33377, 33422,
   33558-9, 33566, 33590-1, 34150, 34246, 34506, 34827, 34866, 34924, 35154,
   35269, 35807, 35827, B35862, 35949, 36060, 36190, B36191, 36445, 36509
   36985, 37112, 37128, 37220-1, 37304, 37331, 37347, 37406-7, 37514, 37699,
   38285, 38357, 38679, 38697, 38710, 38916, 39035-6, 39662, 39971-2, 40336,
   40360

Porphyridium 21720, 22092, 22308, 22623, 25120; 25869, 25882, 26147, 26914, 27616,
   28267; 28960, 29590, 29672, 29955-6, 30877, 30892, 30926, 31709, 32448;
   32557, 32744, 32988, 33613, 33758, 33884, 34007, 34696, 35366-7, 35738, 36279,
   36455; 36772, 36808, 37310, 37455, 37601, 37651, 37705, 37764, 37877, 38295,
   38325, B38475, 38554, 38928, 39516, 39648, 39880, 39883, 39901, 39960

Portulaca 22697-8, 24310; 25360, 26315, 26539 , 26549-50, 26627-9, 27202, 27506,
   27585, 28005; 29865, 29912, 30174, 30272, 30274-5, 30498, 31993, 32430, 32547;
   32766, 32867, 32996, 33080, 33914, 34368, 34980, 35101-3, 35186, 35961; 37022,
   37331, 37344, 37827, 37858, 37863, 38087-8, 38185, 38253, 38385, B38477,
   39237, 39275, 39460, 40358, 40360

Potato see *Solanum*

Prune see   *Prunus*

Prunus (cf. also *Amygdalus, Armeniaca, Cerasus)*
   22284, 22747, B23262, B23264, 23890, B24024, B24176, 24376-8, 24602, 24714;
   25685, 25998, 26476, 26534, 26814, 27197; 29179, 29443, 29467, 29614, 31769,
   31801, 31947, 32526; 32948, 33504, 33602, 34098, 34239, 34511-2, 34581, 34742,
   36478, 36543; 36593, 36780, 36818, 36963, 37293, 37569, 37693, 38285, 38314,
   38365, 38401, 38607, 38679, B39903, 40333, 40355

Pseudotsuga 21633, 22087, 23818, 24336, 24372, 24976-7; 26515, 26952, 28089; 28821,
   29335, 29717, 30262-3, 30476, B30511, 31394, 32051, 32372; 32698, 33422,
   33595, 34287, 34471, 35721, 35808, 36049, 36075, 36332; 36735, 37002, 38285,
   38493, 40022

Pumpkin see *Cucurbita*

Purslane see *Portulaca*

Q

Quercus 21633, 21674, 21705, 21748, 21782, 21808-9, 21854, 22016, 22139, 22165,
   22302, 22402, 22573, 22647, 22752, 22762, 22841, 22893, 22917-8, 23091,
   23232, 23234, 23342, 23534, 23560, 23679, 23804, 23818, 23829, 23962, 24240,
   24567, 24601, 24605, 24670, 24718, 24988
   25186, 25239, 25445, 25578, 25664, 25807, 25897-8, 25989, 25998, B26099,
   26369, 26490, 26499, 26534, 26787-8, 26903, 27242, 27362, 27885, 28052, 28068,
   28125, 28160, 28383, 28460
   28622, 28988, 29071, 29121, 29169, 29206-7, 29209, 29249, 29254, 29320,
   29344, 29372-3, 29394, 29467, 29480, 29489, 29628, 29703, 30096, 30263, 30460,
   B30511, 30787, 31081, 31083, 31090, 31163, 31208, 31270, 31411, 31508, 32064,
   32110-1, 32185, 32194, 32257   (continued)

Quercus (continued)
   32693, 33262, 33345, 33440, 33451, 33477, 33559, 33595, 33632, 33736-7,
   33938, 33984, 34577, 34730, 34924, 35006, 35194, 35233, 35808, 35949,
   35962, 36144, B36191, 36403
   36746, 36748, 36809, 36818, 36973, 37304, 37338, 37407, 37413-5, 37568,
   37693-5, 37874, 38215, 38285, 38293, 38676-7, 38679, 38851, 38886, 38935,
   39002-4, 39077, 39736, 39821, 39991, 40085, 40291

Quince   see   *Cydonia*

R

Radish   see   *Raphanus*

Rape   see   *Brassica*

Raphanus   21719, 21841, 21959, B22052, 22114, 22555, 22571, 22719, 23019, 23125,
   23171, 23399-401, 23657, 24032, 24443, 24485, 24557, 24564, 24685, 24806
   25710, 25835, 26151, 26476, 26587, 26994, 27024, 27439, 27831-2, 27991-2,
   28050, 28130
   28995, 29182-3, 30173, 30199, 30594, 30597, 30992-3, 31291, 31775, 31895,
   31941, 32122, 32133
   32681, 33031, 33062, 33643, 34606-7, 35140, 35310, 35752, 35849, 35983,
   36060, 36137-8, 36142
   36676, 36759, 36827, 37012, 37061-2, 37191,   37739-40, 38158, 38228, B38477,
   38478-9, 39069, 39380, 39582, 39800, 39853, 40201,   40360

Raspberry   see   *Rubus*

Redwood   see   *Metasequoia*

Reed   see   *Phragmites*

Rheum   27673

Rhodopseudomonas   21649-50, 21821, 21993, 22093-4, 22100-1, 22198, 22230, 22259,
   22288-9, 22626, 22733, 22773-4, 22904, 22954, 23047, 23089, 23126-7, 23436,
   23825-6, 23886, 23998, 24028, 24095-7, 24109, 24182, 24205, 24305, 24315-6,
   24329, 24342, 24696, 25076, 25120
   25478, 25555-6, 25645, 25668, 25741, 25746, 25889, 25964, 26205, 26395,
   26406, 26456, 26610, 26631, 26660, 26681, 26696, 26702, 26714, 26731, 26808,
   27174, 27254, 27273, 27447-8, 27504-5, 27517-8, 27525, 27653, 27726, 27871,
   28103, 28259, 28455, 28469
   28530, 28551, 28580,   28640-1,   28692, 28822, 28904,   28949,   29098,
   29148, 29192, 29263-6, 29266, 29339, 29347, 29359, 29431-2, 29474, 29505,
   29610, 29627, 29650-1, 29683, 29712, 29878-9, 29940-1, 29947, 29959-61,
   29969-70, 29995-7, 30050, 30134, 30153, 30265, 30280, 30328, 30340-2, 30382,
   30385, 30421, 30626, 30671-2, 30788-9, 30831, 30915, 30926, 31005, 31033,
   31046, 31059, 31092, 31179, 31225, 31287-8, 31354-5, 31360, 31368, 31371,
   31478, 31490, 31805, 32014, 32017-9, 32045, 32048, 32114, 32123, 32168-9,
   32279-81, 32346, 32367, 32467-8, 32492, 32511
   32619,   32681,   32710-1,   32734, 32744, 32756-7,   32764,   32857-9,
   32915, 32924, 32955, 33063, 33086, 33186-9, 33204, 33241, 33251-2, 33294,
   33337, 33381, 33386, 33400-3, 33448, 33468-9, 33484, 33515-6, 33552, 33581,
   33612, 33640, 33688, 33706-7, 33732, 33735, 33753, 33783, 33786-7, 33838,
   33865, 33964, 33980, 34022, 34024, 34130, 34194, 34237, 34277, 34281, 34297,
   34347, 34356, 34363, 34393, 34522, 34566, 34641-2, 34744, 34789, 34791,
   34842, 34852, 34861-2, 34868, 34910, 34968-9, 35032, 35066, 35085, 35107,
   35167, 35236, 35239, 35246, 35292, 35323, 35376, 35387-8, 35422, 35501, 35568
   35572, 35576, 35651, 35668, 35702, 35724-5, 35758, 35801, 35855-6, 35899,
   35937, 36119, 36189, 36272, 36376, 36482-5, 36513, 36546-7
                                                                  (continued)

Rhodopseudomonas (continued)
      36589, 36620, 36649-51, 36704, 36749, 36796, 36798, 36837, 36861, 36911-2,
      36919, 36943, 36952, 36971, 37109, 37165-7,   37180-1,   37218, 37231,
      37248, 37255, 37350, 37370, 37409, 37446, 37485, 37545-6, 37598, 37611,
      37632, 37675, 37759, 37769, 37800, 37873, 37900, 37954-8, 38025, 38039,
      38041, 38173, 38264, 38305, 38332, 38419, 38424, B38475, 38476, 38538,
      38549, 38565, 38569-70, 38613, 38615, 38637, 38647, 38649, 38654, 38709,
      38714, 38729, 38766, 38796, 38895, 38906, 38918, 38952, 38957, 38972, 39012,
      39023, 39028, 39078, 39095, 39118-21, 39180, 39229-31, 39285, 39296, 39303,
      39315, 39386-8, 39440, 39451-2, 39476, 39491, 39502, 39580, 39613, 39626,
      39685, 39719-20, 39827, 39842, 39845, 39909, 40012, 40031, 40157, 40178,
      40190, 40278

Rhodospirillum 21577, 21651, 21669, 21691, 21777, 21780, 21797, 21855, 21866, 21904,
      22088, 22101, 22126-7, 22253, 22305-6, 22354, 22581, 22612-3, 22648, 22759,
      22824, 22947, 23362-3, 23436, 23470, 23598, 24034, 24121, 24174, 24189, 24333,
      24858-9, 25068, 25104
      25281, 25302, 25323, 25717-8, 25756-7, 25791, 25884, 25914, 25953, 26146,
      26222, 26233, 26328, 26357, 26395, 26486, 26575, 26686, 26969, 27050,
      27262, 27274, 27303-4, 27455, 27517, 27536, 27806, 27869, 27879, 27916,
      27965, 28048, 28072, 28131, 28202, 28237
      28644, 28672, 28701, 28769, 28892, 29038, 29078, 29197-8, 29200, 29272,
      29282, 29371, 29453-4, 29479, 29664, 29666, 29683, 29750-1, 29827-8, 29980-1,
      30567, 30626, 30742, 30926, 30939, 31056, 31093, 31095, 31147, 31295-6,
      31299, 31360, 31370, 31479, 31598, 31660, 31756, 31871, 31892-3, 31958,
      32123, 32149, 32375-6, 32406, 32495, 32549-50
      32585,   32588,   32619,   32670,   32673, 32681,   32710,   32732-4,   32754,
      32764, 32859, 32915, 33237, 33259, 33294, 33401-3, 33515, 33552, 33612,
      33688, 33706-7; 33709-10, 33734, 33779, 33786-7, 33833, 33895, 34010, 34194,
      34217, 34237, 34281-2, 34297, 34299-300, 34356, 34522, 34674-5, 34817, 34852,
      34908, 34965, 35066, 35107, 35130-1, 35167, 35212, 35239, 35323, 35387,
      35409, 35457, 35501, 35692-3, 35702, 35705, 35846-7, 35855, 35863, 35899,
      35934, 35937, 36119, 36189, 36236, 36292, 36359, 36363, 36395-6, 36514
      36605-6, 36624, 36626, 36649-50, 36751, 36764, 36786-7, 36796, 36852, 36890,
      37069, 37149, 37181, 37200, 37231, 37248, 37370, 37438, 37611, 37632, 37642,
      37662-3, 37764, 37773, 37815, 38038, 38063, 38154, 38243, 38247, 38351,
      B38475, 38649, 38796, 38824-5, 38905-6, 38923, 38952, 38972, 38983, 39020,
      39252, 39268, 39293-7, 39315, 39338-9, 39452, 39472, 39483, 39526, 39554,
      39580, 39666, 39673, 39681, 39700, 39704, 39839, 39863, 39909, 40012-3,
      40129, 40157, 40368-9

Rhubarbe see *Rheum*

Ribes 22178, 22747; 25569; 29008; 32948, 34483; 36613, 37027, 38883, 39013

Rice see *Oryza*

Ricinus 22533, 22734, 23927, 24289, 24386; 25373, 25431, 25511, 26142, 27506,
      28267, 28303; 29316-7, 30801, 30840, 31584-5, 32457; 32952, 33068, 33319,
      33880, 35089-90, 35624, 35675, B36191, 36562; 36617, 38003, 38421, B39903,
      39991, 40143, 40261

Robinia 22973, 23551, 23616-7; B26099, 27062; 28919, 28963, 30760, 30764, 30833-4;
      32948, 34285; 36963, 38582, 38695

Rosa   21517, 21809, 22631, 23716, 24063, 24596; 25414-5, 25569, 27362, 27753,
      28436; 29952, 31321; 37263, 37331, 38787, 38893

Rose see *Rosa*

Rubber tree see *Hevea*

Rubus   22948; 26814, 26952; 29467, 29489, 29920, 30760, 31692; 33264, 33845, 33925,
      B33926, 34101, 34485, 34754-5, 35045, 36201; 36593, 36989, 37220, 37489,
      37709, 38607

Rye   see   *Secale*

Ryegrass   see   *Lolium*

S

Saccharum   21950-1, B21979, 21994, 22123, 22249, 22328, 22366, 22532, 22556, 22700,
        22713, 22885, 22965, B23407, 23448, 23586, 23714, B23767, 23814, 23835,
        23877, 24552, 25137
        25190, 25198, 25331, 25431, 25445, 25457, 25482, 25626, 25769, 25913, 26267,
        26539, 26550, 26719, 26806, 26872, 27055, 27242, 27506, 28005, 28247, 28267,
        28301
        28787, 28838, 29018-9, 29050, 29060, 29489, 29812, 29912, 30175, 30677,
        30811, 30951, 31041, 31411
        32611, 32699, 32888, 32919, 33055, 34623, 34949, 35084, 35186, 35341, 35488,
        35744
        36734, 37060, 37701, 37780, 37858, 38185, B38477, 38629, B39903, 40043,
        40250

Safflower   see   *Carthamus*

Sago palm   see   *Cycas*

Salix   21760-2, 22053, 22778, 23829, 24525, 24605, 24624, 24974; 25569, 25581,
        25813, 25822, 27139, 27236, 28158; 29064-5, 29209, 29467, 30393, 31270,
        31595, 32415; 32891, 33264, 33331, 34351, 35006, 35406, 36013, 36132, 36408

Salt marsh and strand plants (*cf.* also Halophilous plants)
        21722, 21903, 21945-6; 25253, 25262, 25543, 25549, 25806, 25886, 26261,
        26663, 27267, 27700, 28192; 29061, 29332-3, 29489, 29865, 30631, 30735,
        30906, 31217, 32544; 32882, 33292, 33304-6, 33694, 34030, 34200, 34444,
        36445; 36679, 36683, 37269, 37348, 37442, 37649, 38358, 38385, 39692, 39992

Sambucus   22614, 24527; 25511, B26099, 26186, 27284, 27566, 27737, 28315; 28543,
        29212, 29467, 29614, 32069; 32948, 35729, 36201; 36963, 39847

Sandal   see   *Santalum*

Santalum   31221

Scenedesmus   21678, 21693, 21706, 21738, 21757, 21770, 21805-7, 21888, 22055, 22121,
        22180, 22248, 22319, 22442, 22454, 22578, 22642, 22658-60, 22673, 22794-5,
        23253, 23272-3, 23311, B23407, 23474, 23614, 23632, 23751, 23824, 23867-8,
        23882, 23888-9, 23954, 24053, 24154, 24169, 24273-4, 24371, 24373, 24401-3,
        24420, 24483, 24493, 24498-9, 24572, 24579, 24608, 24652, 24700, 24723, 24777
        -8, 24854, 24881-2, 24944, 24969, 25001, 25121, 25145
        25163, 25286, 25310, 25339, 25370, 25511, 25633, 25882, 26023-4, 26111,
        B26220, 26270, 26288, 26305, 26350, 26356, 26371, 26402, 26479, 26504,
        26659, 26793-7, 26807, 26822-3, 27086, 27100, 27222, 27227, 27261, 27312,
        27377, 27534, 27548, 27591, 27668, 27681, 27683, 27690, 27762-3, 27795,
        27858, 27875, 27974, 28009, 28074, 28174, 28223
        28814-6, 28845, 28847, 28942, 28944-7, 29014, 29021, 29118, 29127, 29185,
        29208, 29448, 29503-4, 29629, 29653, 29673, 29735, 29802, 29853, 30085,
        30231, 30318, 30381, 30416, 30528, 30594, 30607, 30622, 30624, 30718, 30798,
        30812-3, 30864, 30873, 31031, 31089, 31280, 31349-51, 31409, 31411, 31532,
        31564, 31720, 31723-6, 31738, 31767, 31909, 31998, 32016, 32069, 32089,
        32145, 32207, 32333, 32370, 32448, 32522-4, 32533
        32599-600, 32656, 32814-5, 32817, 32860-1, 32898, 32900, 32909, 32948,
                                                                (continued)

Scenedesmus (continued)
          32599-600, 32656,   32814-5, 32817, 32860-1, 32893, 32900, 32909, 32948,
          32974-5, 33070, 33218, 33265, 33294, 33333, 33352, 33439, 33543, 33545,
          33551, 33706, 33875, 33901, 33904-5, 34149, 34197, 34277, 34303, 34340-2,
          34414, 34417-8, 34468, 34520-1, 34531, 34555, 34607, 34629, 34634, 34681,
          34845, 34858, 34914, 34992, 35043, 35191, 35309, 35418, 35537, 35654, 35729,
          35762-3, 35799, 35995, 36063, 36085, 36182, 36248, 36305, 36385, 36455, 36568
          36592, 36598, 36634-5, 36859, 36998, 37051, 37070, 37194, 37248, 37329,
          37420, 37452, 37481, 37503, 37581, 37630, 37664, 37685, 37718, 37864, 37868,
          38070, 38154, 38225, 38274, 38354-5, 38419, 38451, B38477, 38667-9, 38850,
          38852, 39101, 39171, 39176, 39258, 39308, 39426, 39533, 39538, 39743, 39885,
          39936, 40090, 40270, 40340, 40349, 40358

Secale 22700, 23713-4, 23905, 24508, 24716, 24890-4, 24912, 24943, 25052, 25061,
          25136
          25310, 26003-4, 26063, 26455, 26506, 27239, 27547, 27651, 27678, 27690,
          27739, 27853, 28086, 28139, 28200, 28228, 28252
          29476-7, 30077, 30202, 30568, 30713, 30731, 31055, 31144, 31782, 32266-7
          32644, 32929, 33517-20, 33827, 34071-3, 35029-30, 35117, 35337, 35441,
          35583, 35613, 35729, 35835, 35884, 36047, 36067, 36146, 36261-2
          37192, 37298, 37339, 37342, 37486, 37701, 37946-9, 38285, 38701, 39192,
          40078, 40125

Sedge see *Carex*

Sempervirent plants (*cf.* also *Coffea*, Coniferous plants, *Hedera, Ilex, etc.*)
          21705, 22073, 22284, 22492, 23500, 23693, 24604; 25685, 26095, 26799-801;
          29250, 29587, 30465, 31659, 31947, 32324; 34260, 35354, 35537, 35808, 36543;
          36734, 38215, 38935, 39085, 39500, 39991, 40333

Service-tree see *Sorbus*

Sesamum 22651-3; 25511, 27769, 27794; 29793, 31318, 31697, 32304; 33089, 33504,
          35033; 38285, 39151

Setaria 22630, 24139, 25021; 25913, 26308, 27551, 28099-100; 28875, 29231, 29985,
          30077, 30140-1, 30951, 31418-21, 32047, 32259; 32653, 32766, 32950, 33130,
          34365-6, 34676, 35427-8, 35430-4, 36252; 36692, 37981, B38477, 38793, 38990,
          39238, 40058, 40235

Shrubs see Deciduous trees and shrubs; Sempervirent plants

Sinapis 21933, 22447, 22474, 22588, 22902, 23034-5, 23050, 23528, 23564, 23674,
          23712, 23714, 23920, 24166, 24368, 24716, 24738, 25029
          25235, 25707, 25776, 26077, 26603-4, 27439, 27586, 27623, 27759
          28743, 28842, 29050, 29062, 29169, 29471, 29567-8, 29623, 29871, 30240,
          30841-2, 30893-5, 31096, 32021
          32959, 33062, 33102, 33231, 33551, 33603, 33655, 33692, 34218, 34699, 34920
          -2, 35132-4, 36410
          36618, 36757, 36779, 36925-6, 36956, 37344, 38146, 38903, 38936, 39307,
          39379-80, 40201

Sisal see *Agave*

Solanum (*cf.* also *Lycopersicon*)
          21564, 21639, 21863, 22039, 22123, 22185, 22250, 22281, 22638, 22701, 22719,
          22734, 22829, 22844, 22911, 22999, 23071, 23106, 23226, 23448, 23457, 23544,
          23550, 23625, 23695-6, B23767, 23802, 23846, 24228, 24284, 24424, 24428, 24468
          24505, 24940, 25052, 25142
          25511, 25571, 25693, 25740, 25758, 25772, 25779, 25871, 25935, 25947, 26116,
          26138, 26182, 26267, 26336, 26476, 26733, 26742, 26785, 26853, 27145, 27236,
          27269, 27477-8, 27538, 27564-6, 27592, 27701, 27716, 27740, 27799, 27805,
          28097, 28139, 28175, 28199, 28315
                                                                    (continued)

Solanum (continued)
    28501, 28527, 28731, 28743, 28869, 29008, 29128, 29131-3, 29153, 29158,
    29168, 29430, 29470, 29564, 29615, 30031, 30069, 30080, 30455, 30457, 30571,
    30674, 30802, 30896-7, 30917, 30995, 31034, 31224, 31399, 31654-5, 31765,
    31859, 32181-2, 32265, 32322, 32546
    32678, 32717, 32750, 32874, 32929-31, 32948, 33008, 33034, 33103, 33195-6,
    33291, B33510, 33513, 33645, 33729, 33875, 33897, 33908, 34091-3, 34132,
    34597, 34737, 34808, 34923-5, 34932-3, 34940, 34942, 35006, 35051, 35064,
    35447, 35547, 35704, 35729, 35776-7, 35789, B35862, 35979, 36142, B36191,
    36226, 36262, 36489
    36642, 36716-7, 36734, 36963, 37041, 37150, 37163-4,   37171, 37173,
    37304, 37321, 37339-40, 37533, 37614, 37621, 37701, 37724, 38026, 38062,
    38185, 38260, 38265, 38285, 38312, 38340, 38435, 38464, 38591, 38644, 38657,
    38696, 38803, 38988, 39107, 39123, 39142, 39183, 39247, 39492, B39523, 39548,
    39651, 39677, 39764, 39889, B39903, 39991, 40250, 40344-5, 40360

Sorbus   21809, 23859, 24601; 25569, 27362; 30263, 31090; 32891, 33264; B39951

Sorghum see *Sorgum*

Sorgum   21613, 21705, 21713, 21739, 21822-3, 22028, 22033, 22042, 22123, 22249, 22328,
    22428-30, 22698, 22700, 22762, 22836, 22908-9, 22983, 23007, 23032-3, 23037,
    23219-20, 23231, B23407, 23448, 23586, 23714, B23767, 23782-3, 23814, 23877,
    23945, 24100, 24244, 24323-4, 24468, 24537, 24546, 24623, 24689, 24725, 24808,
    24968, 25037, 25136-7
    25169, 25238, 25331, 25431, 25445, 25454, 25521, 25615, 25705, 25725, 25772,
    25835, 25912-3, 26027-8, 26080, 26141, 26165, 26337, 26372, 26437, 26484,
    26539, 26778, 26872, 26932, 27047, 27199, 27242, 27278, 27382, 27411, 27439,
    27506, 27582, 27789, 27896, 27983, 28098, 28123, 28228, 28231, 28257, 28264,
    28337-8, 28343-4
    28505, 28710, 28787, 28838, 29018, 29050, 29054, B29468, 29486, 29579, 29652,
    29812, 30077, 30082, 30140, 30195, 30217, 30235, B30511, 30808, 30874, 30930,
    30951, 30990, 31348, 31411, 31464, 31710, 31931, 31993-4, 32103, 32183,
    32220, 32243, 32277, 32312, 32434
    32653, 32663, 32765-6, 32803, 32881, 32888, 32918, 32969, 33074, 33150,
    33162, 33192, 33222, 33527, 33547, 33705, 33843, 33914, 34048, 34109-10,
    34147, 34228, 34256, 34295, 34312, 34480, 34506, 34649, 34713, 34743, 34794,
    34942-3, 35186, 35257, 35302, 35341, 35378, 35476, 35893, 35907, 35966,
    36016-7, 36205, 36206-7, 36237, 36267, 36274, 36303, 36325, 36348
    36601, 36703, 36753, 37060, 37334, 37344, 37384, 37401, 37542, 37616, 37827,
    37863, 37866, 37898-9, 37981, 38077, 38082, 38084, 38185, 38285, 38334,
    38584, 38650, 38680, 38743-4, 39069, 39107, 39183, B39537, 39561, 39579,
    39659, 39667, 39678, 39738-9, B39786, 39809, 39812, 39815, B39903, 39991,
    40125, 40250

Soybean see *Glycine*

Sphagnum   23991; 31307, 31559, 31673, 32518; 33264, B33926, 34190, 35014, 35244,
    35463, 35928; 37174, 37497, 37709, 38144, 39639-40, 39665

Spinach see *Spinacia*

Spinach beet see *Beta*

Spinacia   21515-6, 21547-8, 21559, 21567-8, 21573, 21575, 21584, 21588, 21591-2,
    21597, 21606, 21611, 21613-4, 21617, 21620, 21631, 21643-6, 21652, 21655,
    21662, 21672-3, 21675, 21679, 21687-9, 21698, 21717-8, 21720, 21737-8, 21763,
    21766, 21778, 21792, 21816-7, 21819, 21825, 21834, 21839, 21851-2, 21867-8,
    21875, 21898, 21900, 21924-5, 21937, 21965, 21983, 21987-9, 21997, 22017-9,
    22077, 22081, 22098, 22103, 22107, 22136, 22140-2, 22160-1, 22182, 22194-5,
    22203, 22214-7, 22243-4, 22305, 22310, 22320, 22326, 22332, 22345-7, 22355,
    22360, 22370-2, 22374-6, 22397, 22406, 22425-7, 22452, 22456, 22463, 22483,
    22491, 22514, 22527-8, 22530, 22536, 22539, 22553, 22557, 22574, 22576, 22580,
(continued)

Spinacia (continued)

22583, 22586-7, 22615, 22640, 22644, 22656, 22658-60, 22664, 22669, 22671,
22682, 22689-90, 22693-4, 22698, 22705-9, 22719, 22723-4, 22732, 22736, 22738,
22747-8, 22766, 22801-2, 22812, 22820-1, 22835, 22856, 22862, 22886, 22889,
22905-6, 22915, 22922-3, 22934-6, 22940, 22957, 22967, 22970, 22983, 22995,
23050, 23055, 23057, 23062, 23103, 23122, 23183-4, 23192, 23195-8, 23208-9,
23219, 23227, 23253, 23261, 23276, 23292, 23309-10, 23336-7, 23351, 23359,
23370, 23390, 23394, 23397, 23415-6, 23422, 23473, 23484-6, 23492, 23506-7,
23515, 23517-8, 23538, 23541, 23546, 23549, 23555, 23568, 23573, 23577, 23588,
23591, 23600-1, 23632, 23641, 23656, 23691, 23705, 23715-6, 23721-3, 23725-6,
23728-30, 23750, 23761-2, 23837, 23840-2, 23852, 23858-60, 23868, 23873, 23876,
23906, 23916, 23920, 23927, 23948, 23954-7, 23989, 23993, 24035, 24037, 24049,
24054, 24064, 24066, 24081, 24086, 24097, 24103, 24112, 24116, B24176, 24191-
-3, 24222-4, 24237-8, 24262-4, 24281, 24298, 24304, 24306-7, 24311-2, 24327,
24335, 24338, 24344, 24351-2, 24354-5, 24360, 24362, 24364, 24374, 24380,
24384, 24395-6, 24442, 24487, 24492-4, 24497, 24501-2, 24549, 24591, 24637,
24648, 24651, 24663-5, 24667, 24682-3, 24720-1, 24723, 24727, 24750, 24757,
24760, 24791, 24828-9, 24840, 24846, 24856, 24860, 24871, 24874-6, 24878-9,
24881, 24888, 24901-4, 24908, 24910, 24917, 24923, 24925, 24944, 24951, 24999,
25002, 25029, 25057, 25062, B25064, 25081-5, 25092, 25094, 25104-5, 25107,
25113-4, 25131, 25137-8, 25145, 25148, 25152, 25159
25179-80, 25188, 25201-2, 25217, 25222, 25242-3, 25276, 25284, 25290, 25292,
25301, 25304, 25313, 25315-6, 25320-2, 25331, 25347, 25353, 25370, 25372,
25386-7, 25392, 25398-9, 25403, 25416-7, 25421, 25445, 25470, 25487, 25496,
25506, 25511, 25544, 25565, 25570, 25585, 25591, 25636, 25638, 25671-2,
25693, 25712, 25719-20, 25731, 25750, 25762, 25770, 25798, 25820, 25833,
25836, 25861-2, 25879-80, 25890, 25932, 25935, 25942, 25959, 25968, 26036,
26048, 26056, 26059, 26066, 26084, 26100, 26103-5, 26149, 26156, 26172-4,
26181, 26192-4, 26196, 26198, 26210, B26220, 26224, 26231, 26257-8, 26268,
26275, 26288, 26294, 26301, 26305, 26325, 26333-4, 26340, 26342-3, 26352,
26354, 26356, 26366, 26374, 26376, 26378, 26380, 26396, 26418, 26422, 26467-8,
26476, 26485, 26487-8, 26522-3, 26525, 26539, 26553-4, 26559, 26565-6, 26575,
26615, 26622, 26624, 26664, 26668, 26683, 26703-4, 26718, 26726, 26730, 26733,
26760, 26767, 26790-2, 26807, 26809, 26826, 26872, 26888, 26900-2, 26907,
26917, 26946, 26955, 26958-9, 26967-8, 26994, 27004-6, 27031, 27039, 27072,
27093, 27099, 27113, 27131, 27135-6, 27138, 27140, 27186, 27194, 27204, 27222,
27242, 27245, 27263-4, 27297, 27300, 27312, 27324, 27328, 27334-5, 27354-5,
27399, 27420, 27423, 27468-9, 27491-2, 27495, 27497-9, 27528-9, 27546, 27566,
27571, 27617-20, 27639-41, 27655, 27657, 27665, 27680, 27691, 27723-4, 27730,
27742, 27745, 27758, 27772, 27776, 27788, 27816, 27875, 27881, 27901, 27906-
-9, 27934, 27943, 27946, 27966, 27968-9, 27977-8, 28006-7, 28032, 28034,
28072, 28075, 28080-2, 28084, 28104, 28132, 28149, 28165, 28170, 28203, 28225,
28227, 28229, 28235-6, 28258, 28292, 28303, 28305-7, 28315, 28327, 28332-4,
28371, 28386-7, 28396, 28402, 28409, 28411, 28419, 28422, 28437, 28445, 28447-
-8, 28457, 28466, 28488-92
28508, 28513-4, 28538, 28547, 28549-50, 28562, 28564, 28584-5, 28592-4,
28597, 28619, 28634, 28652-3, 28665, 28678, 28682, 28693-6, 28708, 28720,
28730, 28743, 28745, 28747, 28765, 28771, 28774, 28777, 28784, 28794, 28802,
28823, 28839, 28841, 28845-6, 28851-2, 28867-8, 28871, 28875, 28891, 28901,
28920, 28932-3, 28936, 28964, 28966-7, 28969, 29002, 29021, 29027, 29070,
29076, 29114, 29117, 29138, 29161, 29180, 29182-4, 29219-20, 29238, 29245-6,
29262, 29265, 29271, 29279, 29286, 29291, 29304-5, 29308-9, 29353, 29382,
29397, 29405, 29434, 29439, 29442, 29513, 29529, 29533, 29539-41, 29570,
29579-80, 29584, 29595-6, 29598-9, 29618-21, 29635, 29667-8, 29687, 29698,
29718, 29720, 29730-1, 29745, 29756, 29768, 29784-6, 29796, 29807-8, 29814,
29830-1, 29838, 29845, 29853, 29865, 29872, 29877, 29884-5, 29888-9, 29891,
29912, 29932, 29942, 29949, 29974, 29999, 30002, 30036, 30042, 30062, 30085-
-6, 30091-3, 30107-8, 30122, 30125, 30127, 30142, 30154, 30160, 30165, 30177-
-8, 30181-3, 30197, 30210-1, 30213, 30215-6, 30235, 30244-6, 30266-8, 30291,
30306, 30308-9, 30335-6, 30338, 30355, 30406, 30410-1, 30414, 30428-30,
30444, 30449, 30491-3, 30498, 30509, 30518, 30527, 30534, 30539, 30558,
30561, 30604, 30636, 30638-9, 30646, 30667, 30669, 30708, 30717-20, 30744,
30746, 30765, 30777, 30779-82, 30793, 30801, 30806, 30810, 30813, 30854,

(continued)

Spinacia (continued)

30856, 30922, 30937-8, 30952, 30958, 30965, 30969, 30971-4, 30977-8, 30980,
30986, 30989, 30998-9, 31011, 31034-5, 31050, 31080, 31120-1, 31123-4, 31182,
31196, 31207, 31246-8, 31253-4, 31257, 31274, 31291-2, 31320, 31339-9, 31343,
31345, 31365-6, 31413, 31437, 31453, 31461-2, 31469, 31483, 31485, 31488,
31511, 31519-20, 31522-3, 31525, 31573, 31577, 31580, 31612, 31617, 31625,
31630-2, 31639, 31641-2, 31645, 31661, 31665, 31667, 31671-2, 31674, 31677-8,
31687-8, 31692, 31699, 31707-8, 31716-8, 31726, 31755, 31776, 31780, 31810,
31814-6, 31819-20, 31835, 31857-8, 31872-4, 31883, 31897-8, 31923-4, 31930,
31939, 31944-5, 31955, 31959, 31980, 31982, 31985-6, 31992, 32037-40, 32069,
32085, 32101, 32103, 32125, 32128-9, 32141, 32155, 32161, 32178-9, 32204,
32233-5, 32239, 32246-7, 32249, 32254, 32256, 32264, 32270, 32286, 32296,
32301, 32332, 32342-3, 32382, 32384, 32393, 32395, 32407-8, 32420, 32422,
32432, 32437-8, 32453, 32455, 32460, 32462, 32471, 32486-7, 32496-9, 32507-8,
32510, 32535-7, 32547
32594, 32609-9, 32626, B32630, 32631-2, 32636-7, 32641, 32659-60, 32673,
32675, 32679-82, 32687, 32697, 32700, 32729-6, 32744, 32755, 32767, 32769,
32782, 32791, 32828, 32834, 32839, 32841-2, 32845, 32851, 32853, 32889-90,
32894, 32898, 32905-7, 32922, 32932-3, 32944, 32948, 32954, 32968, 32992,
32994, 33005, 33008, 33035, 33037, 33058, 33063-4, 33068, 33100, 33116-7,
33127, 33151, 33171-2, 33212-4, 33256, 33294, 33310, 33341, 33350, 33353-6,
33383, 33387, 33452, 33458-9, 33461-2, 33473-4, 33533, 33553, 33560, 33649,
33690-1, 33712-3, 33743-4, 33748, 33751, 33755, 33772, 33779, 33785, 33801-2,
33805, 33821-2, 33836-7, 33865, 33869, 33875, 33880-1, 33886-7, 33895,
33904, 33910, 33919, 33929-33, 33938-9, 33944-5, 33960, 33986, 33994-5,
33997, 34004, 34032, 34062-3, 34067, 34089, 34093, 34102-6, 34114, 34117,
34126-7, 34138, 34142, 34147, 34165-6, 34168, 34173, 34178, 34180, 34205,
34220, 34231-3, 34277, 34286, 34292, 34328, 34335, 34357-8, 34364, 34367,
34377-8, 34417, 34422-4, 34446, 34438, 34509, 34526, 34530, 34563, 34573-6,
34598, 34659, 34694, 34726, 34728, 34734, 34749, 34769, 34769, 34775, 34802,
34804-5, 34816-7, 34820, 34882, 34889-91, 34956, 34975, 34984, 34996, 35005,
35026, 35086, 35088, 35093, 35100, 35114, 35125, 35140, 35187-8, 35206,
35212, 35214-5, 35231, 35266, 3529', 35345, 35353, 35375, 35378, 35382,
35391, 35402, 35414-5, 35438, 35444, 35456, 35466, 35470, 35479, 35495,
35503, 35510, 35516, 35526, 35540, 35545, 35547-8, 35600, 35615, 35646,
35649-50, 35663, 35665-6, 35669-70, 35681, 35692, 35694, 35716, 35728-9,
35736-7, 35739, 35743, 35752-3, 35816, 35818, 35822-3, 35848, 35853-4,
35858, B35862, 35879, 35889, 35894, 35917-8, 35981, 35983, 35991, 36003,
36022-3, 36025-6, 36035, 36046, 36055, 36063, 36074, 36095, 36102, 36127-8,
36130, 36175-6, 36210, 36239-40, 36244-5, 36263, 36277-8, 36282-3, 36314,
36331, 36355, 36375, 36377, 36389, 36397-2, 36404-5, 36413, 36416, 36430,
36460, 36462-3, 36475, 36488, 36502-3, 36511, 36526, 36555, 36573, 36577
36624-5, 36663, 36668, 36677, 36680, 36710, 36714, 36745, 36755-6, 36782-3,
36808, 36815, 36822, 36833, 36862, 36864, 36871, 36881, 36888, 36892, 36931-
-2, 36935, 36961, 36968-9, 36978, 36990-1, 37019, 37026, 37034, 37038-9,
37042, 37064, 37078, 37089-90, 37115, 37118, 37141, 37146, 37153, 37158,
37175, 37186-7, 37193, 37203, 37222, 37225, 37229, 37245, 37251-3, 37261,
37272-3, 37282, 37335-7, 37339, 37373, 37385, 37419, 37422, 37426, 37494,
37507, 37509, 37534, 37547, 37584, 37626, 37634, 37645, 37653, 37725, 37728,
37745, 37747, 37758, 37763, 37765, 37792, 37806, 37811, 37823, 37830, 37840-
-2, 37849, 37851-2, 37858, 37862-3, 37866, 37875, 37878, 37884, 37891, 37913,
37920-2, 37931-2, 37967-8, 38004, 38006, 38008-9, 38013, 38023-5, 38041,
38049, 38054-6, 38061-2, 38069, 38092-4, 38113-7, B38139, 38185, 38226,
38245, 38247, 38249-50, 38298, 38361, 38373, 38377, 38394, 38408, 38412-3,
38418-9, 38421, 38466, B38475, B38477, 38479, 38481, 38483, 38510, 38525,
38527, 38548, 38560, 38593-4, 38596-7, 38606, 38623, 38626, 38638-9, 38668-
-9, 38671, 38674, 38681, 38698, 38707, 38720, 38733-4, 38748, 38765, 38767,
38773, 38797, 38810, 38812, 38835, 38843-5, 38857, 38864, 38911-2, 38921,
38927, 38932, 38939, 38951, 38953, 38959, 38969, 38971, 39041, 39048, 39051,
39116, 39135, 39154, 39183, 39222, 39225, 39242, 39252, 39283, 39288-9,
39298, 39301, 39305-6, 39318-9, 39324, 39326, 39331, 39333, 39335-6, 39351-3,
39363-5, 39384-5, 39390, 39427, 39432-3, 39435, 39445-6, 39450,
39453, 39462-3, 39466-7, 39471-2, 39476, 39498, 39516, 39518, 39526, 39584,
(continued)

Spinacia (continued)
39590, 39594, 39618-20, 39623, 39627, 39629, 39674-5, 39687. 39694, 39706, 39708, 39745-6, 39772, 39777, 39795-6, 39810-1, 39858, 39840-1, 39880, 39887 39912-3, 39929, 39955, 39957, 39974, 39986, 40021, 40031, 40039, 40050, 40059, 40071, 40108, 40120, 40151, 40181-3, 40185, 40204-6, 40256-7, 40268, 40282, 40294-5, 40306, 40321-2, 40334, 40359, 40371

Spirodela see *Lemnaceae*

Spruce see *Picea*

Squash see *Cucurbita*

Strawberry   see *Fragaria*

Submersed plants see Aquatic macrophytes

Succulents (*cf*. also *Agave, Aloe, Bryophyllum,* Cacti, CAM plants, *etc.*)
23121, 23374, 24287; 25325, 25445-6, 25787, 25807, 25835, 25926, 26282, 26535, 26549-50, 26818, 27292, 27769, 28064-6, 28143, 28308; 28733, 29217, 29281, 29489, 29638-9, 29816-7, 29912, 30332, 30516, 31913; 32555, 33055, 33270, 33858, 33960, 34368-9, 34585, 34924, 35097, 35258, 35442, 35625, 35930, 35958, 36093; 36683, 36713, 36916-8, 36997, 37177, 37399, 37628, 37858, 38088, 38177, 38185, 38235, 38285, 38423, 38540, 38640-3, 38784, 38996-7, 39324, 39500, B39537, 39687, 39724-6, 39830, 39991, 40180, 40273, 40358

Sugar beet see *Beta*

Sugar cane see *Saccharum*

Sunflower see *Helianthus*

Sweet potato see *Ipomoea*

Synechococcus see *Anacystis*

Syringa   22071, 22165, 22614, 23859, 24601, 25018, 25102; 26095, 27370, 28308; 34924

T

Tamarisk see *Tamarix*

Tamarix 23112; 25813; B33281, 36445; B39537

Tapioca see *Manihot*

Taxus 21705, 24674, 24716; 25685, B26099; 32948, 34457, 35006, 35808; 37227, 38194, 38493

Tea see *Thea*

Thea 22647, 24526; 26476, 26798, 27229; 28572, 29524, 30080, 30700, 31782, 31925; 34047, 35017, 35123, 35535, 35619, 35622, 35963; 36686-8, 38847-8, 39402, 39591, 39806

Theobroma 21661-3, 23907-8; 25294, 27220, 27338-9; 28544, 29213, 30994; 35205; 37744, B39903, 40360

Thuja   23745; 25569; 30851, 32324; 33422, 34318, 35949; 38493

Tilia   22168, 23829, B24024, 24240, 24336, 24601, 24817-8, 25018, 25102; B26099,
          26490, 28160, 28485; 29206-7, 29212, 29614, 29703, 31090, 32081; 34924,
          35949, B36191; 36790, 36818, 37230, 37693, 37695, 38285, B38477, 39920,
          39991, 40356

Timothy   see *Phleum*

Tobacco   see *Nicotiana*

Tomato   see *Lycopersicon*

Tradescantia   21840, 22946, 23612, 23873, 25117; 26632-3, 26951, 27109, 27475;
          29777, 30213, 30665, 30707, 31769; 32572, 33615, 34629, 35426, 35571, 35652,
          36326, 36541-2; 36761, 37288, 37756, 37872, 38222, 38289, 39154, 39283,
          B39537, 39753, 39773, 39775, 39991, 40335, 40339

Trifolium   21869, 22123, 22563, 22611, 22668, 23063, 23338, 23588, 23594, 23716,
          23895, 24001, 24427, 24558, 24968, 25103
          25274-5, 25471, 25842, 25965, 26087, 26213, 26267, 26719, 26865, 27054,
          27293-6, 27664, 27808, 27813, 28005, 28102, 28139, 28315, 28343-4
          28566, 28689, 28716, 28757, 28899, 28973, 29456-7, 29537, 29575-6, 30684-6,
          31537, 31570, 31633, 31822
          32948, 33022, 33331, 33338, 33488, 33611, 34086, 34100, 34181, 34541, 34701-2,
          34809-10, 34924, 35158, 35508, B36191, 36434, 36549
          36636, 36656, 36900, 36963, 37836, 37969-70, 38285, 38701, 39244, 39395,
          B39537, 39635, 39818, 39887, 39983, 40044, 40201

Triticum   21529, 21542, 21544, 21561, 21593, 21602, 21634, 21642, 21653-4, 21677,
          21703, 21785, 21800, 21844-5, 21847, 21886, 21947-9, 21967, 22011, 22033,
          22049, B22052, 22064, 22071, 22118, 22123, 22212, 22221-2, 22260, 22281,
          22290, 22295, 22305, 22309, 22331, 22341-2, 22369, 22409-10, 22431, 22441,
          22448, 22467, 22519, 22524, 22639, 22698, 22700, 22719, 22734, 22785-6, 22793,
          22829, 22833, 22835-6, 22840, 22849, 22856-7, 22882-4, 22960, 22983, 23007,
          23049, 23073, 23079-80, 23086, 23095, 23114, 23145, 23149, 23162, 23172,
          23193, 23216, 23231, 23277, 23321, 23339, 23346, B23407, 23417, 23435, 23448,
          23494, 23561, 23586, 23650, 23655, 23665, 23672, 23713-4, 23717, 23748, 23750,
          23763, B23767, 23773, 23778, 23782, 23790, 23805, 23810-2, 23831, 23859,
          23862-3, 23870, 23873, 23899, 23903-5, 23950, 23994, 24022, 24026, 24064,
          24108, 24124, B24132, B24176, 24190, 24196, 24199, 24207, 24252, 24284, 24286,
          24290, 24295, 24309, 24364, 24391, 24424, 24449, 24468, 24504, 24508, 24523,
          24534, 24626, 24633, 24653-4, 24657, 24715-6, 24726, 24752-3, 24775-6, 24781,
          24806, 24808, 24890, 24940-1, 24949-50, 25015, 25037, 25052, 25059, 25134,
          25136-7, 25140, 25143, 25145
          25175, 25224, 25226, 25229-32, 25264, 25268-9, 25277-8, 25287-9, 25311,
          25357-8, 25392, 25413, 25430-1, 25445, 25457, 25480-1, 25496, 25511, 25514,
          25521, 25545, 25681, 25693, 25710, 25765, 25772, 25801, 25828, 25843, 25852,
          25864, 25905-6, 25912-3, 25935, 25937, 25965, 25967, 26001, 26029-31, 26050,
          B26054, 26073, B26099, 26118, 26126, 26131, 26139, 26151, 26204, 26241, 26256,
          26267, 26399-400, 26402, 26450, 26455, 26476, 26481, 26483, 26506, 26522,
          26537, 26539, 26552, 26575, 26735, 26757-8, 26766, 26782, 26827, 26847,
          26863-4, 26876, 26919, 27012, 27052-4, 27070-1, 27149, 27161, 27189-90,
          27200, 27280, 27343-4, 27347, 27405, 27409, 27425, 27439, 27442, 27452, 27460,
          27477, 27584, 27592, 27594, 27623, 27629, 27669-70, 27677, 27685, 27734,
          27778-9, 27783, 27837, 27850, 27873, 27882, 27896, 27982, 27984, 27997, 28005,
          28032, 28086, 28109, 28139, 28176, 28180, 28196, 28199-200, 28212, 28242,
          28262, 28282, 28296, 28300, 28303, 28319-20, 28339, 28379-80, 28432
          28512, 28527,   28533, 28578, 28597, 28605, 28608, 28624, 28637-8, 28650-1,
          28752, 28807, 28809, 28840, 28847, 28862, 28864, 28891, 28898-9, 28915-6,
          28974, 29022, 29043, 29052, 29067, 29102, 29123, 29153, 29221, 29289, 29315,
          29360-1, 29374, 29376, 29410-1, 29422, B29468, 29472-3, 29478, 29489, 29506,
          29550-1, 29586, 29613, 29633, 29637, 29656, 29726, 29783, 29800, 29886, 29912,
          29932-3, 29967, 29977-8, 29983, 29994, 30008, 30070, 30077, 30085, 30087,
                                                                        (continued)

Triticum (continued)
          30098, 30113, 30140, 30162, 30199, 30237, 30267, 30288-90, 30295, 30297,
          30330-1, 30379, 30453, 30455-6, 30468-9, B30477, 30532, 30542, 30548, 30551,
          30554-5, 30602, 30628, 30680, 30689, 30712, 30740, 30743, 30747, 30763, 30804,
          30833-4, 30855, 30914, 30931, 30980, 30995, 31013, 31042, 31156, 31213, 31242,
          31256, 31261, 31263, 31284, 31297, 31306, 31330-1, 31367, 31376, 31384-5,
          31411, 31463-4, 31468, 31494, 31528, 31547, 31552, 31581, 31706, 31769, 31774,
          31842, 31849, 31868, 31875, 31896, 31916, 31935, 31952, 31984, 32057, 32073,
          32104, 32115, 32147, 32164, 32211, 32278, 32307, 32311-2, 32322, 32351, 32424,
          32440, 32514
          32556, 32576,   32604-5, 32634, 32679, 32681, 32689, 32704, 32718, 32761,
          32763, 32786, 32805, 32855, 32893, 32900, 32919, 32948, 32956, 32971, 33000,
          33040, 33085, 33090, 33116, 33128, 33132, 33158, 33162, 33164-5, 33173-5,
          33226-7, 33279, 33339, 33363, 33372, 33374-5, 33405, 33424, 33426-7, 33438,
          33478, 33488, 33519, 33523, 33540, 33547, 33592, 33647, 33655, 33660, 33680,
          33760, 33827, 33835, 33871-2, 33914, 33970, 34014, 34056, 34059, 34066,
          34088, 34090, 34098, 34137, 34228, 34253-4, 34256, 34262, 34308-10, 34312,
          34314, 34316, 34333-4, 34348, 34373-6, 34416, 34437-9, 34443, 34451-2,
          34465, 34467, 34479, 34483, 34499, 34513, 34540, 34542, 34549, 34551-2,
          34554, 34582, 34721, 34817, 34830, 34839, 34879, 34918, 34940, 35029, 35047,
          35064, 35121, 35140, 35142, 35237, 35249, 35257, 35279, 35285, 35294, 35321,
          35327, 35337, 35339, 35341, 35349-52, 35377, 35389, 35394, 35441, 35448,
          35474, 35482-3, 35505, 35517, 35550, 35610, 35616, 35623, 35673, 35676,
          35701, 35704, 35759, 35790-1, 35815, B35862, 35883-4, 35886, 35891, 35893,
          35938, 35952-4, 36010, 36038-9, 36053, 36056, 36058, 36067-8, 36077, 36114,
          36116, 36137-8, 36205, 36281, 36284, 36289, 36296, 36299-301, 36303, 36315-
          -6, 36343, 36406, 36412, 36440, 36450-2, 36556, 36563, 36581
          36618, 36643, 36681, 36692, 36721, 36734, 36747-8, 36816, 36857, 36923,
          36927, 36959, 36973, 37022, 37060, 37066-7, 37082, 37091, 37100, 37108,
          37116, 37150, 37198, 37202, 37239, 37260, 37286, 37294, 37296-8, 37302,
          37325, 37334, 37336-7, 37339-44, 37344, 37363-4, 37372, 37423, 37439-41,
          37451, 37462-3, 37490, 37517, 37588-90, 37592, 37612-3, 37615, 37624, 37636-
          -7, 37701, 37713, 37719, 37768, 37828, 37858, 37866, 37867, 37869-71, 37898,
          37901, 37935, 37960, 37972, 37974, 38064, 38075, 38180, 38185, 38190, 38198,
          38217, 38231, 38252, 38285, 38287, 38294, 38340, 38342, 38387, 38421, 38429,
          38436, 38452, 38463, 38470-1, 38480, 38488, 38539, 38612, 38627, 38695,
          38716-7, 38753, 38769, 38772, 38930, 38933, 38942, 38988, 38994, 39069,
          39076, 39087, 39090, 39094, 39097-8, 39127, 39141, 39143, 39147, 39158,
          39182-3, 39203, 39212, 39271, 39280, 39334, 39335, 39337, 39376, 39401,
          39404, 39408-10, 39464, B39523, 39562, 39579, 39589, 39599, 39601, 39655-6,
          39691, 39695, 39758, 39764, 39779, B39786, 39816, 39843, 39846, 39889,
          B39903, 39905, 39928, 39963, 39980, 39985, 39991, 40009, 40025, 40049, 40060-
          -1, 40078, 40082, 40097-9, 40103, 40118, 40125, 40130, 40194, 40209, 40248,
          40250, 40271, 40306, 40326, 40328, 40342-5, 40350-1, 40360, 40367

Tsuga   22224, 22321, 23616-7, 24213; 27628; 28963, 29300, 29569, 30833-4; 33422,
          35808, 36049; 37002, 38182, 38285, 38493

Tulip see *Tulipa*

Tulipa   29953, 29957, 31442; 33506; B38477, 38540, 39218, B39537, 39899

Tundra plants and ecosystems
          22144, 22314-5, 22543, 23654, 23865, 24301, 24535, 24564, 24634, 24758,
          25024; 26912, 27987-8, 28323; 29290, 31758; 33050, 33052, 33245, 34536,
          34553, 34613, 34888, 35407, 35885, 36014, 36132-3, B36134, 36361; B38211,
          40277

Turnip see *Brassica*

Turpentine tree see *Pistacia*

Typha   23271, 23684; 25254, 26112, 26923, 27710, 28267; 29362, 29605, B31140,
          31239, 31307, 32518; 32867, 33412, 33422, 33536, 33804, 34052, 34351,
                                                                    (continued)

Typha (continued)
34387, 34442, 34486, 34631, 34942, 35006, 35168, 35174, 35244, B35862, 36386-
-7, 36553; 36657, 36869, 37054, 37331, 37817

U

Ulmus 21809, 22190, 22734, 23342; 25283, 25822, B26099, 26534, 27062, 27362;
29207, 29243, 29320, 29467, 31090, 31269, 31270; 33013, 34285; 36818,
37230, 37693, 37695, 38284-5, 39900

Ulva 23552-3, 23876, 23936; 25557-60; 28940, 28950, 28979, 29178, 30026, 31429,
32142-3, 32500; 32681, 33922, 34251, 34385, 34568, 34570, 34626, 34629,
34899, 34912, 35449, 35557, 35735, 36153-4, 36586; 37001, 38499, 38745,
39211, 39349, 40305

V

Vaccinium 21705, 21759, 21762, 22832, 22850, 23865, 24632, 25023; 25659, 26895,
26952; 29704, 30120, 30523, 31670; 33264, B33926, 34291, 35014, 35525,
35691, 35808, 36037, 36201; 37220, 37233, 37250, 38221, 38285, 39821,
39874

Vegetables (*cf.* also *Allium, Asparagus, Bambussa, Beta, Brassica, Capsicum, Cichorium, Cucumis, Cucurbita, Cynara, Daucus, Lactuca, Lycopersicon, Pastinaca,
Petroselinum, Phaseolus, Pisum, Portulaca, Raphanus, Solanum, Spinacia*)
21703, 21705, 21743, 21841, 21948, 21957, 22110, 22131, 22141-2, 22233, 22328,
22483, 22647, 22697, 22747, 22797, 23050, 23101, 23457, 23528, 23577, 23691,
23920, 23972, 24092, 24468, 24564, 24925, 25037, 25158
26218, 26360, 26809, 27024, 27478, 27673, 27799, 28127, 28130, 28227-8,
28315, 28487
28751, 29131-3, 29329, 29603, 29743, 29747, 29979, 30078, B30477, 30718,
31399, 31547, 31586, 31605, 31616, 31657, 31833, 31895, 32069, 32103, 32106,
32122, 32211
32765, 32948, 33790, 33958, 36058, 36137, 36226, 36255
36676, 36963, 37331, 37380, 37548, 37621, 37732, 37735, 38644, 39508, 39678,
40191, 40323

Vetch see *Vicia*

Vicia 21587, 21948, 21974, 22061, 22311, 22711, 23028, 23076, 23263, 23332-3, 23450,
23498, 23547, 23588, 23647, 23716, 23814, 23937-9, 23969, B24024, B24176,
24196, 24268-9, 24284, 24397, 24482, 24504, 24638, 24716, 24754, 24762, 24931,
24933, 24940, 25041, 25145
25193, 25311, 25413, 25511, 25618, 25710, 25770, 25920, 25935, 25943, 25965,
26145, 26252-3, 26398, 26448, 26476, 26561, 26722, 26802, 26813, 26893,
26994, 27047, 27090, 27109, 27195, 27237, 27288, 27364, 27477, 27566, 27573,
28037, 28045-6, 28130, 28158, 28199, 28209, 28268, 28285, 28315, 28321,
28399, 28483
28833, 28875, 29323, 29477, 29616, 29663, 29689, 29740-1, 29758, 29932,
30188, 30212, 30369-70, 30376, B30477, 30659-61, 30801, 30828, 30901, 30909,
30920, 30982-3, 31082, 31155, 31180, 31292, 31494, 31564-9, 31586, 31769,
31826, 31921-2, 31983, 32134, 32147, 32317, 32322, 32462
32610, 32687, 32696, 32739, 32885-6, 32948, 32956, 32969, 33103, 33122,
33318, 33330, 33449, 33623, 33714-5, 33718, 33720, 33791-2, 33820, 33827,
33969, 34046, 34156, 34243, 34311, 34383, 34598, 34659, 34663, 34682, 34704,

(continued)

Vicia (continued)
          34717, 34743, 34924, 34931, 34954, 35139, 35198, 35318, 35461, 35464, 35600,
          35603, 35605, 35729, 35828, 35843, B35862, 35989-90, 36041, 36057, 36077,
          B36191, 36384, 36428
          36673, 36763, 36829, 36907-9, 36963, 37137, 37240, 37303, 37319, 37336-7,
          37340, 37505-6, 37666-71, 37713, 38080, 38175-6, 38186, 38258, 38268-9,
          38421, B38475, 38540, 38555-6, 38656, 38901, 38910, 38912, 39008-10, 39131,
          39218, 39259, 39380, 39391, 39393, B39537, 39773, 39899, B39903, 39920,
          39928, 39991, B40105, 40201, 40360

Vigna   21513, 22352, 23913, 24284, 24849; 25171, 25299, 25431, 25482, 25511, 26465,
          27389, 27941-2; 28866, 29640, 29923, 30173, 30980, 31075, 31570, 31996, 32059;
          32686, 33068, 33668, 33958, 34500, 34513, 34895, 35709, 36036, 36080; 36609,
          36963, 37010, 37298, 37616, 38285, 38431, 38500-1, 38543, 38771, 38859-60,
          39272, 39395, B39537, 39549, 40262-3

Vine   see   *Vitis*

Vitis   21554-5, 21569, 22028, 22300, 23053, 23218-9, B23262, 23526, 23628, 23846,
          23859, 24014, 24089, 24255, 24378, 24619, 24820, 25037; 25203, 25521, 25642-
          -3, 25857, 25947, 26392, 26476, 26498, 26775-6, 27009, 27129, 27134, 27141,
          27388, 28018-20; 28546, 28552, 28761, 29331, 29367, 29490-2, 29944, 29968,
          30480, 30703, 30745, 31187, 31200, 31547, 31827, 31855, 31974, 32011, 32055,
          32201; 32870, 33112, 33208, 33396-7, 34008, 34098, 34370, 34411, 34413,
          34496, 34632-3, 34742, 35010, 35074, 35527, 35563, 35680, 35712, 36009,
          36081; 37088, 37368, 38285, 38465, B38477, 38645, 38897, 39114, B39523,
          B39537, 39784, B39903, 40296

W

Walnut   see   *Juglans*

Watermelon   see   *Citrullus*

Weeds (*cf.* also *Amaranthus, Atriplex, Avena, Bromus, Chenopodium, Digitaria, Seta-
          ria, etc.*)
          21696, 22028, 22209, 23620, 24088, 24231, 24314, 24547; 25978, 26065, 27022,
          27047, 27336, 28092; 28498, 28855, 28875, 28881, 29024, 29433, 30104, 30230,
          30270, 31166, 31201, 31232, 31347-8, 31633, 31884, 32047, 32165, 32519-20;
          32625, 32766, 32875, 32948, 32996, 33216, 35101-3, 35413, 35484, 35529, 36060,
          36424, 36456, 36479; 36659, 36956, 37239, 37244, 37291, 37331, 37344, 37382,
          38082, 38285, 38597, 39043, 39071, 39124, B39537, 39690-1, 39809, 40145,
          40337

Wheat   see   *Triticum*

Whortleberry   see   *Vaccinium*

Willow   see   *Salix*

Wolffia   see   *Lemnaceae*

Y

Yam   see   *Dioscorea*

Yew   see   *Taxus*

Yucca 24872-3; 28065-6; 33440; 38185, 39830

Z

Zea   21507, 21523-6, 21556-7, 21587, 21613, 21637, 21647, 21652, 21657-8, 21676,
21697, 21705, 21708, 21713, 21765, 21782-3, 21790, 21804, 21835, 21871, 21878,
21922, 21948, 21961, 21967-8, 21973, 21994-5, 22011, 22028, 22032-3, 22071,
22123, 22130, 22150, 22156, 22203, 22233, 22249-50, 22263, 22279, 22281-2,
22305, 22311, 22316, 22325, 22343, 22393-5, 22397-9, 22403, 22406, 22421,
22435, 22438-40, 22462, 22498, 22510, 22512-3, 22521, 22530, 22532, 22582,
22601, 22608, 22635, 22641, 22660, 22698, 22700, 22712, 22719, 22734, 22754,
22799, 22806, 22813-6, 22829, 22835, 22840, 22869, 22875, 22958, 22974,
22983-5, 22998, 23002-3, 23021, 23030-3, 23054, 23060, 23067, 23091-2, 23094,
23106, 23108, 23134, 23170, 23231, 23235, 23250-2, 23258, 23343, 23351,
23353, 23376, 23392, 23397, B23407, 23408, 23418-9, 23424, 23426, 23448,
23457, 23461, 23485, 23547, 23586, 23592, 23608, 23619, 23647, 23665, 23671,
23676-7, 23689, 23714, 23716-7, 23732, 23735, 23737, 23739-40, 23746,
B23767, 23782, 23814, 23851, 23859, 23863, 23870, 23877, 23887, 23893, 23919,
23923, 23954, 23969, 23985, 24000-1, 24003, 24005, 24008, 24018, B24024,
24033, 24039, 24047, 24074, 24083, 24100, B24132, 24143, 24157, 24164-5,
24196, 24236, 24244, 24268, 24284, 24318, 24323-4, 24364, 24424, 24468, 24470,
24474, 24504, 24507-8, 24511, 24538, 24543, 24545-8, 24597, 24607, 24615,
24657, 24705, 24723, 24725, 24734, 24737, 24750, 24769, 24772, 24794, 24805-
-6, 24808, 24841, 24850, 24854, 24861, 24887, 24912, 24931, 24933, 24959,
25015, 25037, 25052, 25087, 25091, 25099, 25101, 25134, 25136-8, 25156,
25158
25166, 25221, 25297, 25318-9, 25331, 25354, 25397, 25408, 25431, 25440, 25445,
25449-50, 25457, 25482, 25484, 25511-2, 25521, 25529, 25594, 25662, 25673,
25678, 25711, 25744, 25770, 25772, 25827, 25835, 25837, 25852, 25858, 25877,
25900, 25912-3, 25919-20, 25935, 25947, 25967, 25982, 25987, 26020, 26057,
26063, 26073, 26076, 26091, B26099, 26101, 26117, 26181, 26235, 26237, 26267,
26315-6, 26318, 26327, 26403, 26437, 26444, 26447, 26474, 26476, 26559, 26575,
26598, 26626-7, 26665-7, 26706, 26715, 26744-6, 26779, 26789, 26806, 26812,
26833, 26852, 26862, 26872, 26887, 26890, 26893-4, 26911, 26921, 26932, 26949,
26994, 27007-8, 27036, 27047, 27106, 27131-2, 27167-9, 27176, 27187, 27211,
27228, 27240, 27242, 27255, 27275, 27292, 27296, 27300, 27351-2, 27365, 27402,
27421, 27434-5, 27444, 27468-9,   27506-7,   27532-3, 27546, 27573, 27576-7,
27582, 27592, 27605, 27682, 27687, 27709, 27730, 27792, 27837, 27854, 27862,
27888, 27896, 27920-1, 27925, 27972-3, 27996, 28001, 28003, 28005, 28010-1,
28022, 28058, 28078, 28096, 28109, 28130, 28139, 28146, 28167-9, 28177, 28183,
28196-7, 28201, 28219, 28228, 28265, 28285, 28309-10, 28351, 28353, 28357,
28393, 28420, 28432, B28482, 28489
28505, 28527, 28529, 28532, 28599, 28652, 28655-6, 28688, 28713, 28730,
28743, 28786, 28792, 28807, 28838, 28865, 28896, 28909, 29018, 29021, 29045,
29050-1, 29054, 29090-1, 29103, 29106, 29147, 29153, 29165-6, 29187, 29190,
29262, 29376, 29405-6, 29408, 29430, 29455, 29458, B29468, 29478, 29489,
29501, 29562, 29565-6, 29578-9, 29594, 29603, 29637, 29709, 29826, 29830,
29833-4, 29861, 29865, 29912, 29875-6, 29885, 29912, 29984, 30006-7, 30016, 30049,
30069, 30077, 30080, 30083, 30087, 30097-9, 30140-1, 30150, 30214-5, 30228,
30232-3, 30235, 30243, 30252, 30267, 30272, 30275, 30288, 30296, 30313,
30319-20, 30371, 30438, 30446, 30461-2, B30477, 30498, B30511, 30540-2,
30557-8, 30576, 30643, 30669, 30681, 30705, 30754, 30783-4, 30801, 30819-21,
30859, 30866, 30871, 30887, 30900-1, 30907, 30936, 30951-2, 30959-61,
30980, 30991, 30995, 31015, 31034, 31037, 31053-4, 31075, 31244, 31261, 31265-
-6, 31272-4, 31304-5, 31325, 31337, 31348, 31357, 31359, 31407, 31411, 31443,
31474, 31492, 31494, 31504, 31539, 31551, 31567-8, 31579, 31586, 31611,
31620, 31633, 31674, 31769, 31802, 31821, 31849, 31859, 31903, 31932, 31966-7,
31977, 31984, 31993, 32047, 32065, 32076, 32084, 32096, 32103, 32118, 32147,
32163, 32172, 32192, 32220, 32223, 32226, 32261-2, 32292, 32294, 32300,
32304, 32312, 32317, 32322-3, 32382, 32391, 32399, 32429-30, 32462, 32484,
(continued)

Zea (continued)
        32489, 32503, 32545
        32566, 32600-1, 32633, 32640, 32647, 32653, 32662, 32681, 32686, 32700,
        32709, 32742, 32765, 32850, 32859, 32867-8, 32874, 32888, 32902, 32919,
        32948, 32964, 32969, 32986-7, 33027, 33062, 33097, 33103, 33108, 33143,
        33150, 33158, 33202, 33205, 33254-5, 33339, 33388, 33391, 33431, 33488,
        33493, 33495, 33498-9, 33501-2, 33505-6, 33527, 33539, 33541, 33557, 33571,
        33577, 33579, 33630, 33641, 33654, 33742, 33748-9, 33769, 33773, 33805,
        33808, 33817, 33827, 33843, 33853-4, 33869, 33875-6, 33914, 33935, 33975,
        33998, 34013, 34043-5, 34052, 34095-8, 34116, 34123, 34132, 34155, 34206,
        34228, 34252-6, 34259, 34286, 34322, 34390, 34418, 34451, 34454, 34456,
        34492, 34500-3, 34513, 34533, 34535, 34550, 34585, 34616, 34618-20, 34622,
        34625, 34630, 34693, 34695, 34705, 34708, 34711, 34713, 34720, 34758, 34770-
        -1, 34857, 34873-4, 34876, 34896-7, 34927, 34930-1, 34940, 34942, 34958-9,
        34983, 35025, 35049, 35055, 35059, 35064, 35103-5, 35140, 35160, 35163,
        35186, 35249-50, 35257, 35286, 35303, 35321, 35341, 35346, 35398, 35410,
        35444, 35454, 35460, 35489, 35492, 35504, 35558, 35564, 35582, 35603-5,
        35631, 35636-8, 35641, 35652, 35656-8, 35662, 35701, 35704, 35730, 35806,
        35835, B35862, 35870, 35877, 35891-3, 35964, 35970, 35986, 35993, 35996,
        36004, 36019, 36021, 36056-7, 36060, 36069, 36077, 36146, 36148, 36163,
        36178, B36191, 36205, 36212, 36225, 36230, 36237, 36269, 36275, 36303,
        36348, 36412, 36425-6, 36440, 36497, 36520, 36533, 36537, 36548, 36579-80
        36590, 36595, 36601, 36630, 36636, 36640, 36663, 36671, 36684, 36692, 36719-
        -20, 36732, 36760, 36763, 36898, 36933, 36960, 36973, 37060, 37072,
        37084, 37098-9, 37119-20, 37147-8, 37150, 37186, 37199, 37207, 37212-3,
        37215-6, 37260, 37266, 37270-1, 37280, 37294-5, 37298, 37311, 37334, 37336-7,
        37339, 37344, 37359, 37361, 37389-91, 37412, 37464-5, 37470, 37472, 37479,
        37516, 37518, 37539, 37541, 37559, 37602, 37623, 37643-4, 37691-2, 37696,
        37701-2, 37711, 37716, 37723, 37733, 37742, 37747, 37752-3, 37755, 37768,
        37771, 37827, 37829, 37858, 37866, 37872, 37896, 37898, 37919, 37944, 37976,
        37981-3, 37992, 37995, 38020, 38062, 38082, 38163, 38170, 38185, B38211,
        38214, 38219, 38256, 38260-1, 38285-6, 38294, 38340, 38390, 38407, 38421,
        38425, 38429, 38433, 38440-1, B38477, 38530, 38540, 38552, 38562-4, 38590,
        38617-8, 38636, 38644, 38670, 38701, 38724, 38738, 38757, 38768, 38783,
        38786, 38798, 38814, 38841-2, 38874, 38962, 38988, 39019, 39037, 39069,
        39088, 39093, 39097, 39107, 39129, 39139-40, 39144, 39166-7, 39183, 39201-2,
        39259, 39275, 39277, 39292, 39318, 39350, 39355, 39380, 39385, 39403, B39523,
        B39537, 39556-8, 39560-1, 39572, 39593, 39606, 39609, 39632, 39652, 39667,
        39678, 39710, B39711, 39747, 39752, 39764, 39767, B39786, 39788, 39799,
        39832, 39855, 39888-9, 39900, B39903, 39940, 39945-6, 39958, 39991, 39995,
        40018, 40045, 40075, 40077, B40105, 40113-4, 40121, 40124-5, 40132, 40250,
        40264-5, 40297, 40299, 40302, 40344-5, 40352-4, 40358, 40360, 40365

Zebrina   see   *Tradescantia*

Zingiber   28267; 35426